Engineered Wood Products for Construction

Edited by Meng Gong

Published in London, United Kingdom

IntechOpen

Supporting open minds since 2005

Engineered Wood Products for Construction
http://dx.doi.org/10.5772/intechopen.92960
Edited by Meng Gong

Contributors
Maryam Shirmohammadi, William Leggate, Wan Mohd Nazri Wan Abdul Rahman, Nur Sakinah Mohamed Tamat, Nor Yuziah Mohd Yunus, Jamaludin Kasim, Mario Ramos-Maldonado, Cristhian Aguilera-Carrasco, Kaimeng Xu, Guanben Du, Siqun Wang, Anita Ogrin, Tomaž Hozjan, Zhiqiang Wang, Tianxiao Yin, Hongmei Gu, Prakash Nepal, Matthew Arvanitis, Delton Alderman, Jing Li, Jingxiao Li, Shamsul Haq, Hüseyin Emre Ilgın, Markku Karjalainen, Lassi Tulonen, Zizhen Gao, Meng Gong, Elena Vladimirova, Olli-Paavo Koponen, Kelly Bossardi Bossardi Dias, Ricardo Marques Barreiros, Halil Turgut Sahin, Yasemin Simsek, Ranjana Yadav, Jitendra Kumar, Anu Soikkeli

First published in London, United Kingdom, 2022 by IntechOpen
IntechOpen is the global imprint of INTECHOPEN LIMITED, registered in England and Wales, registration number: 11086078, 5 Princes Gate Court, London, SW7 2QJ, United Kingdom
Printed in Croatia

British Library Cataloguing-in-Publication Data
A catalogue record for this book is available from the British Library

Additional hard and PDF copies can be obtained from orders@intechopen.com

Engineered Wood Products for Construction
Edited by Meng Gong
p. cm.
Print ISBN 978-1-83962-771-2
Online ISBN 978-1-83962-772-9
eBook (PDF) ISBN 978-1-83962-790-3

For EU product safety concerns:
IN TECH d.o.o., Prolaz Marije Krucifikse Kozulić 3, 51000 Rijeka, Croatia, info@intechopen.com or visit our website at intechopen.com.

Meet the editor

Meng Gong is a leading researcher and the director of the Wood Science and Technology Centre, University of New Brunswick (UNB), Canada. He recently completed a five-year project on increasing the use of wood for structural and non-structural applications as the New Brunswick Innovative Research Chair in Advanced Wood and Construction, Canada. He received his master's degree in Wood Technology and Ph.D. in Wood Engineering from Nanjing Forestry University, China, and UNB, respectively. He worked as a post-doctoral fellow in the Wood Research Institute, Kyoto University, Japan. He has published more than 120 refereed journal and conference papers, and more than 50 technical reports. Currently, he is conducting research and development on mass timber products using underutilized wood species.

Contents

Preface

I spent my simple and happy childhood in Wuxi, a beautiful city by Lake Tai in China. I admired my father for being a wood science professor, and for his ability to make beautiful wooden furniture. Very often, I helped hold a wood board when he drilled a hole in it and passed nails to him. I found that wood is such an easy material to play with. My father shared many stories with me about wood when he discovered my curiosities and interest in the matter. Ginkgo is one of the most suitable wood species for use as cutting boards. Ash can be readily bent after being soaked in hot water, good for making curved arms for chairs. Chinese fir is a smart choice for building a house. My father always encouraged me to explore the science behind these stories. In the summer before I started my university life, I assisted my father to use bamboo fibers and phenol resin to make the shuttles for knitting cloth, which required a super impact strength and minimal friction resistance. During my undergraduate studies, I majored in wood-based composites, understanding that wood can be modified to produce various types of products to meet the needs of everyday life. Back in the 1990s, and likely nowadays as well, wood and wood products were mainly used for making furniture and decorating interiors. With an aim to broaden my horizons, I went to Canada to pursue my doctoral degree in 1995.

Canada is a forest-rich country that utilizes lumber to construct more than 90% of its residential houses and buildings. With the change in raw materials, many engineered wood products (EWPs) such as parallel strand lumber, laminated veneer lumber, and cross-laminated timber have been invented as a substitution for conventional lumber products. These EWPs are fabricated with wood elements (such as strands, veneer, and lumber) and adhesives and/or fasteners (such as nails and dowels) for use in structural applications. Use of adhesives, dry wood elements, heat, and pressure in the manufacturing of EWPs provides them with two great advantages over sawn timber: stable dimensions and uniform properties. Moreover, the designability of EWPs makes it much easier for these products to meet market requirements than it is for solid wood products to do so. These merits of EWPs have been attracting architects, engineers, and builders to adopt EWPs in their design and construction of buildings. The current National Building Code of Canada allows for the construction of wood buildings up to 12 stories. Canadian research scientists and code developers have been working hard towards a very ambiguous but feasible and promising goal, which is to allow people to select EWPs in the same way as they choose steel or concrete in the construction of tall buildings in 2025.

Unusual weather, such as flooding and snowstorms, has been occurring worldwide for the last few decades. It is widely recognized that forests and wood products play a critical role in the reduction of greenhouse gas emissions. Forests can sequester carbon and wood products can continue to store carbon over their lifespan, particularly structural wood building products that can last 50 to 100 years. This is a benefit of using wood rather than fossil fuel-intensive materials such as steel and concrete. Furthermore, we can recycle and reuse wood products at the end of their lifespan, which can help store carbon for an extended time. Therefore, increasing the use of forest wood can help mitigate climate change.

This book is a product 'fabricated' by authors from Australia, Brazil, Canada, Chile, China, Finland, India, Malaysia, Slovenia, Turkey, and the United States. It discusses EWPs, their uses in construction, and their contributions to reducing the construction industry's carbon footprint. The book is divided into two sections that address general overviews and applications of EWPs and recent research and development of EWPs.

The first section contains eight chapters discussing EWPs in terms of definition, classification, design principle, basic manufacturing processes, and mechanical properties. They also examine the perspectives of using EWPs from architects and builders, design considerations, and life cycle assessment on carbon.

The second section includes twelve chapters that address the development, improvement, and prospect of EWPs from material science to structural performance, including element size effect, adhesive bonding evaluation, preservative treatment, Industry 4.0 manufacturing, dovetail joint technique, architectural modelling, and protection from lightning.

This book provides a general picture of the state of art of EWPs in construction, making it a valuable reference for manufacturers, engineers, architects, builders, researchers, and students in the field of construction. I am grateful to all the chapter authors for their contributions. I also appreciate the support provided by IntechOpen and its author service managers and editorial and publication staff throughout all stages of production.

Wood is a gift from nature. The use of wood and EWPs helps reduce adverse environmental effects and thus improves human wellbeing. My love for wood was triggered by my father and encouraged by my mother. There will never be enough words to express my gratitude towards my parents. I sincerely wish they lead a healthy life and enjoy their retirement. I also express my thanks to my wife and my son for their understanding and support.

Meng Gong
Wood Science and Technology Centre,
University of New Brunswick,
Fredericton, Canada

Section 1

General Overviews and Applications of EWPs

Chapter 1

Wood and Engineered Wood Products: Stress and Deformation

Meng Gong

Abstract

Wood, as a natural, sustainable, and renewable bio-composite material, has a long history of serving humanity as construction materials. With the advance in technologies, many modern engineered wood products (EWPs) have been invented, produced, and used in construction, such as laminated veneer lumber, oriented strand board, and cross laminated timber. This chapter first introduces the classification, rationales, and pros and cons of EWPs. Secondly, it discusses the stress-related topics, including growth stresses in living trees, the evolution of wood strength from the molecular level to the actual design implementation. Thirdly, this chapter discusses moisture-induced deformation with examples. Finally, it mentions the benefits of using EWPs and their market shares.

Keywords: wood, engineered wood products, classification, growth stresses, strength evolution, design principle, moisture-induced deformation, market shares

1. Introduction

Wood is a gift from nature. Wood is a sustainable and renewable bio-composite material, which has a long history of serving human beings in the form of fuel, construction materials, furniture, paper, sports equipment, musical instruments, and transportation components. Wood is "manufactured" in a living tree with aims to grow it by transporting water and minerals and providing strength and rigidity to anchor it to the ground. A tree is optimally "designed" to resist the loads created by gravity, wind, snow, and others, rather than produce lumber and boards. The anatomical structure of wood is adapted to generate maximum strength in the stressed directions; yet in other directions, the strength is quite low [1]. This results in the anisotropic nature of wood, i.e., the properties of wood in a given direction are different from those in another. In addition, wood, as a biomaterial, maintains its fairly high variability of anatomical structures and physical properties; therefore, it requires a very large sampling size in research practice.

The dimensions of solid wood are entirely dependent on the dimensions of trees. The largest tree in the world is, in terms of the overall volume of its trunk, reported to be the Giant Sequoia (*Sequoiadendron giganteum*), which takes about 2300–2700 years to form its current dimensions, roughly 84 m in height and 11 m in diameter at the base [2]. On the other hand, the free span of composite glue laminated timber arch beams used in the Richmond Olympic Oval, Vancouver, Canada reaches 100 m with a depth of 1.6 m [3]. Undoubtedly, natural wood fails to meet the requirements for constructing modern timber structures, suggesting a need for "man-made" wood products. Furthermore, to address climate change and protect the earth's environment, logging in old-growth forests has been, in the last several decades, restricted or almost banned in most of the countries in the world. Consequently, available trees are largely from faster-growth plantations, which usually produce small-diameter logs of low-density wood and large-percentage juvenile wood. Traditional large-diameter solid timber, which was often used for long-span wood buildings in the past, has been phasing out. However, there is an increasing demand for using wood, as a green building material, to construct large and tall residential, commercial, and industrial buildings. In order to address the foregoing challenges of sourcing raw wood materials and catering to the market demands, an ever-growing number of value-added wood-based commodities and building materials have been created through the advances of technology. Contemporarily, modern engineered wood products (EWPs) have been widely used in construction [4].

An EWP is a product fabricated with wood materials and adhesives and/or fasteners (such as nails) targeted mainly for structural applications. An EWP has gone through an engineering design, which is often inspired by nature, and innovative technology. With the great efforts made by scientists and engineers in the last century, EWPs have grown into an extended family. **Figure 1** illustrates commonly used EWPs and their respective abbreviations.

EWPs offer many advantages over traditional solid timber products [5, 6]: (1) EWPs can reach a size that is not confined by the tree dimension. Theoretically, EWPs are only limited in width and length under transportation considerations; (2) EWPs accommodate a wide spectrum of species and sizes of trees, allowing more efficient

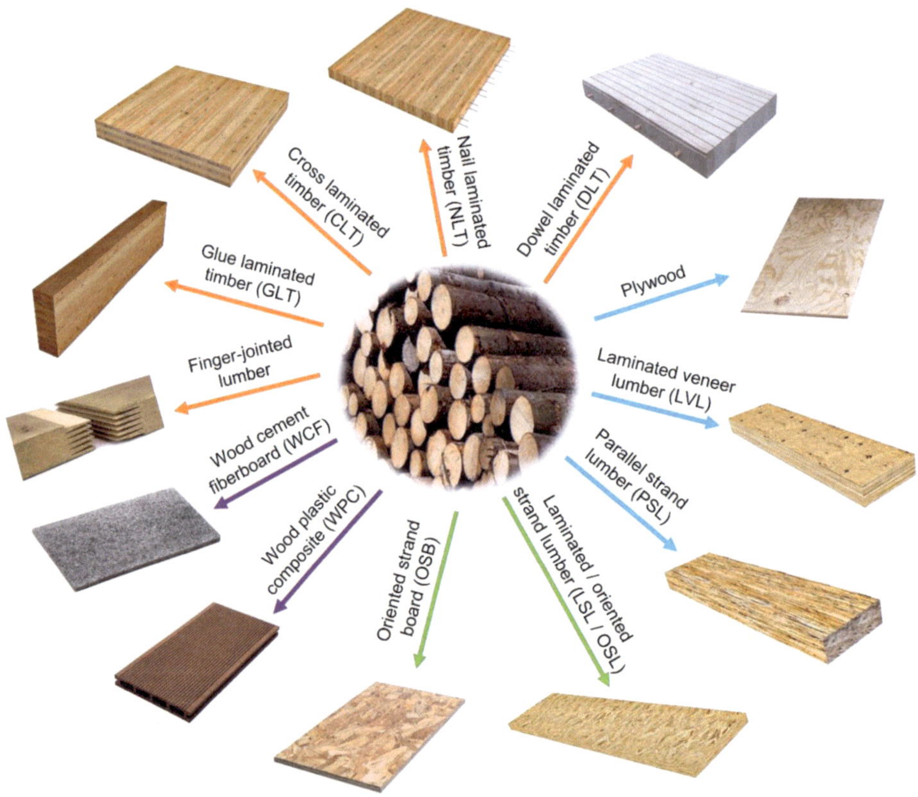

Figure 1.
Major types of engineered wood products and their abbreviations (source: images obtained from archiproducts.com, canac.ca, diy.com, globalsources.com, leben.co.in, nrcan.gc.ca, and structurecraft.com).

utilization of raw wood materials in the form of fibers, strands, veneers, and lumber; (3) EWPs have more uniform and reliable properties than solid wood, since the strength-reducing defects present in solid wood can be removed to a large degree or placed in a less critical zone(s) in the products; (4) EWPs exhibit greater dimensional stability and tolerances than sawn timber, due to the use of adhesives, dry wood elements, heat and pressure during their manufacturing processes; and (5) EWPs can make themselves much easier to adapt to market requirements than solid wood due to their designability. **Figure 2** illustrates the yields of raw wood material usage from logs to various EWPs. It can be found that laminated strand lumber (LSL) and oriented strand board (OSB) have a higher yield (larger than 75%) than plywood, laminated veneer lumber (LVL), and parallel strand lumber (PSL) (less than 65%). It can also be reasonably estimated that finger-jointed lumber, glue laminated timber (GLT), cross laminated timber (CLT), nail laminated timber (NLT), or dowel laminated timber (DLT) has a yield being less than that of sawn lumber due to the loss of wood materials during their manufacturing. A higher raw wood material yield means less waste and lower production cost, suggesting that EWPs are a great solution to the utilization of wood resources.

However, there are some disadvantages associated with EWPs [5, 6], one of which is that the process of manufacturing of an EWP requires more variables to manipulate than that of sawn lumber. Thus, highly automated equipment and technologically intense processes are essential in the production of an EWP, which significantly increases the capital cost of establishing an EWP mill. Therefore, the production of an EWP with the existing technologies is very costly compared to that of sawn lumber. Another shortcoming of most EWPs is that the use of adhesives in those glue-bonded EWPs causes a negative impact on the ecological image of wood as a natural biomaterial [5].

From **Figure 1**, it can be easily distinguished that there exist two groups, i.e., beam-like and panel-like EWPs. The beam-like EWPs is a group of relatively large length and depth compared to width, which is commonly used for beams and columns. The beam-like EWPs include finger-jointed lumber, GLT, LVL, PSL, LSL, and oriented strand lumber (OSL). The panel-like EWPs are of relatively small thickness and large width and length, which are usually used for floors, walls, and roofs. The panel-like EWPs can be further classified into two sub-groups, in terms of thickness: (1) thick-panel-like EWPs, containing CLT, NLT, and DLT, and (2) thin-panel-like EWPs, consisting of plywood and OSB. However, there is not a widely accepted criterion for sorting EWPs according to their dimensions and shape. For example, GLT can be used flat as panels for decking like NLT. For another

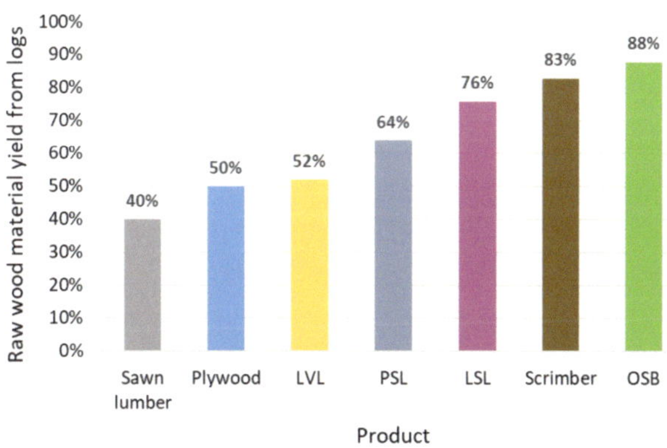

Figure 2.
Raw wood material yields of sawn lumber and EWPs from logs (data from [7, 8]).

example, LSL can be manufactured in the form of panels at the beginning and then ripped along the panel length direction to make beam-like or lumber-like products. In addition to those EWPs listed in **Figure 1**, there are many other types of EWPs such as I-joist (or I-beam), timber concrete composite (TCC), and fiberglass-reinforced GLT. It is noteworthy that **Figure 1** also lists two types of fiber-based EWPs, which are made of wood fibers as "fibers" and plastic or cement as "matrix". These two fiber-based EWPs are not commonly used for structural components in wood buildings nowadays, but they can be used for ceilings and decking. Wood plastic composite (WPC) has found its applications in the automotive, marine, and construction industries. Wood cement fiberboard (WCF) has also been steadily invading the housing and construction market.

With advancing wood-based nanotechnology, nanomaterials, such as nanocrystalline cellulose (NCC), have been derived from woody biomass and other cellulose sources such as straws. NCC is the celluloses in their crystalline form, which can be extracted by removing the amorphous sections from the celluloses and processed into solid flake, liquid, and gel forms. NCC can be employed, for instance, to reinforce the adhesive bond in EWPs. There is another innovative EWP called scrimber that is made of "scrims", a kind of interconnected loose webs. The steps of manufacturing scrims include crushing small diameter logs into webs, drying the webs, applying an adhesive into the webs, cutting the webs to the required length and width, laying up the webs into a mat, and pressing the mat into a billet using a radio frequency heating press [5]. However, scrimber has not been well commercialized because of the application of large quantity of adhesive (causing the high cost of production), and damaged wood generated while preparing scrims (reducing the strength of wood).

EWPs (usually excluding fiber-based ones) can be also classified into parallel and cross-laminated groups. The parallel-laminated EWPs include LSL, OSL, LVL, PSL, GLT, NLT, and DLT; meanwhile, cross-laminated EWPs contain OSB, plywood, and CLT. The parallel-laminated EWPs are usually used for load-carrying members such as beams and headers; while the cross-laminated EWPs are used for floor plates and sheathing sheets. The way of lamination inspires a philosophy of designing EWPs, which will be briefly outlined in Section 2 "Stress" of this chapter.

Another way of grouping EWPs is rooted in the wood elements that make them, which include fiber, strand, veneer, and lumber-based EWPs, shown in **Figure 3**. It

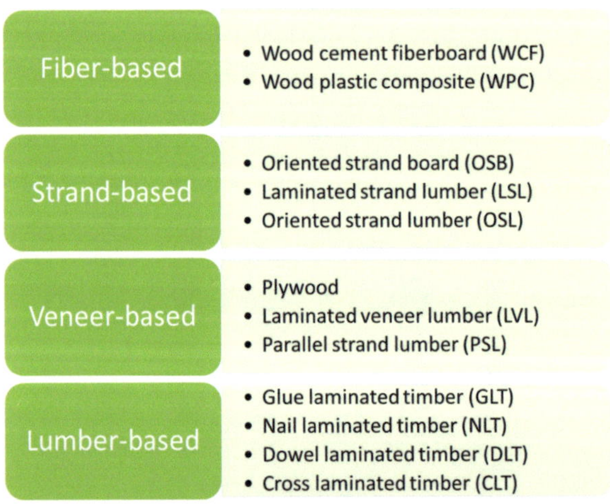

Figure 3.
Classification of EWPs in terms of the wood elements used for making them.

should be noted that PSL is made of long veneer strands (also called veneer strips). Thus, PSL is classified into the group of veneer-based EWPs rather than the group of strand-based EWPs albeit it has "strand" in its name. The wood elements largely govern, in terms of their dimensions, shape, moisture content (MC), species, the physical and mechanical properties of the product made from them. For instance, the density of an LVL made of yellow poplar is just slightly greater than that of yellow poplar veneer, which is attributed to the use of an adhesive(s) and slightly densified veneer during manufacturing. For another example, the equilibrium moisture content (*EMC*) of plywood is usually lower than the wood from which it is made due to the use of adhesive, dried veneer, and heat and pressure in the course of manufacturing. However, the *EMC* of GLT is very similar to that of the lumber used for making it, since only a relatively small amount of adhesive is applied in comparison to the volume of GLT itself, and the room temperature is applied during its manufacturing process.

In the course of discussing EWPs, it is worth introducing another two terms: structural composite lumber (SCL) and mass timber products (MTPs). SCL refers to those products that combine dried strands, veneer, or other small wood elements bonded with an exterior structural adhesive(s) to form thick-panel-like or beam-like EWPs [6, 9]. SCL basically includes LVL, LSL, OSL, and PSL. One outstanding characteristic of SCL is that the grain of the wood elements used is essentially aligned parallel to the length direction of its products with an aim to maximize its structural properties in this direction. Thus, SCL products are broadly used for beams and columns in wood buildings. MTPs connote a family of EWPs of a large section size, which can be employed to make strong but light load-bearing components for structural applications such as floors and walls [9]. MTPs basically include lumber-based EWPs, i.e., CLT, GLT, NLT, and DLT. However, MTPs also include SCL, TCC, and other large-size EWPs such as fiberglass-reinforced GLT, see **Figure 4**. Among EWPs, the lumber-based MTPs also possess other unique features, such as wood-look appearance, environmental friendliness, and low carbon emission. MTPs have been, since the mid-1990s, attracting architects, engineers, and builders to employ them in their design and construction of tall and large buildings with an aim to compete with or even substitute steel and concrete.

The philosophy of designing a timber structure/component/connection is largely rooted in two aspects: safety, limiting the maximum load-carrying capacity (i.e., strength) and serviceability, restricting excessive deflection (i.e., deformation) [10, 11]. The reaction of a material (such as wood) or a product (such as EWP) to the action of external forces is indicated by its mechanical properties, or

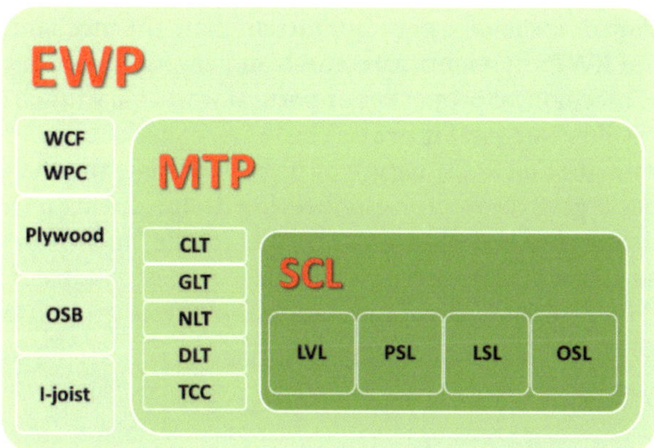

Figure 4.
Classification of commonly used EWPs.

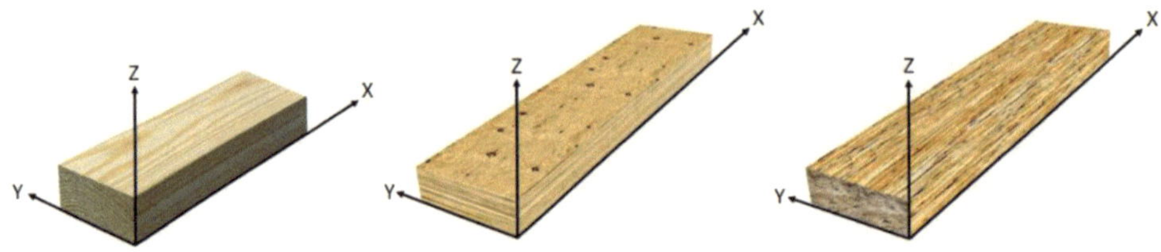

Figure 5.
Cartesian coordinate system for wood/lumber (left), LVL (middle) or PSL (right).

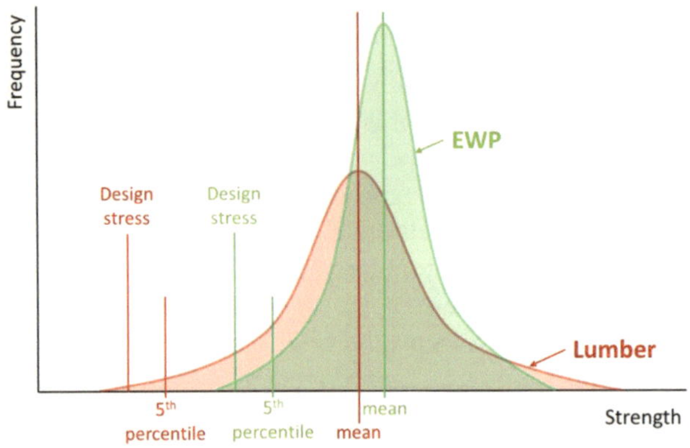

Figure 6.
Probability density functions for strength of lumber and EWP.

otherwise known as engineering properties, including tensile strength, compressive strength, shear strength, bending capacity, ductility, and creep. When an external load is applied on a wood component/connection/structure, it causes the stresses inside the wood component/connection/structure, generating deformations and eventually leading to failure. Failure can be, from an engineering point of view, defined as a fracture, when stress exceeds the strength of the wood component/connection/structure or failure as deformation exceeds the design value.

Under certain assumptions, such as ignorance of natural growth characteristics (e.g., knots and eccentricity) and growth ring curvature, a three-dimensional Cartesian coordinate system can be introduced in **Figure 5**, representing the wood/lumber longitudinal or EWP major direction (X), wood/lumber tangential or EWP minor direction (Y), and wood/lumber radial or EWP thickness direction (Z). On this basis, an orthotropic model can be built to simulate the mechanical behavior of wood, lumber, and EWPs in scientific research and engineering design. As mentioned above, the mechanical properties of natural wood vary much more than those of EWPs, which is illustrated in **Figure 6**. The lumber product has a relatively wide distribution of strength due to the nature of its biomaterial, and the EWP has a fairly narrow range since it goes through an engineering design during its manufacturing. Overall, the EWP has greater design stress, 5th percentile, and mean values than sawn lumber. The following two sections will discuss these two basic mechanical terms (stress and deformation) from the standpoint of tree growth and wood uses.

2. Stress

The strength of wood is a measure of its resistance to failure. If the stress applied to the wood exceeds the strength of the wood, will break. Stress accompanies wood during its formation in a living tree and its services over its life span. As mentioned, a tree is subjected to various types of stresses during its growth. Mattheck and Kubler depicted external loading and internal stresses distributed in a tree [1], **Figure 7**. By neglecting the weights of the stem and branches, and loads generated by wind and snow, they simplified their model by only considering the weights of the upper and lateral crowns (F_1 and F_2). The compressive stresses (σ_1) produced by F_1 act on the area (A_1) of the stem above the lateral branch, equaling to F_1/A_1. The bending moment generated by F_2 is applied on the lateral branches, which increases linearly towards the stem. Thus, the stem below the branch joint bears both bending moment (tensile and compressive stresses) and axial compressive stress (σ_1).

These stresses refer to the mechanical stresses permanently supported by wood in a living tree during its growth, which are called tree growth stresses. The tree growth stresses result from the combined effects of the increase of dead weight and maturation of cell walls [12]. There is an interesting phenomenon that can be viewed in a leaning stem when it is subjected to a bending moment. In this situation, the stem tries to resume its original, usually upright, position, thus, it needs to counteract the bending moment. In such a stem, the growth stresses often differ on its two opposite sides, resulting in abnormally wide growth rings appearing in the upper or lower side of a leaning stem, **Figure 8**. The wood of such abnormally wide growth rings is called reaction wood. Reaction wood in hardwoods or softwoods is named tension wood or compression wood, respectively. Understanding compression wood is of great importance since softwood lumber is commonly used in construction of wood buildings. Compression wood has a relatively large longitudinal

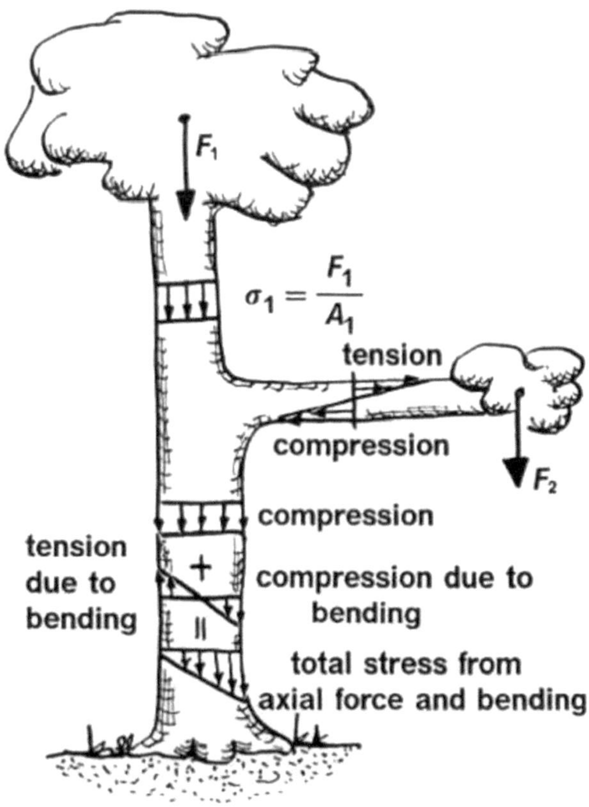

Figure 7.
Combined axial stresses and bending stresses in a tree generated by its crown weights [1].

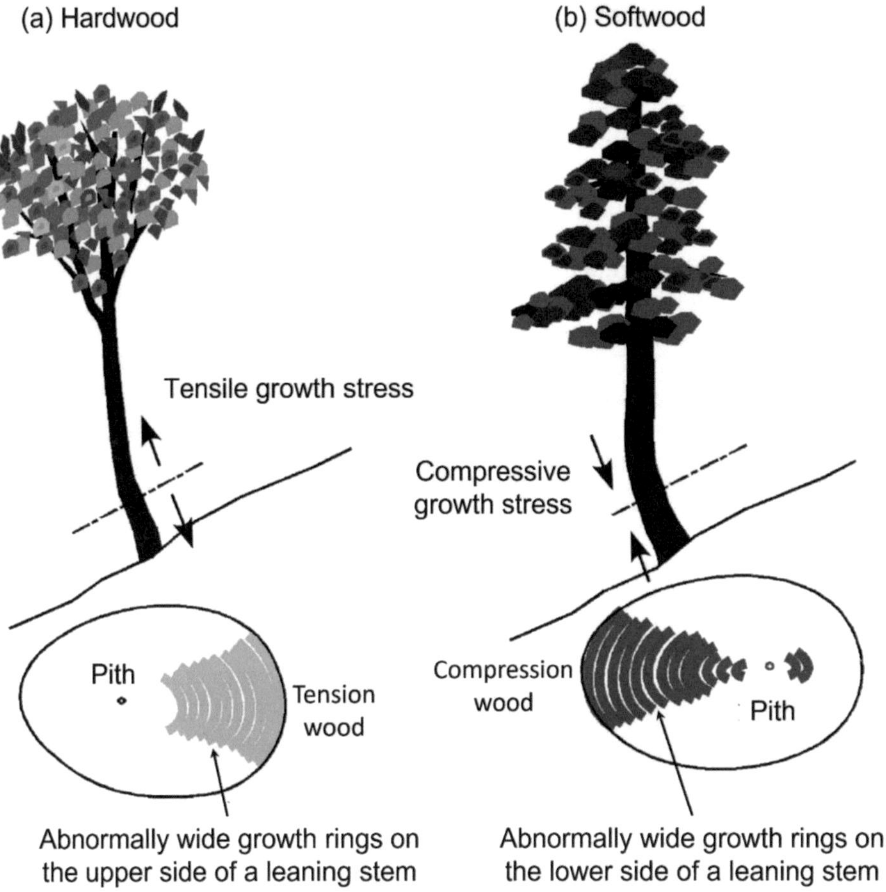

Figure 8.
Reaction wood in a leaning stem: (a) tension wood in hardwood, and (b) compression wood in softwood (modified from [12]).

shrinkage compared to normal wood, which can reach 1–2% [13], around 10 times as large as that of normal wood. As a result, warping and even cracks, often emerge in softwood lumber. In comparison to normal wood, compression wood has a relatively high density but similar strength, resulting in a low strength-to-weight ratio [13].

Wood is a complicated hollow structure consisting of substances and voids. The substance is the basic building materials constructing cell walls made of an ordered association of cellulose, hemicellulose, and lignin, with an average density of 1.5 g/cm^3 under oven-dry conditions [13]. The voids in wood appear in the form of cell lumens, pit openings, pit cavities, and intercellular spaces. The density of wood is largely governed by these voids, i.e., if a wood species has a larger volume of voids, its wood has a lower value of density. **Figure 9** illustrates the tensile strength values of wood at various levels. At the molecular level, the strength of wood is extremely high in the longitudinal direction, with estimates exceeding 7000 MPa [14]. Wood strength is about 15 times larger than that of structural steel, which has a strength of 400–550 MPa. Bundles of cellulose molecules form so-called microfibrils, the basic cell wall elements constituting cell walls in association with hemicellulose and lignin. From the composite theory point of view, wood is a natural composite, i.e., nature's fiberglass, in which celluloses are the "fibers" and hemicellulose and lignin are the "matrix". Microfibrils have a strength of about 480 MPa and individual cells are estimated to have a strength of about 140 MPa [14]. The tensile strength of clear

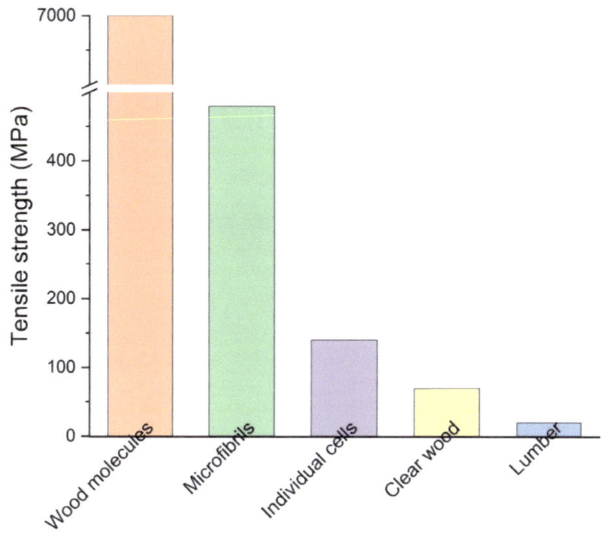

Figure 9.
Tensile strength of wood at various levels.

softwoods in the longitudinal direction has values ranging from 40 to 200 MPa [5]. However, lumber has, in the same longitudinal direction, a tensile strength of only 15–40 MPa [5] due to the existence of many strength-reducing characteristics such as knots, the slope of grain, checks, and splits. In the derivation of the design value of lumber, its characteristic strength properties, such as characteristic tensile strength in the longitudinal direction, are determined using the lower 5th percentile value of the Weibull distribution, which is much lower than the mean. From here, an allowable property value can be calculated by dividing the 5th percentile with a property reduction factor (n), or a so-called adjustment factor. For example, allowable tensile strength is equal to the 5th percentile divided by n, where n = 2.1 [15]. Finally, design value can be derived by multiplying the allowable property value with modification factors such as load duration, service condition, size, treatment, system, and other factors. For instance, the tensile strength of wood in the longitudinal direction can be as low as 2.5 MPa in the design of a structural component, which is fully attributed to the biomaterial nature of wood and its service conditions.

The above discussion suggests that human beings have not fully utilized the strength of wood. Contributed by the recent technological progress, the optimized use of wood has been improved to a certain degree in terms of strength. Song et al. selected three hardwood species (basswood, oak, and poplar) and two softwood species (western red cedar and eastern white pine) as test materials, and took two steps to transform bulk natural wood directly into super strong and tough densified wood [16], **Figure 10**. Step 1 used a chemical treatment to partially remove lignin/hemicelluloses and Step 2 mechanically hot-pressed the chemically treated wood at 100°C to reduce its thickness by about 80%. They discovered that the tensile strength of densified wood reached about 550 MPa, which was 12 times as large as that of natural wood. This value is higher than that of microfibrils (about 480 MPa). They indicated that most of the densified wood consisted of well-aligned cellulose nanofibers, greatly enhanced hydrogen bond formation among neighboring nanofibers. Their research provides a promising method of maximizing the use of wood strength by removing most voids and some lignin/hemicellulose. As mentioned in Section 1, NCC derived from wood attracts increasing attention due to the non-renewability of petroleum and the global promotion of green materials and products. NCC made from bleached softwood kraft pulp exhibits a diameter of 10 nm and a length of 150 nm [17].

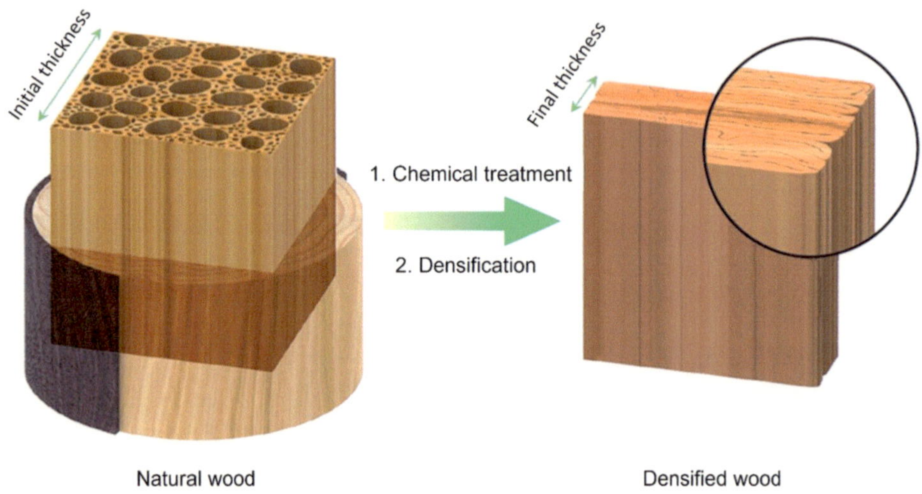

Figure 10.
Two-step processing approach for making densified wood (modified from [16]).

The Young's modulus of NCC can reach about 137 GPa [18]; while that of Douglas fir clear wood, one of the strongest softwoods, reaches only about 14.5 GPa [19]. This is one of the key reasons for applying NCC as high-performance "fibers" to "matrix" to produce composite materials of increased stiffness and tensile strength.

EWPs are fabricated through an engineering process, the basic principle of which is to place stronger materials at the most stressed zones [5, 7, 9]. Each lamination, such as lumber and veneer, must be graded visually or mechanically. **Figure 11** illustrates two lumber-based EWPs, GLT, and CLT. GLT here is made of sawn lumber of various visual grades, notably select structural (SS, the highest grade), No. 1, No. 2, and No.3 (the lowest grade here). The GLT in **Figure 11** (left) is intended to be used as a beam subject to the bending of a simple span. Therefore, an unbalanced layup is designed with an aim to optimally and efficiently use lumber by locating SS-grade lumber on the bottom face and No. 3 lumber in the core layers. **Figure 11** (right) depicts the basic idea of designing CLT, in which lumber grades in its major strength axis are required to be at least 1200f-1.2E MSR or visually graded No. 2, where No. 3 is the minimum lumber grade required in the minor strength axis [20]. This design gives CLT two-way action capacities, suitable for floor uses.

Figure 12 further justifies the principles of engineering design in wood in terms of two basic veneer-based EWPs, namely plywood and LVL. Veneer can be, according to the size and number of defects (such as knots), sorted into three grades (high, mid, and low). In the construction of plywood and LVL, the best quality

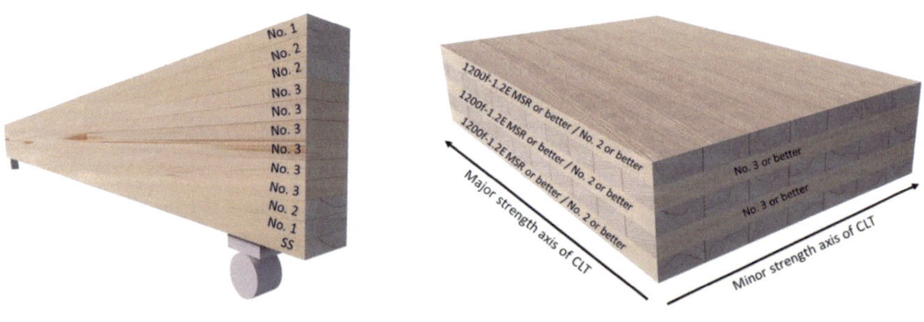

Figure 11.
Design of GLT (left) and CLT (right) [9].

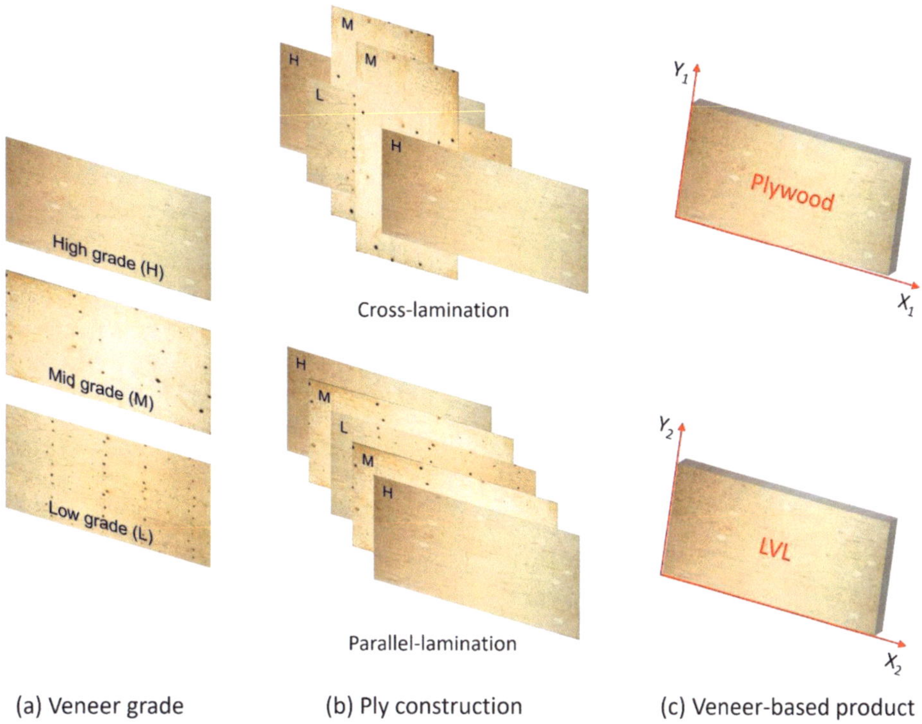

Figure 12.
Design of veneer-based products.

veneer is used as top and bottom laminations; however, the laminating approach differs. The cross-lamination of veneer enables plywood to be relatively strong in both the length (X1) and width (Y1) directions when loaded on its face, enabling it be widely used as sheathing materials. In LVL, the grain of each layer of veneer runs in the same direction, i.e., parallel-lamination. As a result, the strength in the length direction (X2) of LVL is much stronger than that in the width direction (Y2), making it a more suitable material for beam and column uses.

Assuming the same wood species, veneer quality, adhesive type, and other manufacturing parameters (except the laminating approach) are used in the fabrication of plywood and LVL, it can be reasonably predicted from **Figure 12** that the modulus of elasticity (MOE) and modulus of rupture (MOR) are the largest in the direction X_2, the second largest in X_1, the second smallest in Y_1, and the smallest in Y_2. This is verified by a group of undergraduate students at the University of New Brunswick, Canada, who made 5-layer poplar plywood and LVL panels in their laboratory using a phenol-formaldehyde adhesive at two levels of pressure. Each panel they made had a width of about 150 mm and a length of about 300 mm, from which small specimens were cut for bending tests. **Figure 13** summarizes their results indicating that there is a notable difference in MOE and MOR between plywood and LVL in either the major (X_1 or X_2) or minor (Y_1 or Y_2) direction. Plywood has fewer degrees of difference in MOR and MOE than LVL between the major and minor directions, suggesting a more uniform structure in plywood. However, LVL has a much higher MOR and MOE in the major direction (X_1) than that in the minor one (X_2), indicating its one-way strength capacities.

As discussed above, one of the advantages of EWPs over sawn lumber is that they can be engineering-designed. For instance, a cylindrical LVL was inspired by the hierarchical structure of a wood cell wall. A cell wall consists of three major layers in its secondary wall, and each layer has many lamellae containing microfibrils at

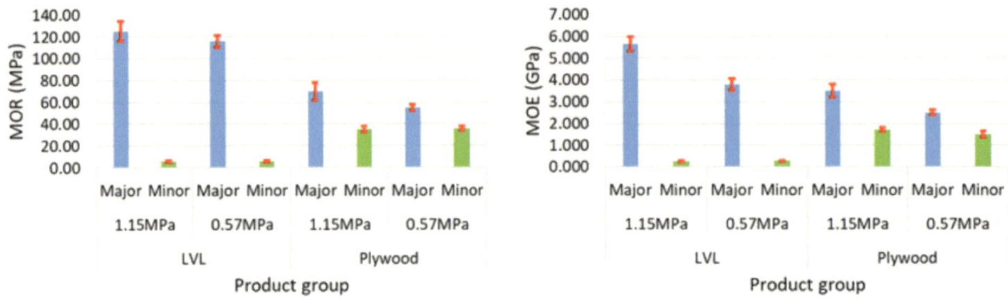

Figure 13.
The means and standard deviations of MOR and MOE of plywood and LVL.

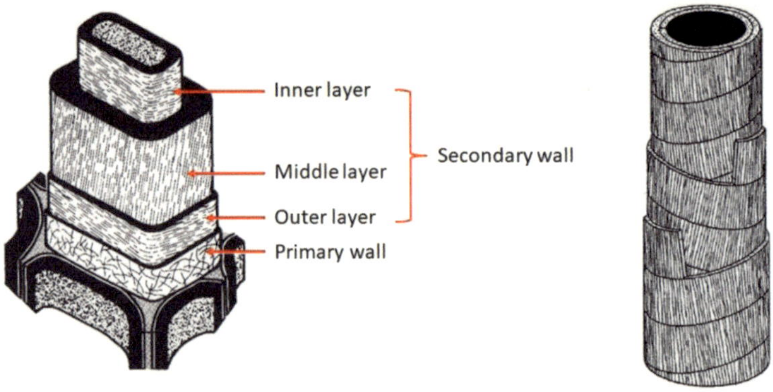

Figure 14.
The hierarchical structure of a mature cell wall (left) (adapted from [21]) and a cylindrical LVL (right) (adapted from [22]).

different angles [21], **Figure 14** (left). The middle layer in the secondary wall has 30–50 lamellae, occupying about 75% of the thickness of the cell wall. The microfibrillar angle in the middle layer ranges from 10 to 30 degrees relative to that of the longitudinal axis of the cell (almost parallel to the longitudinal direction of wood). Therefore, the middle layer governs the properties of the cell wall and furthermore the properties of wood, and provides considerably more strength and stiffness parallel to its axis than perpendicular. Yamauchi et al. used 2.5-mm-thick Japanese cedar veneer and a resorcinol resin to make cylindrical LVLs using the spiral-winding method, i.e., by laying neighboring veneer sheets at ±10 degrees to form an interlocked grain structure [22], **Figure 14** (right). The cylindrical LVL specimens they made were about 300 mm in the outer diameter, 25 mm in the wall thickness, and 3600 mm in the specimen length. They discovered that (1) the MOE of cylindrical LVLs was the same as that of solid lumber of the same species; and (2) as the number of veneer plies used in cylindrical LVLs increased from 6 to 10, its MOR increased; yet MOE remained almost unchanged. Yamauchi et al. concluded that the interlocked grain structure they applied could effectively prevent a decrease in MOE and indicated that cylindrical LVL was suitable for structural uses, especially for posts in construction [22].

3. Deformation

Wood and EWPs may undergo dimensional changes due to variation in ambient temperature and relative humidity, and stresses caused by external loads. The interaction of the surrounding atmosphere and loading conditions can create an enhanced level of deformation in wood. This section briefly explains how the change in moisture content of wood affects the shrinkage or swelling of wood and aims to increase the awareness of how the moisture-induced deformation impacts the structural performance of a timber building.

As a hygroscopic material, wood and its products can absorb and release moisture, resulting in dimensional changes. **Figure 15** illustrates the possible changes in moisture content (MC) of wood during the construction and use of wood buildings. The initial MC is the MC at the time of the manufacturing of a wood product, which is usually less than 19% for lumber and 4–15% for EWPs [10, 23]. The initial MC of strand-based or veneer-based EWPs, such as OSB and LVL, ranges from 4–6% [10]. However, the initial MC of lumber-based EWPs, such as GLT and CLT, varies from 11–15% [10]. The difference in initial MC between strand/veneer-based and lumber-based EWPs can be attributed to the dimension and shape of wood elements, amount and type of adhesive, and heat and pressure applied during manufacturing. The MC can significantly increase during construction if the wood components are not well protected from moisture/water, which can cause a large change in dimension and shape as well. Proper cautions and measures must be used to minimize such a large dimensional change. After a wood building is completed and occupied, its in-service MC may vary from 7–15% [23], depending on the surrounding temperature and relative humidity, before eventually reaching equilibrium moisture content (EMC). The EMC fluctuates between the low and high in-service MC, resulting in some dimensional changes in wood from time to time.

As discussed above, the wood components of a low *MC* at the time of delivery may get wet on a construction site, generating a swelling value that cannot be ignored. **Figure 16** illustrates a floor joist that is supported by a wood frame wall on the right end and by a masonry block on the left [10]. During the service of the floor joist, differential wood movements may occur between the two ends, **Figure 16** (upper). To ensure that the floor is at a horizontal level, the movements at the two ends must be the same. To address this, a wooden sill beam, just like the interior beam, can be added to the concrete wall, as shown in **Figure 16** (lower), which provides an equal amount of wood movement in the vertical direction. This example provides a hint to designers, i.e., it is important to identify, in the course of designing a wood building, the locations where potential differential wood movements could affect structural integrity and serviceability.

Wood is indeed an anisotropic material. Thus, its dimensional change varies from one direction to another. The dimensional change in the longitudinal direction is as low as 0.1–0.2% for mature wood [13]; thus, it can be ignored. However, the dimensional change in the transverse (i.e., radial, tangential or in-between) direction can reach as high as 12% or so [13]; therefore, it must be taken into account during the design of wood buildings. However, it is not practical and sometimes

Figure 15.
Variation of moisture content of wood during the construction and use of wood buildings [23].

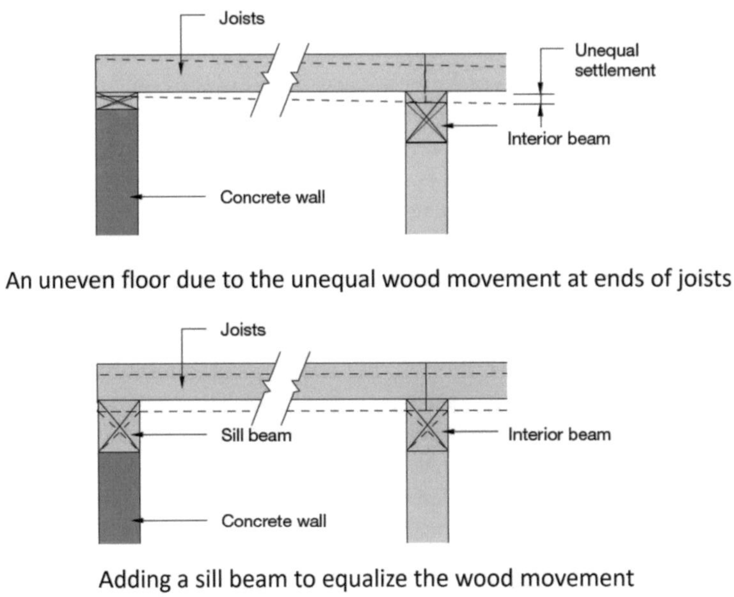

Figure 16.
Detailing to account for vertical movement due to the change in moisture content [10].

impossible to estimate the dimensional change in the transverse direction because the grain orientation in the cross-section of lumber cannot be predicted in the construction of a wood building. To resolve this problem, the Canadian standard CSA O86 "Engineering design in wood" specifies Eq. (1) to estimate the dimensional change of a member made of sawn lumber or lumber-based EWPs such as GLT and CLT [11]:

$$S = D \times \left(M_i - M_f\right) \times c \tag{1}$$

where, S is the dimensional change (mm) due to moisture; D is the actual dressed dimension (mm) (i.e., thickness, width, or length); M_i is the lesser of the initial moisture content of the fiber saturation point (28%); M_f is the final moisture content; and c is a coefficient. As for lumber, c = 0.002 or 0.00005 for the dimensional change perpendicular to the grain (i.e., the transverse direction) or parallel to grain (i.e., the longitudinal direction), respectively.

Scientists at FPInnovations, Canada, conducted a study by monitoring the vertical movement in a 4-storey wood-frame building over 22 months [24]. The floors consist of 38 mm by 240 mm "S-Dry" dimension lumber joists with a concrete topping. The walls consist of 38 mm by 140 mm "S-Dry" solid sawn plates and studs. Double top plates and double bottom plates are used in all storeys. The stud length of all storeys is 2.44 m. The joist spacing is 400 mm; joist spans are 3.75 m; and stud spacing is 400 mm. The scientists calculated the vertical wood movement including shrinkage from an initial *MC* of 19% to the final *MC* of 8%, using the above equation and deformation by assuming the specified roof and floor dead loads to be 0.5 kPa and 1.3 kPa, respectively. Deformation generated by stress includes instantaneous deformation and creep deformation. The equations for calculating these deformations are provided in the report by Doudak et al. [24]. **Figure 17** summarizes the estimated vertical movement values at each storey, indicating that the accumulated shrinkage over 4 stories accounts for about 90% of the total vertical movement. This suggests the vertical deformation generated due to the change in the moisture content of wood is critical and must be considered in the design of a wood building. The actual vertical movement measured in this 4-storey building

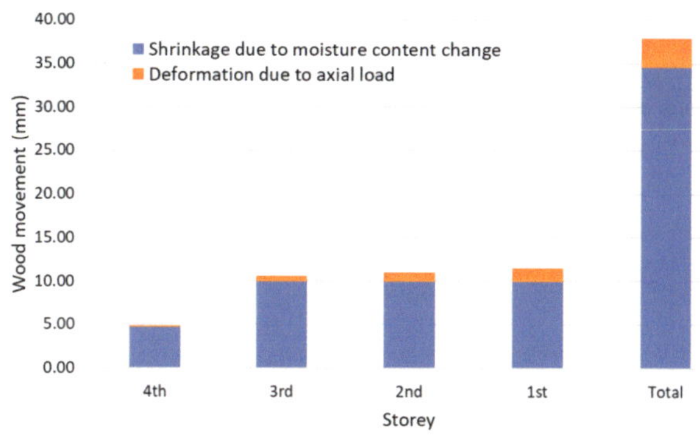

Figure 17.
Estimated vertical movement due to changes in MC and axial compressive load (source: data from [24]).

was about 40 mm, which is in good agreement with the estimated value (about 38 mm). They concluded that it was possible to make a good estimation of vertical movement to avoid the potential problems of structural integrity, serviceability, and building envelope over the lifespan of this wood building. The scientists also found that the use of EWPs could reduce the accumulated shrinkage to about 80% of the total vertical movement, based on their monitoring of another 5-storey building with floor joists

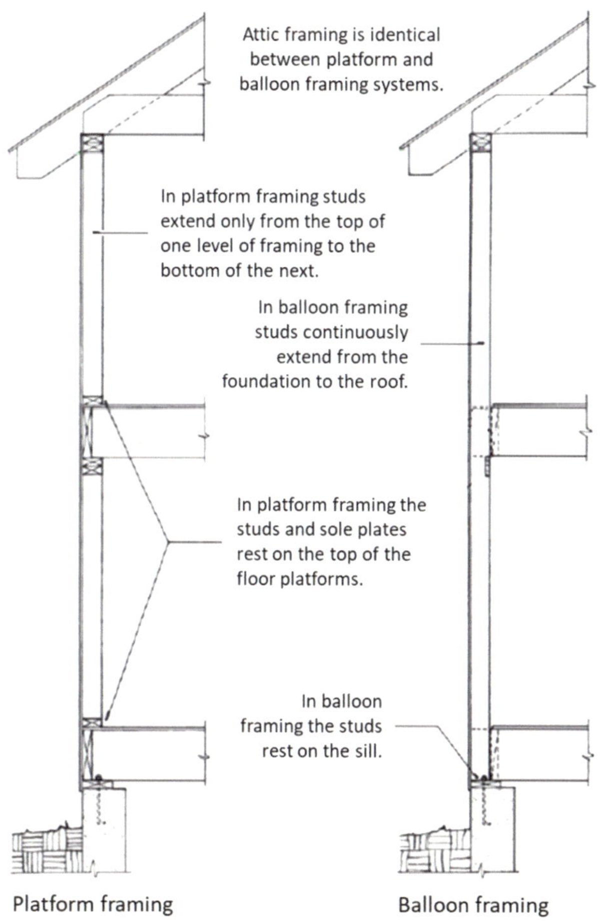

Figure 18.
Light frame construction methods: platform framing and balloon framing (modified from [25]).

made of LVL flanges and OSB webs. The total vertical movement of this 5-storey building was about 30 mm, which was 75% of that of the 4-storey building they studied.

There are two basic wood framing construction methods, notably platform framing and balloon framing [25], **Figure 18**. In the platform framing, all vertical structural elements of the exterior bearing walls and partitions consist of single studs extending the full height of the frame. Meanwhile, in the balloon framing, the studs of the exterior walls and some of the interior walls are continuous from the foundation sill plate to the top plate below the roof framing. In comparison to balloon framing, platform framing is more commonly used for modern structures due to its simplicity and ease of erection; but its vertical movement due to MC changes is much larger. Balloon framing is rarely used nowadays since the length of sawn lumber available is not sufficient due to changes in forest resources with the production of smaller trees compared to older times. However, with advancing technologies, many modern EWPs have been invented and manufactured, providing sufficiently sized materials with less moisture-induced wood movements. This gives an opportunity for people to rethink the use of the balloon framing method in building construction.

CLT panels can, attributed by their inherent two-way spanning capabilities, eliminate the necessity of placing beams underneath the panels, as with other MTPs. This facilitates the emergence and application of a post-and-panel mass timber construction system [9]. Therefore, this system significantly reduces the building height and construction time, as well as overall costs [26]. In this type of modern construction system, the CLT floor plates are point-supported by GLT and PSL columns. This may cause two potential issues, i.e., excessive vertical wood

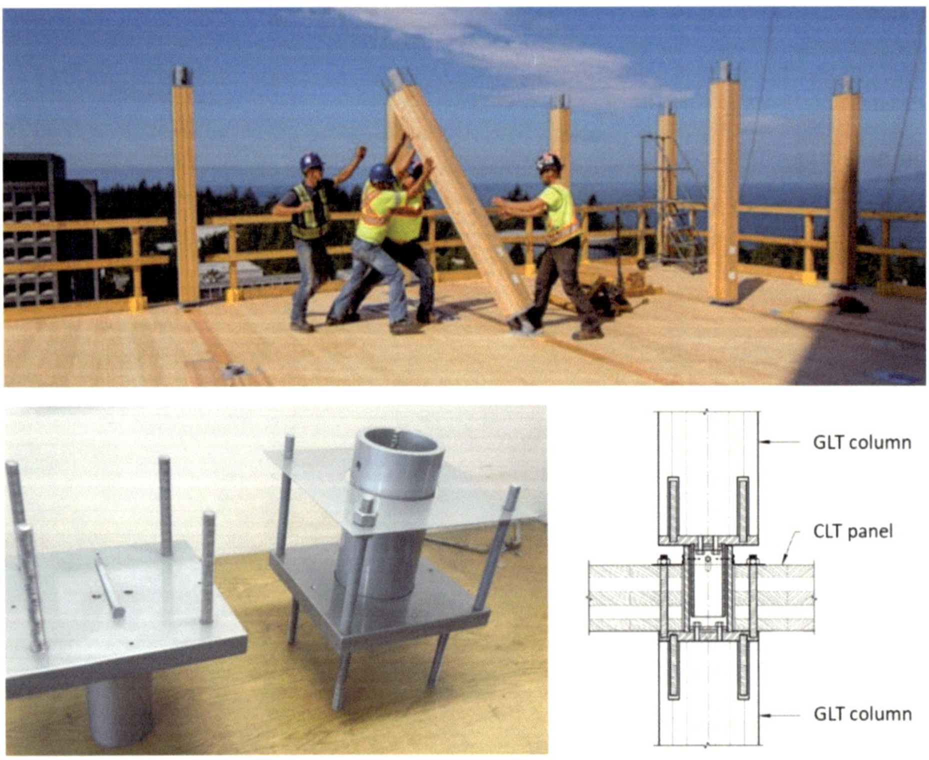

Figure 19.
Column-to-column connection used in the Brock Commons Tallwood House in the University of British Columbia, Vancouver, Canada (upper: on-site installation of columns connected to CLT panels (source: www.fastepp.com); lower: HSS steel connectors [26]).

movement because of MC change and crushing of CLT panels due to axial loads, if there are no proper connectors joining CLT plates and columns. These issues occur because the vertical direction of the building is the thickness direction of CLT (i.e., the transverse direction of lumber), along which larger accumulated wood movements and lower compressive strengths exist in CLT. To address these issues, a so-called HSS steel connector was developed and employed, **Figure 19** (lower), which can directly transfer load from upper columns to lower columns and provide some tolerance for dimensional change. The development of various types of metal connectors is of great importance in the design and construction of wood buildings with EWPs.

4. Endnotes

With climate change being an inevitable and urgent global challenge, it is essential to address such issues with real-life content, for instance, the global warming impacts caused by buildings and constructions. The World Green Building Council reports that building construction and operation account for 39% of greenhouse gas (GHG) emissions annually [27]. Among this 39%, 11% is from embodied carbon and 28% from operational carbon. The embodied carbon of a building is defined as the amount of carbon emitted during its construction. Whereas, operational carbon is defined as the amount of carbon emitted during the operation of a building. As innovative building technologies continue to develop, operational carbon will be significantly reduced, and embodied carbon can be responsible for almost 50% of total new construction emissions from now to 2050 [28]. In the next 40 years, with a doubled urban population, a building area of 2.48 trillion square feet is required to fit the needs of urban population growth [29]. The combination of considerable global CO_2 emissions from the building sector and the increasing demand on buildings reveals that actions should be taken immediately to mitigate emissions from the embodied carbon. One of the answers to this global challenge is to increase the use of wood and wood-based products in the construction sector. As a biomaterial, wood possesses its natural ability to mitigate carbon dioxide (CO_2). During the growth of trees, wood is produced, sequestering carbon. After trees are harvested, wood can be processed into various products. These products can be therefore used in construction of buildings, which store carbon over their lifespan. The recently released report "The state of mass timber in Canada 2021" from the Government of Canada [30] indicates "As high-value wood products, mass timber can play an instrumental role in the circular economy by providing a renewable source of building materials and contributing to a lower carbon footprint for the construction sector." In the last two decades, the development of mass timber was rapid in Canada, which can be viewed through the number of completed mass timber projects, **Figure 20**, with 10 projects in 2007 and upwards of 60 projects in 2018. It should be noted that each project listed in this figure must meet two criteria, i.e., a minimum floor area of 300 m^2 and structural use of MTPs.

Figure 21 illustrates the percentage share of each EWP in major markets over a time horizon [31]. The data in this figure are outdated, only depicting the status of EWPs in 2006, but it still provides some insight into the market shares of each EWP. The sheathing and industry plywood markets are in the stages of "decline" and "maturity", respectively. This is mainly due to the decreasing volume of veneer quality logs and the rapid development of OSB. Sheathing OSB is in "rapid growth", which has taken a big market share from plywood because of its high yield of using raw wood materials. Framing lumber takes 90% or more market shares since SCL is still much more expansive than sawn lumber, resulting in its failure to completely

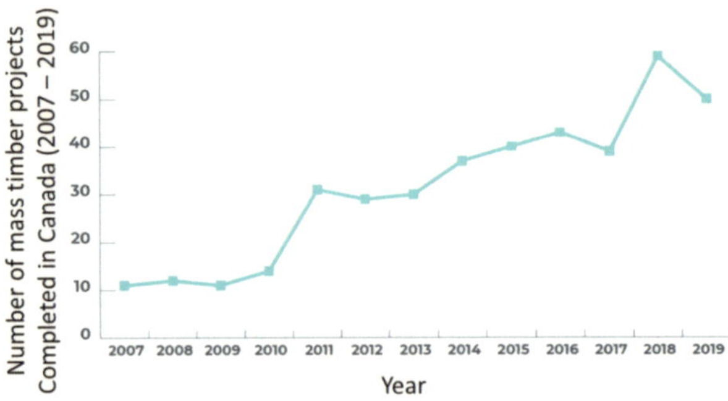

Figure 20.
The number of completed mass timber projects per year in Canada [30].

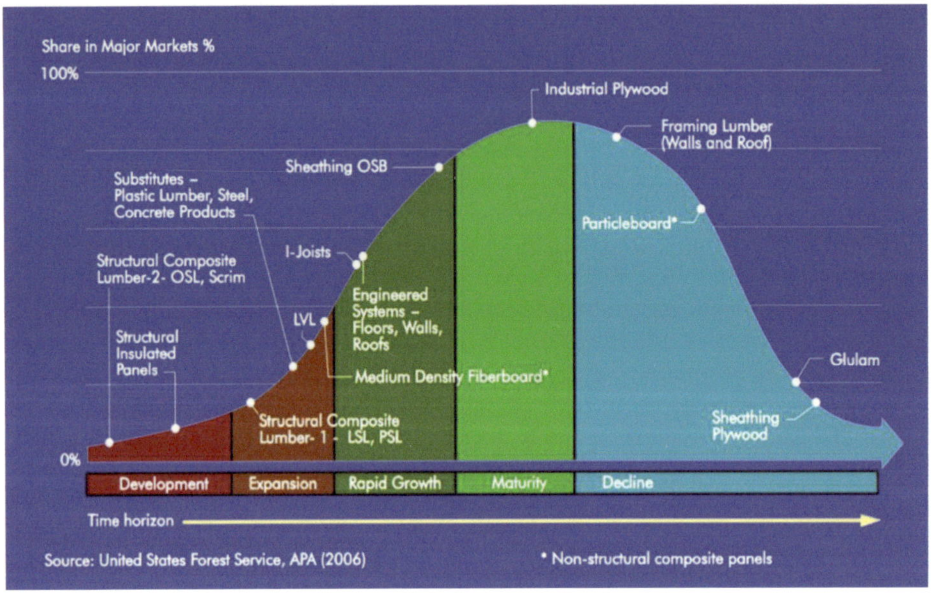

Figure 21.
Percentage share in major markets of EWPs over time horizon (source: photos obtained from [31]).

replace lumber [6]. Glulam, a commonly used name for GLT, has passed its plateau of market demand, moving into the "decline" stage. The main reason for this decline could be the emergence of other EWPs such as LVL, LSL, and PSL. Albeit LVL stays at the end of the "expansion", it could have reached its maximum market share. This implies that the cost of producing LVL will, just like manufacturing plywood, definitely increase, since premium logs for producing veneer are becoming scarce [6]. LSL and PSL are in "expansion" due to their high raw material yield from logs to the final products. **Figure 21** does not show the market shares of CLT, NLT, and DLT, but it can be reasonably speculated that these lumber-based EWPs are in the "development" and "expansion phase", and will enter "rapid growth" quickly, particularly for CLT. As technologies advance, the cost of manufacturing EWPs will be further reduced. For example, adoption of the artificial intelligence in production can lead to an increased yield of raw material usage from logs and reduced labor costs. Therefore, EWPs will be more competitive to sawn lumber and make inroads into more market shares.

It can be well foreseen that EWPs will have a bright future in construction because (1) EWPs are designable, producing an optimal structural performance

DOI: http://dx.doi.org/10.5772/intechopen.101199

for construction; (2) EWPs have more uniform strength properties and fewer changes in dimensions and shape, making them more suitable building materials; (3) EWPs provide a wide selection of dimensions, allowing designers and builders to design and build tall timber, wood-concrete, and wood-steel hybrid structures; and (4) EWPs fall into the category of environmentally friendly and recyclable products, contributing to a lower carbon footprint for the construction sector.

Acknowledgements

This piece of work was financially supported by the New Brunswick Innovation Foundation (Canada) under its New Brunswick Innovation Research Chair Initiative Program and the Natural Sciences and Engineering Research Council of Canada under its Discovery Grant Program (RGPIN 2017-03994).

Author details

Meng Gong
University of New Brunswick, Canada

*Address all correspondence to: meng.gong@unb.ca

References

[1] Mattheck C, Kubler H. Wood—The Internal Optimization of Trees. Berlin, Germany: Springer-Verlag; 1997

[2] National Parks Conservation Association. Introduction to Giant Sequoia, Sequoia National Park, California, USA; 2019.

[3] Canadian Wood Council. The Richmond Olympic Oval. Canadian Wood Council, Ottawa, ON, Canada; 2019. Available from: https://cwc.ca/case-studies [Accessed: June 30, 2021]

[4] McKeever DR. Engineered wood products: A response to the changing timber resource. In: Pacific Rim Wood Market Report. Madison, Wisconsin, USA: USDA Forest Products Laboratory; 1997

[5] Thelandersson S, Larsen HJ. Timber Engineering. Chichester, West Sussex, UK: John Wiley & Sons Ltd; 2003

[6] Hsu WE. Structural Composite Lumber Manufacturing. CreateSpace Independent Publishing Platform, Scotts Valley, California, USA; 2016

[7] Smulski S. Engineered Wood Products: A Guide for Specifies, Designers and Users. Madison, Wisconsin, USA: PFS Research Foundation, Inc.; 1997

[8] Schuler A. Innovative use of wood promotes market development and supports forest sustainability: A win-win situation for society, forest products industry, and forest owners. UNECE Timber Committee Sixtieth session, Item 4 "Innovative uses of wood", Economic Commission for Europe, Economic and Social Council, United Nations, September 24-27, Geneva, Switzerland. 2002

[9] Gong M. Lumber-based mass timber products in construction. In: Timber Buildings and Sustainability. London, UK: IntechOpen; 2019

[10] Canadian Wood Council. Introduction to Wood Design. Canadian Wood Council, Ottawa, ON, Canada; 2011

[11] CSA Group. Engineering Design in Wood. CSA O86-14. Canadian Standards Association, Mississauga, ON, Canada; 2014

[12] Gril J, Jullien D, Bardet S, Yamamoto H. Tree growth stress and related problems. Journal of Wood Science. 2017;**63**:411-432

[13] Shmulsky R, Jones PD. Forest Products and Wood Science: An Introduction. 6th ed. West Sussex, UK: John Wiley & Sons Ltd.; 2011

[14] Canadian Wood Council. Structure and Properties of Wood. Slide-Cassette Presentation. Ottawa, Canada: Canadian Wood Council and Forintek Canada Corp; n.d.

[15] ASTM International. Designation D1990-19. Standard Practice for Establishing Allowable Properties for Visually Graded Dimension Lumber from In-Grade Tests of Full-Size Specimens. West Conshohocken, PA, USA: ASTM International; 2019

[16] Song JW, Chen CJ, Zhu SZ, et al. Processing bulk natural wood into a high-performance structural material. Nature. 2018;**554**:224-228. DOI: 10.1038/nature25476

[17] Landry V, Alemdar A, Blanchet P. Nanocrystalline cellulose: morphological, physical, and mechanical properties. Forest Products Journal. 2011;**61**(2): 104-112

[18] Julkapli NM, Bagheri S. Progress on nanocrystalline cellulose biocomposites.

Reactive and Functional Polymers. 2017;**112**:9-12

[19] Bodig J, Jayne BA. Mechanics of Wood and Wood Composites. New York, NY, USA: Van Nostrand Reinhold Company Inc.; 1982

[20] Karacabeyli E, Gagnon S. Canadian CLT Handbook, 2019 Edition, Special Publication SP-532E. Pointe-Claire, QC, Canada: FPInnovations; 2019

[21] Côté WA Jr. Wood Ultrastructure—An Atlas of Electron Micrographs. Seattle, WA, USA: University of Washington Press; 1967

[22] Yamauchi H, Miura I, Sasaki T, Koizumi A, Kawai S, Sasaki H. Manufacturing condition and mechanical properties of cylindrical LVL manufactured by the spiral-winding method. Journal of the Society of Materials Science. 1998;**47**(4):350-355

[23] McLain R, Steimle D. WW-WSP-10. Accommodating Shrinkage in Multi-Story Wood-Frame Structures. Washington DC, USA: WoodWorksTM, Wood Products Council; 2019

[24] Doudak G, Lepper P, Ni C, Wang J. Vertical Movement in Wood Platform Frame Structures: Movement Prediction. Vancouver, BC, Canada: FPInnovations; 2013

[25] Allen E, Iano J. Fundamentals of Building Construction: Materials and Methods. 6th ed. Hoboken, New Jersey, USA: John Wiley & Sons, Inc.; 2014

[26] Canadian Wood Council. Brock Commons Tallwood House: A Case Study. Canadian Wood Council, Ottawa, ON, Canada; 2019. Available from: https://cwc.ca/case-studies [Accessed: July 8, 2021]

[27] World Green Building Council. Global Status Report for Buildings and Construction. 2019. Available from: https://www.worldgbc.org/news-media/2019-global-status-report-buildings-and-construction [Accessed: February 16, 2021]

[28] Architecture 2030. New buildings: Embodied carbon. 2020. https://architecture2030.org/new-buildings-embodied [Accessed: November 18, 2020]

[29] Green M. How timber will define the skylines of our future. "Online Education Series "The future of Sustainable Buildings", October 27" Soprema, Drummodville, QC, Canada; 2020.

[30] Natural Resources Canada. The State of Mass Timber in Canada 2021. Ottawa, ON, Canada; 2021

[31] UNECE/FAO. Forest Products Annual Market Review 2019-2020. Geneva, Switzerland: United Nations Economic Commission for Europe (UNECE) and the Food and Agriculture Organization (FAO); 2020

Chapter 2

Engineered Wood Products as a Sustainable Construction Material: A Review

Ranjana Yadav and Jitendra Kumar

Abstract

Engineered wood products are considered as best building materials due to environmentally friendly. Huge change to the way in which wood has been utilized in primary application of construction in the course of the most recent 25 years are in light of decreased admittance to high strength timber from growth forests, and the turn of events and creation of various new design of manufactured wood products. Engineered wood products are available in different variety of sizes and measurements like laminated veneer lumber, glued laminated timber, finger jointed lumber, oriental strand board etc. It is utilized for rooftop and floor sheathing, solid structure, beams and the hull of boats. This review objectively explores not only the environmental aspects of the use of different engineered wood composites as a building material, but also their economic aspects, to understand their effect on sustainability.

Keywords: engineered wood product, plywood, cross laminated timber, laminated veneer lumber, oriented strand board etc.

1. Introduction

Wood is one of the world's most promptly accessible and adaptable building materials, and recent advances in engineered wood products innovation have permitted these to be utilized in a basically load-bearing segment [1, 2]. Wood has assumed a huge part in the existences of people from ancient occasions to the advanced time. It is richly found in an assortment of normal settings and effectively retrievable in its raw form. Wood has been a fundamental component of development for quite a long time, with saved instances of such constructions dated to millennia B.C. Engineered wood products (EWPs) are among the most excellent and harmless to the ecosystem building materials. In manufacture, they are produced efficiently from a renewable resource. In development, the way that engineered wood products is accessible in wide assortment of sizes and measurements [3].

EWPs are chiefly laminated veneer lumber [4], wood I Beam, glued laminated lumber [5], cross laminated timber [6, 7], finger jointed lumber, oriented strand lumber, oriented stranded board [8, 9], medium density fiber board (MDF) and Particle Board (PB). These EWPs are regularly created from the adhesive bonding of wood chips, pieces or veneers, as well as the mechanical securing of timber segments to frame bigger segments, beams, boards or other structural components [10–14].

The advantages of EWPs incorporate upgraded dimensional stability, the development of bigger and more complex structural segments, decreased impact of common imperfections (for example knots), more noteworthy toughness and more homogenous mechanical properties [15, 16]. These properties of wood can be improved through controlled changes, and this is the establishment of designed wood items [17–20]. This builds the overall performance of structural wood composites, prompting a more viable structure material, accordingly growing potential end uses [21–23]. Engineered wood products have permitted wood to be utilized in circumstances where solid timber is incapable, prompting specific items to help a more different exhibit of uses. This has extended market openings prompting the positive financial development of this industry [24].

As well as being utilized as a substitute for more "traditional" engineering materials, engineered wood products have demonstrated to beat conventional sawn lumber in the structural applications for which they are intended for [25, 26]. Extra advantages of working with engineered wood products include: lower building costs with less expensive materials/speedier construction time, lower ozone depleting substance outflows by not utilizing energy escalated materials, unrivaled adaptability under seismic burdens and better energy execution/effectiveness [27].

As per estimates of Forest survey of India (2017), while the annual production from the natural forests is quite low, the production from the tree outside forest is much higher. Most of the Industrial wood in India is produced from outside government forests and agroforestry/ farm forestry in the country. India is one of the emerging markets in Asia pacific engineered wood industry, currently accounting for 10% of the Asia pacific engineered wood market share. Cross laminated timber market is rising construction industry. Laminated veneer lumber is one of the most popular EWPs.

Engineered wood products are making it conceivable to build taller and bigger wood structures, and there is innovative work on this theme. Hence, this review chapter focusses on various engineered wood products, which are more economical. This review chapter of the current writing offers bits of knowledge. This chapter takes a gander at another age of wood products, made from the sustainable structural material. These products can assume an undeniably significant part in a naturally concerned world.

2. Types of engineered wood product

2.1 Glued-laminated lumber (glulam)

Glued laminated lumber, or Glulam, is the most essential and oldest member of the engineered wood products family, which has helped with growing the structural uses of wood and conventional sawn wood development [28, 29]. Solid sawn substantial timber is in restricted accessibly for extremely huge sizes, and is not basically proficient because of deformities like bunches and checking. Glulam has disposed of the limitations of utilization of huge sawn wood concerning size of the stem cross-area, the length of the stem, and the structural deformities present. Production of this product started in Europe at when the new century rolled over, in the U.S. in the 1940's, and in Canada in 1952. Glulam individuals comprise of various wood overlays (or "lams") that are fortified together using glue [30–34]. The boards are pressed with hydraulic equipment in the process to ensure tight bonds. Dimensional softwood lumber is regularly picked for the lamination, and care is taken to guarantee that the grain of the boards runs parallel to the longitudinal axis of glulam member (**Figure 1**). Boards utilized in the lamination process may differ in sizes but do not exceed two inches in thickness. Notwithstanding the benefits related with huge

Figure 1.
Glulam [35].

cross-sectional zones, lamination boards are regularly joined at the ends to develop glulam individuals that increased the lengths of stock lumber. Recently, fiber-reinforced polymers have been included in the manufacture of some glulam production [36]. This component enhances the tensile performance of the member and is said to offer economic benefit in some applications [37–42]. Glulam is often used as straight beams, including lintels, purlins, ridge beams and floor beams, Columns including round, square and complex section, curved beams and roofs.

2.2 Cross laminated timber (CLT)

The construction industry is starting to use new enormous scope engineered wood composites known as mass lumber items. CLT is a moderately new wood product that holds extraordinary potential for significantly expanding the utilization of wood products in construction [43–45]. CLT as enormous boards built through the overlay of various layers of structural grade softwood boards. Each layer of boards is usually oriented perpendicular to adjacent layers and glued on the wide faces of each board, usually in a symmetric way so that the outer layers have the same orientation (**Figure 2**). The products are discovering use in building projects as floor slabs, load bearing wall and shear wall [47–49].

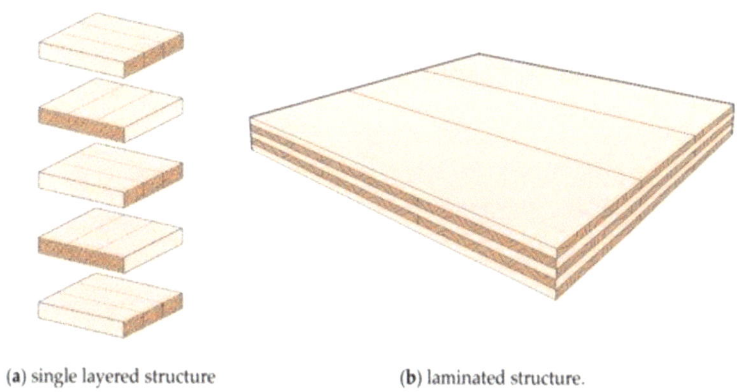

Figure 2.
Cross-laminated timber [46].

Being a suitable choice for heavy frameworks manufacturer expanding interest in CLT and giving a green choice to steel and cement. CLT offers high strength and the structural simplicity needed for cost effective buildings, as well as a lighter environmental footprint than concrete or steel. CLT board manufacture considers a wide scope of board sizes and thicknesses. Architect use CLT boards as burden bearing plate components in construction projects, working as floor slabs, rooftops and walls. The substantial idea of CLT and its great strength properties give the likelihood to use in multistory structure. CLT can also be processed as a "ready-made" building material, tailoring its process to the precise required measurement to reduce wastage of building material.

2.3 Laminated veneer lumber (LVL)

While glulam and CLT boards are contained wooden boards, various engineered wood products are made with wood veneer. LVL is perhaps the most generally utilized engineered wood products for constructional applications. It is a composite board made from various thin layers of veneers that are lined up with the length of the finished lumber [50–53]. The product was invented in the last part of the 1960's and has gotten grounded as a high strength pillar and header segment in both residential con commercial constructions. Since it is made from veneers, LVL makes up to 35% more powerful utilization of logs than is conceivable with solid lumber. At fabricate, veneers are dried to 8% moisture content, and reviewed for uniform strength and width before lay-up. Adhesive is applied and the board is pressed under heat and consistent pressure until cure (**Figure 3**). Laminated veneer lumber is planned for use as high strength, load carrying beams to help the heaviness of development over window and entryway, and in floor and rooftop frameworks of residential and light commercial wood frame development. It can give the both boards and beam/column components [55].

2.4 Laminated strand lumber (LSL)

LSL is usually realized Timber strand. As of now, LSL is being made from excess, over develop aspen trees that normally are not huge, solid, or sufficiently straight to develop conventionally wood products. In this cycle, the debarked logs are utilized to give the material to chipped strands, which can be up to 300 mm long. These strands are then dried, coated with adhesive, and pressed into huge billets by a process which incorporates steam injection (**Figure 4**). The billet might be up to 140 mm thick, 2.4 m wide and 10metres long. Subsequent to sanding, countless sizes are sliced to suit applications like headers, edge joists for floor frameworks,

Figure 3.
Laminated veneer lumber [54].

Figure 4.
Laminated Strand lumber [56].

columns, joists and studs. It is utilized for a wide scope of millwork, like doors, windows, and practically any item that require high grade lumber. It is additionally utilized for truck decks, fabricated housing, and some structural lumber such as window and door headers [57, 58].

2.5 Parallel strand lumber (PSL)

PSL, ordinarily known Parallam, is intended to supplant huge dimension lumber (beams, planks, and posts). Parallel strand lumber was developed in Canada, advanced onto the market in the last part of the 1980's [59]. PSL comes in numerous thicknesses and widths and is fabricated up to 66 feet in length (**Figure 5**). The strands are for the most part taken from veneers peeled from the outermost part of the logs, where more grain is found. Veneers e are dried to 11% moisture content and reviewed for strength prior to chopping into strands [61]. They are then adjusted parallel to each other, coated with waterproof glue, then pressed and cured [62, 63]. It is utilized for enormous individuals in residential construction and as middle and huge individuals in commercial building construction.

Figure 5.
Parallel Strand lumber [60].

2.6 Structural plywood

One of the most important well-known building materials constructed with veneer is plywood. It is handily sourced from everywhere the world and has exhibited fruitful. Plywood is utilized for some light duty building materials. It is likewise utilized for rooftop and floor sheathing, concrete formwork, webs of wood beams, and surprisingly the frames of boats. It very well may be utilized to oppose gravity loads or to oppose horizontal burdens as in plywood diaphragms and shear walls. Plywood is fabricated from stacked veneers which are organized in an odd number of layers, the grain of the face layers arranged to the long dimension of the board (**Figure 6**). The cross-overlaid lay-up of the veneers gives strength, stiffness and dimensional strength [65].

2.7 Oriented strand board (OSB)

OSB was first produced in Canada in 1964. Since the mid-1980s, OSB has been one of the most commonly used engineered wood-based panels for structural construction in residential sectors due to excellent properties, especially due to the increasingly competitive price [64, 66]. OSB is an engineered structural panel made from strands of wood sliced from small diameter timber logs and bonded together with an exterior grade adhesive, under heat and pressure [67–70]. OSB is manufactured in various grades with improving resistance to the effects of moisture (**Figure** 7). OSB is extensively used for wall sheathing, floor underlayment, roof cover and I-joist in both commercial and residential building. OSB also is used in furniture, reels, trailer liners and recreational vehicle floors [72–74].

2.8 Wood I beam

Wood I beam are engineered wood products which have great strength in respect to its size and weight. Wood I beam is a light beam support assembled by gluing together wooden flanges and fiber board and plywood beams. The flanges of beam made of laminated veneer lumber or finger jointed solid wood lumber. The web of beam made of plywood, laminated veneer lumber or oriented strand board. Wood I beam are available up to 80 feet long (**Figure 8**). It has been used in residential and commercial construction as floor, rood structure of structure and external wall frames. I beam are best for the structure which required rigidity, heat insulation and

Figure 6.
Structural plywood [64].

Figure 7.
Oriented Strand board [71].

Figure 8.
Wood I beam [75].

economy. Nowadays, wood based I beams are becoming popular. Beams allow easy execution of installation penetration. Their 'I' configuration provides high strength and stiffness.

3. Advantages of engineered wood products (EWPs)

The term engineered wood products covers a wide cluster of wood-based items produced using veneers or strands peeled, chipped or sliced. These items are appropriate for construction reason as structural materials [76]. EWPs tend to perform better as they have higher load-carrying abilities compared to solid wood of the same dimension. This is because EWPs have more uniform and predictable structural properties, as the usual deficiencies in the wood (like knots and cracks) are either removed or offset by the manufacturing process. The low cost of engineered wood is the most immediate benefits and the reason that it is so widely used.

The utilization of glue laminated timber offers various benefits including structural opportunity, proficient utilization of timberland asset like the utilization of merchantable and non-merchantable wood species and mixed species application. CLT gives various ecological advantages notwithstanding its excellent thermal performance. CLT boards are produced for explicit end use applications, which bring about almost no waste. (CLT's thick cross segment gives significant imperviousness to fire since boards char gradually [77]. Probably the greatest advantage of utilizing CLT is that the design can be fabricated rapidly and productively. The size of LVL is not restricted by log size, because of its assembling strategy [78]. Since, it is fabricated with homogenous quality that has least number of defects. The advantage of I-joists/beam is they are less likely to bow, crown, twist, up or split as would a dimensional piece of lumber.

Engineered wood product lie plywood permits developers to diminish the quantity of trees required for building a home. Plywood sheathing for walls and rooftops definitely diminishes home construction cost compared with utilizing boards of solid wood. Some engineered products, for example, MDF and PB can be produced using sawmill scraps, wood chips and even saw dust reutilizing wood waste.

Engineered wood product is likewise useful in that it assists with cultivating up the interaction of development. Perhaps the most appealing highlights of these engineered components and assemblies are that they can be manufactured to longer lengths than their sawn lumber partners. Additionally, engineered products preserve or extend the use of the forest resource by using a higher percentage of fiber, which previously was burned or left to rot. The use of wood from residual sources, plantations and second-growth forests reduces the pressure to harvest more forest area. Waste timber can be recycled and turned into strands and fibers and reconstituted into engineered wood products.

3.1 Comparing material options

In considering the utilization of engineered wood products for a construction, the manufacturer is given a stable of material choices which permit extensive potential for the declaration of sculptural structure. This opportunity to make a masterpiece, when taken, regularly brings about a wood structure whose elegance and excellence is genuinely a supplement to function. It is significant that the wood industry keeps on advancing the advantages of the unique character and warmth of the uncovered structural products. Engineered wood products can be considered as a reasonable substitute or complement for concrete, steel and brick in large building projects.

Simultaneously, it is similarly imperative to know about the premium related with the stock of regularly elaborate shapes and treatments. These may draw structural components as architectural components in a way that utilizes essentially more material that is needed to help a given burden. To appropriately evaluate the expenses of wood versus steel, one should perceive such contemplations. Really frequently, a wood alternative is saved due to such defective examinations. For instance, a school whose structural framework, in the brain of the designer, will be alright served by a structural steel frame and open web steel joist framework ought not just be assembled that way in light of the fact that a weighty wood post and beams framework is costlier. The Engineered Wood Products explained in this chapter currently permit the wood industry to contend with practically identical items to the steel alternative. Cross laminated lumber is able to replace concrete slabs in the frames of multistory buildings [79]. The weight CLT is about 4 times less than concrete which reduces foundation loads and transportation cost. Glulam is two third the weight of steel and one sixth the weight of concrete. The high strength of laminated timbers enables glulam beams to span large distances without intermediate columns, allowing maximum design flexibility than traditional timber construction.

The different sections of this chapter which describe various EWPs refer to the enhanced utilization of logs due to the use of small sections, in many cases veneers or strands. Engineered wood products bring huge advantages compared to competing products such as steel, concrete and aluminum in terms of embodied energy, and emissions of carbon dioxide and other pollutants during manufacture and extraction.

4. Conclusions

This chapter presents an overview of the EWPs. The advancement of new designs of structural wood products – called EWPs- throughout the most recent 25 years has utilized wood fiber and allowed the wood industry to rival other structure materials in more construction applications. Based on the properties and end uses engineered wood products will see an expanding market share within the wood products manufacturing industry. There are various variables impacting this, for example, the advantages of wood building construction, improved fire execution over dimensional timber products. CLT offers high strength and the structural simplicity needed for cost-effective buildings, as well as a lighter environmental footprint than concrete or steel. Current experience shows that engineered wood products are excellent structural wood building material for future.

Conflict of interest

The authors declare no conflict of interest.

Author details

Ranjana Yadav* and Jitendra Kumar[2]

1 Indian Plywood Industries Research and Training Institute, Centre, Mohali, Punjab, India

2 Department of Biosciences, University Institute of Biotechnology, Chandigarh University, Mohali Punjab, India

*Address all correspondence to: ranjanay7@gmail.com

References

[1] Maxim P, Darryl P, Len G, Joseph C. Taller Wood Buildings and Fire Safety. University of the Fraser Valley; 2013.

[2] Smulski S. (Ed) Engineered wood products. PFS Research Foundation: Madison, Wisconsin; 1997.

[3] Milner HR. Sustainability of engineered wood products in construction. Sustainability of construction materials. wood head publishing series in civil and structural engineering, Cambridge UK; 2009. p. 184-212.

[4] Kurt R, Hayrettin M, Aslan K, Cil M. Laminated Veneer Lumber (LVL) Manufacturing Using Three Hybrid Poplar Clones. Turkish Journal of Agriculture and Forestry. 2012; 36 (2): 237-245.

[5] Moody RC, Hernandez R. Glued-Laminated Timber. In: Engineered wood Products: A guide for specifiers, Designers and Users, Smulski S. (Ed.) PFS Research Foundation, Wisconsin; 1997. p. 1-39. ISBN:096567360X.

[6] Gagnon S, Pirvu C. (eds.) CLT handbook: Cross-laminated timber. FP Innovations, Quebec, Canada; 2011.

[7] Divekar N. Introduction to New material- cross laminated timber. International Journal of Engineering research. 2016; 5 (3): 675-679.

[8] Lowood J. Oriented strand board and waferboard. In: Smulski S (ed) Engineered wood products. Miller Freeman, Madison; 1997. p. 5.123-5.145.

[9] William G. Davids, Habib J. Daugher, Eric D. Cassidy, Douglas G. FRP-reinforced oriented strand board panels for disaster-resistant construction, 2003.

[10] Hemmila V, Adamopoulos S, Karlsson O, Kumar A. Development of sustainable bio-adhesives for engineered wood panels–A Review. RSC Advances. 2017; 7 (61): 38604-38630.

[11] Baldwin R. Plywood and Veneers Based Products, Manufacturing Practices, Miller Freeman: San Francisco, California; 1995.

[12] Barbu MC, Reh R, Irle M. Wood-based composites. In Research Development in Wood Engineering Technology, IGI Global. 2014. p. 1-45.

[13] Nelson SPE. Structural composite lumber. In: Engineered wood products a guide for specifiers, designers and users, Sumulski S. (ed.), PFS, Research Foundation, Madison; 1997.p. 147-172.

[14] Shukla SR, Rao RV, Sharma SN. Evaluation of parallel strand lumber (PSL) and 144 WOOD RESEARCH its comparison with laminated veneer lumber (LVL), rubber wood and teak. Holz als Rohund Werkstoff. 1999; 57(4): 267-270.

[15] Lam F, Prion HG. Engineered wood products for structural. Timber Engineering, Wiley, New York; 2003.p. 81.

[16] Nicole MS, Cai Z, Carll C. Wood-Based Composite Materials Panel Products, Glued-Laminated Timber, Structural Composite Lumber, and Wood–Nonwood Composite Materials, Chapter 11, General Technical Report FPL–GTR–190; 2010. p. 11.1-11.28

[17] *Gaff* M, Ruman D, Gašparík M, Štícha V, Boška P. Tensile-Shear Strength of Glued Line of Laminated Veneer Lumber. BioResources. *2016;* 11: 1382-1392.

[18] Gary C W. Engineered wood products - experience and opportunities. NZ Timber Design Journal. 2020; 8 (2):6-13.

[19] Sotayo A, Bradley D, Bather M, Sareh P, Oudjene M, El-Houjeyri I, Harte AM., Mehra S, O'Ceallaigh C, Haller P, Namari S, Makradi A, Belouettar S, Bouhala L, Deneufbourg F, Guan Z. Review of state of the art of dowel laminated timber members and densified wood materials as sustainable engineered wood products for construction and building applications. Developments in the Built Environment. 2020; 1: 100004.

[20] Wang B, Chui YH, He G, Xie D. Development of high performance and durable engineered wood products from mountain pine beetle veneers using novel resin impregnation technologies. Natural Resources Canada; 2009. p. 40.

[21] Smith I, Snow M, Asiz A. Design methods for connections in engineered wood structures, Final Report UNB2, Natural Resources Canada, Ottawa; 2006.

[22] Smith I, Asiz A, Snow M. Performance of mechanical fasteners used with engineered wood products, Final Report UNB75, Natural Resources Canada, Ottawa; 2007.

[23] Snow M. Fracture development in engineered wood product bolted connections [PhD thesis]. University of New Brunswick: Fredericton; 2006.

[24] William C. A Review of Engineered Wood Products in Canada [An undergraduate thesis]. Lakehead University Thunder Bay: Ontario; 2018.

[25] Fadai A, Winter W, Gruber M. Wood based construction for multi-storey buildings. The potential of cement bonded wood composites as structural sandwich panels. World Conference on Timber Engineering (WCTE 2012), 2012 ;1. WCTE 2012 Committee. Aukland.

[26] Stark NM, Cai Z, Carll C. Wood-based composite materials: panel products, glued-laminated timber, structural composite lumber, and wood-nonwood composite materials. In: Ross R. (Ed.), Wood Handbook - Wood as an Engineering Material. U.S. Department of Agriculture Forest Service, Forest Products Laboratory, Madison, WI; 2010. p. 11.1-11.28.

[27] Adair C. Regional production and market outlook tor structural panels and engineered wood products 2001-2006. APA Economics Report E67 Table 28, APA, Tacoma, WA, 2001. p. 51.

[28] Brandner R, Schickhofer G. Glued laminated timber in bending: new aspects concerning modelling. Wood Science & Technology. 2008; 42(5):401-425

[29] Younquist JA. Wood based composites and panel products. In wood handbook: forest product laboratory. United States, department of agriculture forest service, Madison, Wisconsin; 1999. P. 10.1-10.31-113.

[30] Anshari B, Guan ZW, Kitamori A, Jung K, Komatsu K. Structural behavior of glued laminated timber beams pre-stressed by compressed wood. Construction Building Materials. 2012; 29: 24-32.

[31] Anshari B, Guan ZW, Wang QY. Modelling of Glulam beams pre-stressed by compressed wood. Composite Structures. 2017; 165: 160-170.

[32] Correal JF, Echeverry JS, Ramirez F, Yamin LE. Experimental evaluation of physical and mechanical properties of glued laminated guaduangustifolia Kunth. Construction Building Materials. 2014; 73:105-112.

[33] Herawati E, Massijaya MY, Nugroho N. Performance of Glued-Laminated Beams Made from Small Diameter Fast-Growing Tree Species. Journal of Biological Sciences. 2010; 10: 37-42.

[34] Herawati E. The characteristics of glued-laminated beams made from small diameter fast growing species [MS Thesis]. Postgraduate School, Bogor Agricultural University; 2007.

[35] GLULAM. Versatile engineered wood product delivers strength, beauty and reliability, https://ecohousemart.com

[36] Castro G, Paganini F. Mixed glued laminated timber of poplar and Eucalyptus grandis clones. Holz Als Roh-Und Werkstoff. 2003; 61(4): 291-298.

[37] Nadir Y, Nagarajan P. The behavior of horizontally glued laminated beams using rubber wood. Construction Building Materials. 2014; 55: 398-405.

[38] Ezeagu C, Eromosele A, Okoro H, Chukwujekwy U, Emetomo T. Flexural Strength of solid and glue laminated timber beams. American Journal of Engineering Science & Technology Research. 2015; 3 (1): 1-14.

[39] Nagarajan P. The behavior of horizontally glued laminated beams using rubber wood. Construction Building Materials. 2014; 55: 398-405.

[40] Sreeshna KS. Study on Flexural Behavior of Glued Laminated Timber. International Journal of Innovative Research in Science, Engineering and Technology. 2016; 5 (7): 13373-13381.

[41] Vratuša S, Kuzman MK, Kilar V. Structural particulars of glued laminated beams of variable height. Drewno. Pr. Nauk. Donies. Komunik. 2011;54: 185.

[42] Xiao Y, Yang RZ, Shan B. Production, environmental impact and mechanical properties of glubam. Construction and Building Materials. 2013;44: 765-773.

[43] Lehmann S. Developing a prefabricated low-carbon construction system using cross-laminated timber (CLT) panels for multistorey inner-city infill housing in Australia. Journal of Green Building, 2012; 7(3): 131-150.

[44] Liao Y, Tu D, Zhou J, Zhou H. Feasibility of manufacturing cross-laminated timber using fast-grown small diameter eucalyptus lumbers. Construction Building Materials. 2017; 132: 508-515.

[45] Mohammad M, Gagnon S, Douglas BK, Podesto L. Introduction to Cross Laminated Timber. Wood Design Focus. 2012;22: 2.

[46] Dong Y, Cui X, Yin X, Chen Y, Guo H. Assessment of Energy Saving Potential by Replacing Conventional Materials by Cross Laminated Timber (CLT)—A Case Study of Office Buildings in China. Applied Science. 2019; 9: 858.

[47] Arul Mozhi Bharathi DV, Dhanalakshmi E, Manickam K. Study on cross laminated timber. International Research Journal of Engineering and Technology (IRJET). 2020; 7 (9): 1196-1199.

[48] Brandner R, Flatscher G, Ringhofer A, Schickhofer G, Thiel A. Cross laminated timber (CLT): overview and development. European Journal of Wood and Wood Products. 2016; 74: 331-351.

[49] Frangi A, Fontana M, Hugi E, Jubstl R. Experimental analysis of cross-laminated timber panels in fire. Fire Safety Journal. 2009; 44(8): 1078.

[50] Bal BC. Some technological properties of laminated veneer lumber produced with fast-growing Poplar and Eucalyptus. Maderas. Ciencia y Tecnología. 2016; 18(3): 413-424.

[51] Kunesh RH. Micro = Lam: structural laminated veneer lumber. Forest Product Journal. 1978; 28(7): 41-44.

[52] Stump JP, Smith LA, Gray RL. Laminated veneer lumber made from plantation-grown conifers. Forest Product Journal. 1981;31(4):34-40.

[53] Hesterman ND, Gormon TM. Mechanical properties of laminated veneer lumber made from interior Douglas-fir and lodgepole pine. Forest Product Journal. 1992; 42(11/12): 69-73.

[54] Hakkarainen J. LVL Handbook EUROPE. Federation of the Finnish Woodworking Industries, Snellmaninkatu, Finland 2019.

[55] Neuvonen E, Minna S, Jani H, Micha H, Matthias W. Laminated Veneer Lumber - Wood-Based Panels Technology. Laminated Veneer Lumber - Wood-Based Panels Technology. N.p., n.d. Web; 2013.

[56] Laminated strand Lumber. https://images.app.goo.gl/1GiWYAXXKKmk94919.

[57] Wang Z, Gong M, Chui YH. Mechanical properties of laminated strand lumber and hybrid cross-laminated timber. Construction Building Materials. 2015; 101:622-627.

[58] Weight S, Yadama V. Manufacture of laminated strand veneer (LSV) composite. Part 1: Optimization and characterization of thin strand veneers. Holzforschung. 2008; 62(6):718-724.

[59] Ahmad M, Kamke FA. Properties of parallel strand lumber from Calcutta bamboo (*Dendrocalamus Strictus*). Wood Science and Technology. 2011; 45:63-72.

[60] Andre M Iskandar. Potential Market of Parallel Strand Lumber (PSL) and Laminated Veneer Lumber (LVL) in Indonesia. A Report Submitted in Partial Fulfillment of the Requirements for the Degree of Bachelor of Science in Wood Products Processing in The Faculty of Forestry, 2013.

[61] Yihai L, Le W. Selected properties of parallel strand lumber made from southern pine and yellow poplar. Holzforschung. 2003;57(2): 207-212.

[62] Fan MZ, Xu RX, Yang AH. Study on parallel strand lumber made of waste veneer/bamboo strips. Wood Industry (Beijing). 1995; 9(1): 10-13.

[63] Kurt R, Cavus V. Manufacturing of parallel strand lumber (PSL) from rotary peeled hybrid poplar I-214 veneers with phenol formaldehyde and urea formaldehyde adhesives. Wood research. 2011; 56 (1): 137-144.

[64] Structural Plywood. https://images.app.goo.gl/8T8kocPbFfHevuCy5.

[65] Bekhta P, Hiziroglu S, Shepelyuk O. Properties of plywood manufactured from compressed veneer as building material. Materials and Design. 2009; 30 (2009): 947-953.

[66] Mirski R, Dziurka D, Derkowski A. Properties of oriented strand boards with external layers made of non-strand chips. BioResource. 2016; 11: 8344-8354.

[67] Fiorelli J, Soriano J, Lahr FAR. Roof modular system in wood and particle board (OSB) to rural construction. Scientia Agricola. 2012; 6 (3): 189-193.

[68] Barbuta C, Blanchet P, Cloutier A, Yadama V, Lowell E. OSB as substrate for engineered wood flooring. European Journal of Wood and Wood Products. 2012; 70: 37-43.

[69] Dos Santos RBG, Da Costa DF, Bezerra CMVO, Bezerra JFO. Constructive-process of using plates OSB (oriented strand lumber) in sustainable structural systems in a residential building in the city of Manars-AM. International Journal of Advanced Engineering Research & Science (IJAERS). 2019; 6 (10): 85-91.

[70] Cassidy E. Development and Structural Testing of FRP Reinforced

OSB Panels for Disaster Resistant Construction. MS Thesis, Department of Civil and Environmental Engineering, University of Maine, Orono, Maine; 2002.

[71] Oriented strand board, Wikipedia. https://images.app.goo.gl/3KA6wHxsw9MqGg5M8.

[72] Gu H, Wang S, Neimsuwan T, Wang S. Comparison study of thickness swell performance of commercial oriented strandboard flooring products. Forest Product Journal. 2005; 55(12):239-245.

[73] Lunguleasa A, Dumitrascu A, Ciobanu V. Comparative Studies on Two Types of OSB Boards Obtained from Mixed Resinous and Fast-growing Hard Wood. Applied Science. 2020; 10: 6634.

[74] Hiziroglu S. Oriented strand board as a building material. Oklahoma State University; 2006.

[75] Voth C, Yadama V. Sustainable Lightweight Wood-Strand Panels for Building Construction, Proceedings of the International Convention of Society of Wood Science and Technology and United Nations Economic Commission for Europe – Timber Committee, Geneva, Switzerland; 2010.

[76] Abeysuriya N. Effective use of structural timber for sustainable construction. http://www.academia.edu.

[77] Yu D, Xue C, Yin X, Chen Y, Guo H. Assessment of Energy Saving Potential by Replacing Conventional Materials by Cross Laminated Timber (CLT)—A Case Study of Office Buildings in China. Applied Science. 2019; 9: 858.

[78] Tellnes LGF, Eide S, Kristjansdottir TF, Kron M. Assessment of carbon footprint of laminated veneer lumber elements in a six-story housing – comparison to a steel and concrete solution, Chapter 11 - LCA of Sustainable Materials and Technologies; 2011.

[79] Mallo MFL, Espinoza O. Cross-Laminated Timber Vs. Concrete/Steel: Cost Comparison Using A Case Study. World Conference on Timber Engineering (WCTE 2016), Vienna, Austria; 2016.

Chapter 3

Veneer-Based Engineered Wood Products in Construction

Elena Vladimirova and Meng Gong

Abstract

Veneer-based engineered wood products (EWPs) are widely used in construction. Veneer-based EWPs are made of thin veneer sheets or veneer strands with adhesives, mainly including plywood, laminated veneer lumber (LVL), and parallel strand lumber (PSL). This chapter first discusses veneer-based EWPs in terms of their manufacturing, properties, and applications. Secondly, it introduces how veneer sheets or veneer strands intersect with each other in these products, providing additional strength and stable dimensions. Thirdly, this chapter overviews the effects of element dimensions, basic structure, veneer grade, adhesive type, and processing parameters on the properties of these products. Finally, it illustrates the uses of veneer-based EWPs through examples with a focus on their construction applications.

Keywords: veneer-based engineered wood products, plywood, laminated veneer lumber, parallel strand lumber, adhesives, engineering properties, structural uses

1. Introduction

Due to the significant population growth and the rising housing standards, the need to use structural wood products has been increasing [1]. At the same time, the timber industry must come up with solutions for ensuring the preservation of natural resources because of the growing demand for lumber and decreasing availability of large-diameter old-growth trees [2, 3]. Previously sawn from massive logs, structural lumber is now made from reconstituted wood in various shapes and sizes, which is classified as engineered wood products (EWPs). EWPs can maximize the use of wood and utilize small-diameter logs in comparison with conventional lumber [3, 4]. There are several types of EWPs in terms of the elements used, such as veneer-, strand-, fiber- and lumber-based EWPs, among which the veneer-based group is the oldest but still widely used.

The veneer-based EWPs, or called layered wood composites, are made of veneer sheets or veneer strands bonded with an adhesive [2], mainly including plywood, laminated veneer lumber (LVL), and parallel strand lumber (PSL), **Figure 1**. These products are largely made from peeled logs and reconstituted wood, which can then be fabricated into large sheets known as engineered panels [7]. The significant advantage of using veneer, as opposed to sawn lumber, is that it can increase the yield of wood materials from logs, particularly from small-diameter logs [8]. Veneer-based EWPs have a more homogeneous structure and uniform mechanical properties than solid lumber, making them a good candidate for building materials in construction.

Figure 1.
Veneer-based EWPs. Top left – plywood (source: photos obtained from Indiamart [5]) top right – thick plywood. Bottom left – LVL, bottom right – PSL (source: photos obtained from think wood [6]).

Veneer-based EWPs differ by wood species, adhesive type, as well as by layup structure. **Figure 2** shows the cross-sections (i.e., the width-thickness plane, or x-y plane named here) of four widely used wood products in construction, i.e., solid wood/lumber, plywood, LVL, and PSL. In the y-axis, the dimensional change is similar between solid wood, plywood, and LVL due to limited efficacy of adhesive bonds in this direction, i.e., the radial direction of the wood. However, the dimensional change of PSL in the y-axis is smaller than that of the other three products because of its irregular arrangement of veneer strands in the x-y plane and application of an adhesive. On the x-axis, the dimensional change is largest in solid wood (Note: the x-axis is the tangential direction of the wood.) and smallest in plywood

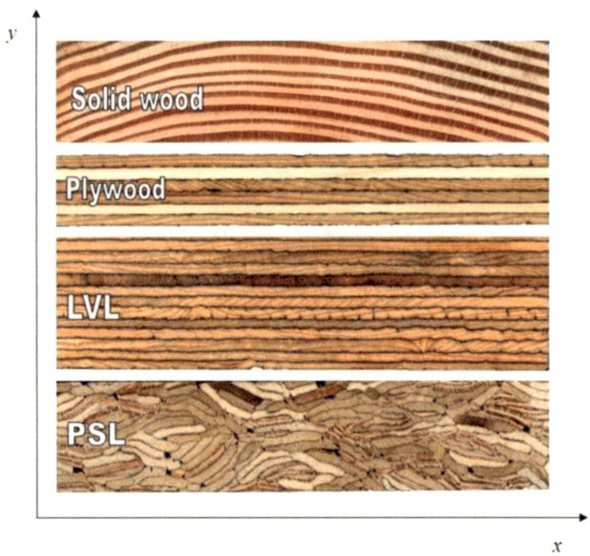

Figure 2.
Cross-sections of solid wood and veneer-based EWPs.

and PSL, with LVL being in between. In other words, solid wood has the largest variability in both x- and y-axes; plywood and LVL have the reduced variability in x-axes, and PSL has the smallest variability in both x- and y-axis.

The first type of veneer-based EWPs invented is plywood [9]. Later, modifications applied to the veneer layups resulted in LVL, and afterward, the long veneer strands were used to make PSL. The veneer-based EWPs have been widely used in construction nowadays [8]. Plywood is usually used as the sheathing material for walls, floors, and roofs, and the web stock for I-joists. LVL is commonly used as beams, columns, and the flange stock of I-joists. PSL is mainly used as columns and beams.

2. Plywood

2.1 Introduction

Plywood is a glued wood panel consisting of several thin layers of veneer with wood fibers in adjacent layers at right angles in most cases. Usually, a plywood sheet consists of an odd number of veneer layers [2, 3, 10]. Each layer is called ply, so the plywood can be deemed as a wood sandwich [7]. The cross lamination of adjacent plies in plywood contributes to improved mechanical properties and dimensional stability in both length and width directions [10].

Plywood is one of the oldest veneer-based EWPs. More than 3500 years ago, a type of plywood was found in ancient Egypt, which is part of the coffin, dating back to the third Egyptian dynasty [8]. Later, around 1500 BC, some images were discovered in which workers cut plywood with an axe-like tool. These images also show that the glue, apparently of animal origin, was prepared in a pot on fire [6]. Furniture constructed from overlapping sheets of wood and inlay had been discovered in Egyptian tombs. Hardwood veneer was preferred due to its attractive texture and shades [11]. The introduction of plywood was linked to the high cost of wood. Due to the shortage of available wood than supply, Egyptians had to import, by sea, ebony and mahogany from East Africa and cedar and pine from Lebanon at a very high price [12]. Later, the ancient Greeks and Romans started producing plywood. Plywood was primarily used for the manufacturing of furniture and household items [11]. Plywood production took off in the 1850s thanks to the Swedish inventor Emmanuel Nobel, who created a model of a rotary lathe [8]. This model made it possible to remove the veneer in a certain and constant thickness from a wooden block. It gave the plywood a uniform thickness and structure [8].

Despite the fact that plywood is now widely used for sheathing in residential and commercial construction, early builders were hesitant to use the newly-born plywood panels because the blood and soybean protein-based glues used were not waterproof, and some panels delaminated when they got wet [13]. In 1934, waterproof synthetic wood adhesives were introduced, which solved the problem and eased builders' concerns [8, 13]. During World War II, the use of plywood was exploded in many industries such as boats, aircrafts, footlockers, crates, and buildings [13]. It led to the post-war boom in plywood production [8], which was adopted for structural and exterior applications. One notable example of using plywood is the construction of the legendary bomber Mosquito [14]. This aircraft was introduced during the World War II. Spruce wood, birch plywood, and balsa wood were used in the construction of aircraft, which made it possible to achieve the necessary strength with a low weight structure [15]. Plywood and other structural panels have changed the way of constructing light wood-frame houses and buildings [11, 16]. Since the middle of the past century, usage of structural panels has expanded from a few niche applications to a popular commodity such as subflooring, roof and wall sheathing, corner bracing, and concrete forming [16]. Initially concentrated in the Pacific Northwest of the United States, where old-growth, large-diameter Douglas-fir was mostly used the plywood business therefrom expanded into the southeastern regions in the 1970s as the technological barrier of bonding southern yellow pine veneer was removed [13]. As seen from **Figure 3**, plywood consumption in Canada was rather stable in the last 15 years or so despite the emergence of other new types of building materials. However, Canada also imports plywood from other countries to meet its increasing demand in construction and other industries such as furniture [17].

2.2 Manufacturing

Figure 4 illustrates the key processes of manufacturing three major veneer-based EWPs, i.e., plywood, LVL, and PSL. An example of manufacturing Canadian softwood plywood is given below, which is used for structural applications. Specially chosen peeler logs are transported to a barker, where they are rotated against a steel claw, which removes the bark [18]. Then debarked logs are cut into peeler blocks. A block is placed on a massive lathe, rotating against a sharp knife. When the block turns, a continuous thin layer of wood, i.e., veneer is peeled off, similar to how paper unwinds from a roll.

The whole block is tried to use with an aim to generate a high yield of good quality wood material. The leftover small spindles are used to make other wood products. The long ribbon of the veneer is then cut with clippers into desired widths and sorted. It is also possible to remove defective pieces of veneer. Subsequentially, the veneer is dried to a moisture content of 5% or so in steam- or gas-heated ovens [18]. Depending on its intended use, the veneer may range in thickness from 0.3 mm (0.01 in) to 6.3 mm (0.25 in) [11]. After drying and sorting, the veneer is fed by glue spreaders, which apply an adhesive layer of uniform thickness. Phenol-formaldehyde (PF) adhesives are usually used in the manufacturing of plywood for structural and outdoor applications when exposed to the weather in its service [3]. Veneer sandwiches are sent to the hot press, which is a key step in the production process of curing the adhesive, subjected to a temperature of 150°C (300°F), and a pressure of 1.38 MPa (200 psi). After the press panels are cut to required dimensions, sanded, and graded [18].

In the fabrication of plywood for non-structural uses, such as furniture, cabinets, and indoor decoration, water-resistant urea-formaldehyde (UF) adhesives are used. The UF adhesives can be cured at a temperature of about 120°C (250°F) during hot-pressing, which can also be cured with high-frequency heating system with an aim to reduce the hot-pressing time and increase the production efficiency [3].

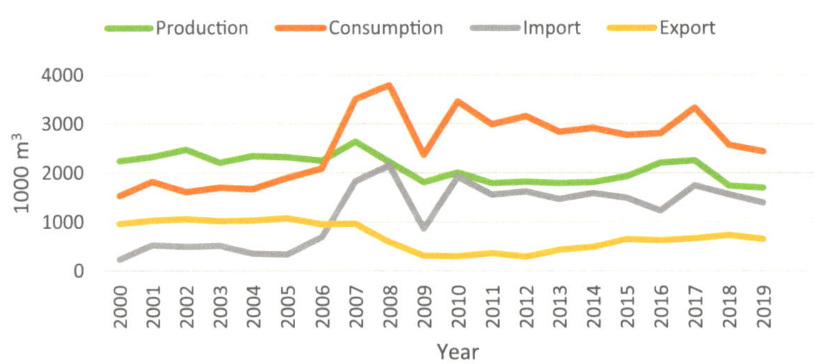

Figure 3.
Plywood production and consumption in Canada [17].

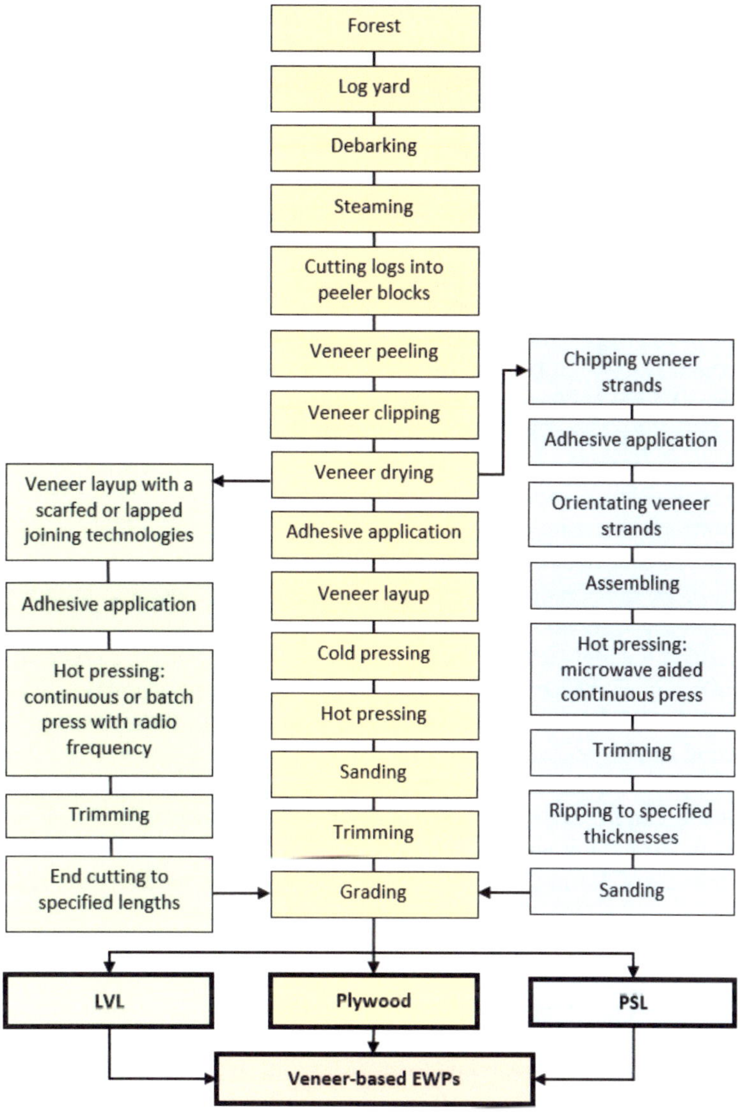

Figure 4.
Processes of manufacturing veneer-based EWPs.

Quality control, which includes incoming management of raw materials, such as wood and glue, and manufacturing parameters at all stages of the production, must be applied in order to produce good quality plywood products. Acceptance quality control is the final stage of the manufacturing process. Many plywood manufacturers in western Canada produce structural plywood under the supervision of the British Columbia Council of Forest Industries (COFI), which constantly checks glue bond strength and other properties to guarantee that the products satisfy the Canadian Standards Association (CSA) standard [18].

2.3 Typical species and sizes

Plywood can be made from various types of wood. Softwoods are commonly used to make veneer for plywood in North America, containing Douglas fir, western hemlock, spruces, pines, and firs [14]. These wood species can be divided into various categories based on their strength and use within the plywood structure. Spruce is used to make the majority of construction-grade softwood plywood in Canada [7]. More discussion on softwood plywood is given through the text in following sections.

Of hardwoods, birch, alder, linden, and lauan ("Philippine mahogany") are most popular for veneer production [7, 10]. These species do not have distinguished early-wood and latewood zones, which are characterized by uniform density and structure, making them easy to be peeled to produce thin and durable veneer.Beautifully grained hardwoods are often combined in several ways to make a unique face pattern [7].

The first standard plywood sheet had a width of 1.22 m (4 ft) and a length of 2.44 m (8 ft), which appeared in 1928 [19]. Such a standard size for plywood sheets has, since then, almost not been changed. The common thickness of plywood varies from 3.2 mm (1/8 in) to 76 mm (3 in) [10]. It depends on the thickness of the veneer and the number of layers. The most common plywood contains 3, 5, or multiple layers. With a three-layer, the plywood is 2–3 mm (0.08–0.12 in) in thickness, which can be used as an underlayment between the subfloor and the tile. Hardwood decorative plywood is often uniformly selected for grain texture, which is widely employed for indoor uses. The universal hardwood plywood has five layers, resulting in a thickness of 4 mm (0.16 in) or so, which can be used for multiple outdoor and indoor applications. Multiple layer plywood with more than seven layers can be classified as thick plywood, which is widely used for structural purposes, requiring acceptable strength and durability under the loading condition [11]. The thick plywood needs a sub-floor or structural sheathing attached to the framing elements of a new canopy.

Hardwood can be peeled or sliced for the production of decorative veneer for making furniture, cabinets, and interior decoration. Slicing results in more loss in raw materials and more intensiveness in labor [16]. Hardwood veneer, such as birch, usually has a thickness of 1.5 mm (0.06 in), whereas softwood veneer is often cut to a thickness of 3 mm (0.12 in) for plywood and LVL production [8].

2.4 Grading

Plywood comes in a range of appearance grades, from flat natural surfaces suitable for finishing to cost-effective unsanded grades suitable for sheathing. More than a dozen typical thicknesses and over twenty different grades of plywood are available [14]. The plywood is usually graded based on the appearance quality of veneer in North America. There are commonly two classes of plywood, each of which has its own set of standards: (a) construction and industrial plywood and (b) hardwood and decorative plywood [3].

In Canada, the two most popular types of softwood plywoods are unsanded sheathing grade Douglas Fir Plywood (DFP), which conforms to CSA O121 "Douglas fir plywood", and Canadian softwood plywood (CSP), which conforms to CSA O151 "Canadian softwood plywood". The poplar plywood, which conforms to CSA O153 "Poplar plywood", is also designated but less uses in construction [14]. The group of DFP can include other species in addition to Douglas fir. For example, the front and back faces are made of Douglas fir, but the inner plies can be made from any of the specified species, including Douglas fir, western hemlock, and the majority of spruce, pine, and fir species in Canada [14]. Plywood that contains other selected Canadian wood species in the face and back plies is labeled CSP. Most species that are only allowed as inner plies for DFP may also be used as the face or back plies for CSP. Three hardwood species, i.e., balsam poplar, trembling aspen, and cottonwood, are restricted to use as inner plies in DFP and CSP [14]. The sizes, grades, specialty panels, manufacturing tolerances, and glue bond quality of plywood are all stipulated in the standards CSA O121, CSA O151, and CSA O153.

The structural plywood is put with a legible and durable stamp showing the manufacturer, the bond style (EXTERIOR), the species (DFP or CSP), and the grade [14]. DFP and CSP are both made in a variety of grades based on the appearance and quality of the veneer used for making the outer plies.

Many plywood mills are members of the associations, which are responsible for inspecting, testing, and certifying the products with stamps. These stamps indicate that the stamped products meet the standards accepted by the associations. One of the largest associations in North America is APA – The Engineered Wood Association (formerly American Plywood Association) [20]. There are usually two letters on a stamp, the first indicating the quality of one surface, while the second showing the quality of the opposite surface, **Figure 5** [7]. This stamp ensures the customer that this product has followed the association's stringent quality and efficiency standards [3]. In Canada, the CertiWood™ Technical Center (formerly CANPLY– the Canadian Plywood Association), a non-profit, industry-funded association, represents manufacturers of EWPs [21]. Those mills, being the members of CertiWood™ Technical Center, can put the stamp with the trademark CANPLY on their products [7, 21].

The vast majority of construction and industrial plywood is used in applications where structural performance surpasses appearance. Some construction and industrial plywood are manufactured with faces chosen mainly for appearance of either plain natural finishes or lightly pigmented finishes [3]. Structural plywood is available in two exposure durability classes: interior and exterior [13]. INTERIOR plywood is only intended for use in dry indoor applications where the panels should be protected from moisture permanently; which is even glued with a water-resistant interior-use adhesive [13]. EXTERIOR plywood is the only panels suitable for outdoor exposure. They are bonded with a waterproof exterior-use adhesive [13], including EXPOSURE 1 and EXPOSURE 2. EXPOSURE 1 panels are waterproof and designed for applications where long construction delays or exposure to high moisture in service are possible [13]. EXPOSURE 2 panels are water-resistant and designed for protected applications, where only minor construction delays are expected since they are mainly developed for interior use [13]. Sheathing grades that are not listed for appearance usually have the grading stamp on one of the faces, whereas grades such as Good Two

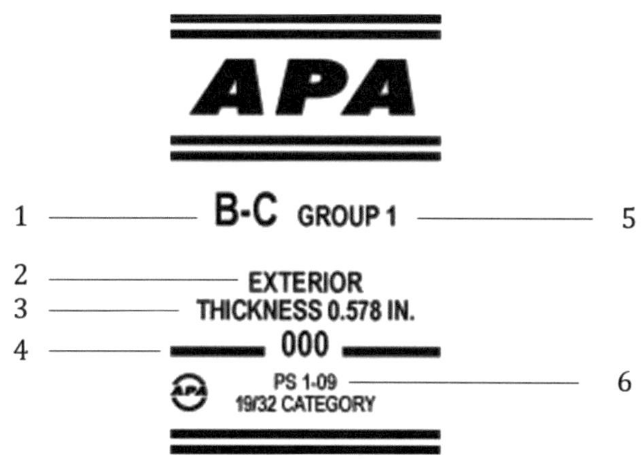

Figure 5.
A sample stamp on plywood: 1- panel grade - panel grades are generally identified in terms of the veneer grade used on the face and back of a panel (e.g., A-B, B-C); 2- bond classification - exposure ratings for APA wood structural panels may be exterior or exposure 1; 3 - decimal thickness declaration; 4 - mill number - manufacturing mill identification number; 5- species group number - classified according to strength and stiffness under manufacturing standard; 6 - product standard [20] (source: photos obtained from APA – the Engineered Wood Association [20]).

Sides are stamped on the edge to avoid affecting the appearance. The strength values stipulated in CSA O86 "Engineering design in wood" [22] are used for Sheathing grade panels based on layups containing only C-grade veneer. Typical DFP and CSP grades include Sanded grades, primarily used in concrete formwork or non-structural applications, and Select and Select Tight Face grades, which are primarily used in floor underlayment applications requiring a smooth and solid surface [14].

Chemical treatments can be applied to plywood to increase its resistance to decay and fire. In Canada, the preservative-treated plywood must be made following CSA O80 "Wood preservation" [23]. To assess the effects of fire retardants or some other potentially strength-reducing compounds, plywood producers shall conduct tests following ASTM D5516 "Standard test method for evaluating the flexural properties of fire-retardant-treated softwood plywood exposed to elevated temperatures" [24] and ASTM D6305 "Standard practice for calculating bending strength design adjustment factors for fire-retardant-treated plywood roof sheathing" [14].

2.5 Properties

The density of plywood depends on the wood species and thickness used, which varies from 400 kg/m^3 (25 lb./ft^3) to 800 kg/m^3 (50 lb./ft^3) [3]. This is compared to the density of oven-dry wood, ranging from approximately 320 kg/m^3 (20 lb./ft^3) to 720 kg/m^3 (45 lb./ft^3) [3]. Plywood has good machining properties; thus, it is possible to work with it just like with ordinary wood, such as sawing, nailing, and gluing. However, the cross-lamination design of plywood, in contrast to wood that is broken down the grain, prevents it from splitting readily in the grain direction. As a result, screws and nails can be used in structural applications near the edges of plywood panels.

Plywood has exceptional built-in resistance to raking, twisting, or distortion, which is especially crucial when care is taken for transferring large shear stresses generated by powerful winds or earthquakes [11]. Many strength properties are equalized by changing the direction at 90 degrees to the grain with each consecutive wood layer of veneer. This provides plywood with a two-way capacity, i.e., the properties in the width direction are approximately equal to those in the length direction. For example, 6 mm (1/4 in) sheathing plywood on a typical framed construction wall with doors and windows delivers double the rigidity and strength furnished by 19 mm (3/4 in) thick boards laid diagonally. When glued to the framework, the strength values for plywood walls are raised even further [11].

Because structural plywood uses waterproof resins, a weather-resistant panel can be obtained if the edges are properly sealed [10]. Awareness of the allowable design values of a plywood panel is not required except in special engineering applications such as diaphragms and earthquake-resistant shear walls. When properly fastened to framing at the correct spacing, the span ratings alone ensure that the panels can work well under the roof and floor loadings stipulated in the building code. In North America, the design values can be found in CSA O86 "Engineering design in wood" [22], Wood Design Manual [25], and APA – Plywood Design Specification [20] or in its Design Capacities of APA Performance-rated Structural-Use Panels Technical Note N375 [26]. For engineering applications, STRUCTURAL I panels are typically the best choice. The typical values of sheathing grades are listed in **Table 1.** The properties of plywood vary with the quality of the constituent layers.

2.6 Applications

Plywood is widely employed in structural and non-structural applications [3], which can be an ideal option for use in both wet and dry environments [14]. It was

Mechanical properties	Metric	Imperial	Remarks
Tensile Strength	27.6–34.5 MPa	4000–5000 psi	Parallel to face; ASTM D3500 [28]
Modulus of Rupture	48.30-68.90 MPa	7000-10000 psi	Parallel to face; ASTM D3043 [29]
Modulus of Elasticity	8.200-10.300 GPa	1190–1490 ksi	Parallel to face; ASTM D3043 [29]
Compressive Strength	31.00–41.40 MPa	4500–6000 psi	Parallel to face; ASTM D3501 [30]
Shear Modulus	0.138–0.207 GPa	20–30 ksi	In-plane (rolling shear) ASTM D2718 [31]
	0.586–0.758 GPa	85-110 ksi	Through thickness (edgewise shear) ASTM D2719 [32]
Shear Strength	1.72–2.07 MPa	250–300 psi	In-plane (rolling shear) ASTM D2718 [31]
	5.52–6.89 MPa	800–1000 psi	Through thickness (edgewise shear) ASTM D2719 [32]

Table 1.
Mechanical properties and testing standards of plywood (source: Wood Engineering Handbook [27]).

reported that plywood took about 54.8% of the market share in 2017 in the construction sector in North America **Figure 6** [21].

In construction, plywood is mainly used as a load-bearing element in platform-frame structures, including single-family and multi-family housing, such as sheathing and underlayment, since it has good dimensional stability and does not crack, cup, or twist [18]. Plywood panels are used as wall sheathing materials, providing high lateral resistance to shear walls and high racking strength, and assisting in achieving the overall thermal efficiency of walls [16, 18]. Roof sheathing is frequently made of plywood. The stiffness of which constitutes diaphragm action when using prescribed framing and nailing patterns [18]. Also, plywood often finds its uses in the fabrication of I-joists as web stocks, marine applications, pallets, industrial containers, and furniture, **Figure 7**. Extra thick plywood with special surface treatment can be used for facing concrete formwork in concrete structures [7, 10, 14].

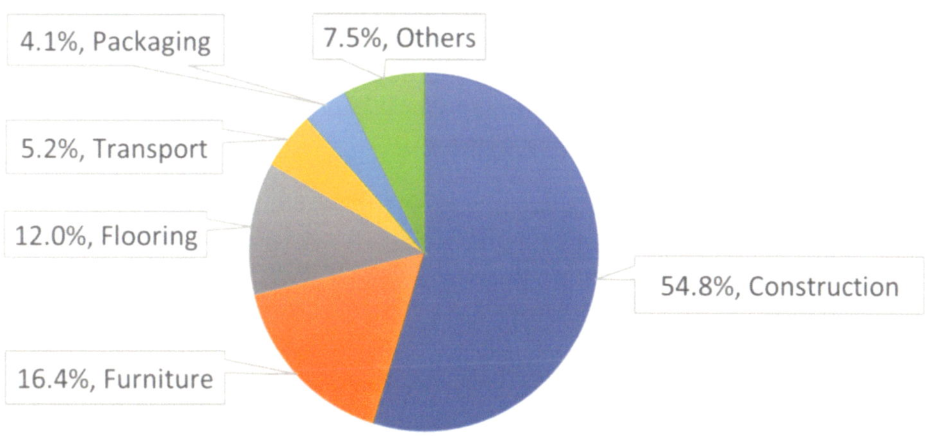

Figure 6.
North America plywood market share by segment in 2017 (source: Figure obtained from BCC research, FAO, RAUTE, IMF, National Statistics Offices [33]).

Floor sheathing and underlayment
(Source: Photos obtained from Georgia-Pacific [34]*)*

Roof sheathing
(Source: Photos obtained from Georgia-Pacific [35]*)*

Plywood webs making I-joists
(Source: Photo obtained from Woodware [36]*)*

Concrete Forming
(Source: Photos obtained from APA – The Engineered Wood Association [20]*)*

Figure 7.
Plywood applications [20, 34–36].

3. Laminated veneer lumber (LVL)

3.1 Introduction

Laminated veneer lumber (LVL) is a type of structural composite lumber (SCL) made by gluing several layers of veneer in the longitudinal direction of the wood, which differs from plywood that has the veneer layers cross-laminated. LVL is one of the most important members in the family of veneer-based EWPs [8]. This material was initially used to produce aircraft propellers and other high-strength aircraft components during World War II [18, 37]. The research and development of LVL can be dated back to the 1940s with an aim at making high-strength parts for aircraft structures out of Sitka spruce veneer [3]. LVL was used as a building material since the mid-1970s [18, 37, 38] when the research was focused on examining the effects of manufacturing variables on LVL with a thickness being up to 12.7 mm (1/2 in) [3]. LVL is now widely used as building and packaging materials [18].

3.2 Manufacturing

The veneer manufacturing and drying processes are almost the same as those used in making plywood. **Figure 4** illustrates the different processes in the manufacturing of LVL from plywood, largely including veneer orientation during layup, hot press type, and end cutting to produce the length required.

To make veneer sheets, the logs are usually peeled in a lathe. The thickness of veneer sheets in 1.5 mm (0.06 in) up to 6.4 mm (0.25 in) [16, 38], the length is 2640 mm (104 in), and the width is 1320 mm (52 in) or 660 mm (26 in) [18]. The veneer is dried to a moisture content of 6–10%, ideally 6–8% [38]. The veneer is clipped to remove any strength-reducing defects and graded. The veneer sheets are cut to the desired width for billet production [18]. The individual veneer sheets are then joined, with the grain of all veneers running in the direction of a billet's length direction. End joints between different veneer pieces are staggered along the length of the billet to distribute any defects that could reduce strength. To effectively transmit load, the joints might be scarf joined or overlapped for some distance [18]. Then the veneer sheets are covered with a waterproof phenol-formaldehyde adhesive [18, 37, 38] or phenol-resorcinol-formaldehyde or polyurethane adhesives [39].

The veneer layup of LVL differs from that of plywood. In the production of LVL, the veneer is oriented in the same direction, i.e., the longitudinal grain direction of the wood, providing the super strength in this direction, which is similar to or larger than solid lumber, **Figure 8**. Thus, LVL is commonly used for beams and columns in the construction of buildings [8, 10]. Veneer sheets in plywood are cross-laminated, making it possess two-way properties, i.e., similar properties in both major and minor directions, as mentioned in Section 2, suitable for sheathing materials.

The pre-pressing of LVL billets could be carried out in a single-opening cold press or a short continuous cold press [38]. The completed billets are simultaneously exposed to pressure to consolidate the veneer and heat to accelerate the curing of the glue [18]. In general, the press temperature used to produce LVL is rarely higher than 175°C (350°F). For the batch type presses, it is usually 160°C (320°F). For the continuous presses, the temperature might be significantly higher since there is a pre-heating zone. As veneer sheets are relatively low in permeability, it is recommended to avoid using a high press temperature, especially when combined with a long press time [38]. This process is similar to used in manufacturing of plywood, except that instead of being formed into thin flat panels, the veneer sheets for making LVL is formed into long billets up to 25 m (80 in) in length. After curing, the billets are sawn to specific lengths and widths for the target application(s) of a LVL product [18].

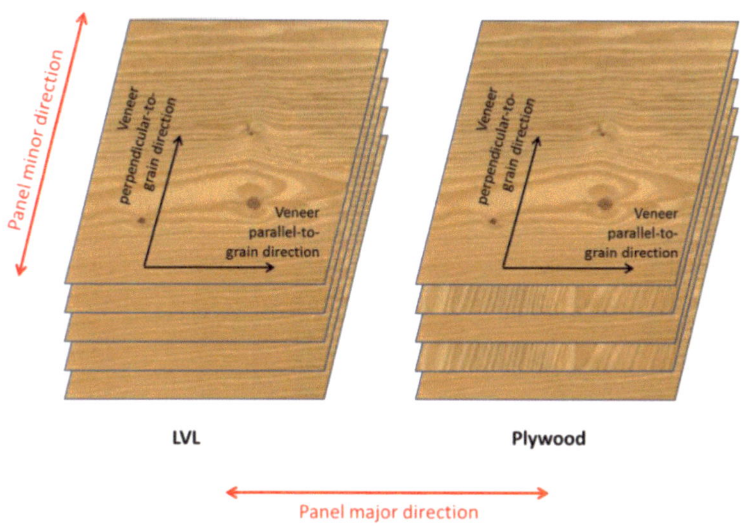

Figure 8.
Layups of LVL and plywood (source: image obtained from Gong [39]).

During the manufacturing of LVL, the selection of veneer is, in terms of thickness and grade, of great importance. The right veneer thickness can help balance LVL properties and manufacturing costs in the production. The veneer used in the manufacture of LVL must be carefully selected in order to obtain the desired engineering properties. Ultrasonic scanning is often used to sort veneer sheets to ensure that the final product has the desired engineering properties [37]. The individual veneer is typically graded so that the strength characteristics of each LVL can be customized [10]. For esthetic reasons and superior flatwise bending properties, the best veneer sheets are usually used as surface plies, while lower-grade sheets are used for the inner plies [38]. For example, if the final use of LVL is scaffold planks, the higher grade veneer will be put on the plank's outer sides [18].

With decreasing veneer thickness, the number of veneer sheets required increases for the same density, thickness, and layout technique of an LVL product. As a result, defects in LVL with thinner veneer will disperse more defects than in LVL with thicker veneer. Because of this, as the veneer thickness decreases, the variation diminishes, and the strength values increase. However, as veneer thickness decreases, resin content, press cycle time, and production cost increase [38].

The strength properties of veneer are more critical for LVL than those in plywood in general. As a result, it is highly desirable for LVL producers to avoid using the veneer sheets with deep lathe checks that cause a reduction in veneer strength. Deep lathe checks can decrease LVL's shear strength and stiffness while having little effect on its MOE. Low shear rigidity can reduce LVL's MOE rating [38].

3.3 Typical species and sizes

LVL can be made from different softwood and hardwood species; however, in North America, Douglas-fir, larch, southern yellow pine, hemlock, aspen, and yellow poplar are the most widely used wood species for producing LVL [18, 37, 40].

LVL is available in thicknesses ranging from 19 mm (3/4 in) to 89 mm (3–1/2 in) and likely to 178 mm (7 in) [18, 41]. The most typical thickness of LVL used in construction are 38 mm (1–1/2 in) [3] and 45 mm (1–3/4 in) [18], from which broader beams can be conveniently assembled on a job site by fastening several LVL plies [18]. The typical depth is from 140 mm (5–1/2 in) to 508 mm (20 in). Different manufacturers can also provide different widths and depths. At the job site, LVL can easily be cut to a length required [37]. Typical lengths of LVL are 14.6 m (48 in), 17 m (56 in), 18.3 m (60 in), 20.1 m (66 in), and 24.4 m (80 in) [10, 37, 41]. LVL is manufactured in the form of billets with widths of 610 mm (24 in) or 1220 mm (48 in). The required depth of LVL can be cut from these billets [18].

3.4 Grading

LVL is a proprietary product; therefore, its engineering properties and sizes can differ from one manufacturer to another. As a result, there is no general production standard or design values in the LVL industry [37]. However, the Canadian Construction Materials Centre (CCMC) reviews and approves the design values, which are derived from test results following CSA O86 "Engineering design in wood" and ASTM D5456 "Standard specification for evaluation of structural composite lumber products" [37]. Each manufacturer develops the characteristic properties of its LVL products by in-grade testing. The manufacturer is also responsible for checking the properties of its products by constant monitoring and quality management. Each manufacturer publishes its own list of design properties, resulting in a unique grade for a given LVL product [42]. Products that satisfy the CCMC criteria are assigned an Evaluation Number and an Evaluation Report that describes the design strengths. They are then entered into the CCMC's Registry of Product Evaluations. The manufacturer's name or product marking, as well as the stress grade, are stamped on the material at different intervals, although this may not be present on every piece due to end cutting [37].

Figure 9 presents a stamp of LVL from APA – The Engineered Wood Association, which shows a qualified LVL grade (e.g., 3100F_b-2.0E), product evaluation reports, the treatment facility, and standard specifications for SCL.

3.5 Properties

The density of LVL is about 480–510 kg/m^3 (30-32 lb./ft^3) [43], which is similar to that of the wood made from. Compared to solid wood, LVL has more stable characteristics than solid timber. This is due to the fact that natural defects, such as knots, splits and slope of grain, are dispersed throughout the material or completely removed during the manufacturing, and dried veneer and adhesives are employed [37].

LVL can easily absorb water, resulting in the change in dimensions, in particular in the thickness direction since there are almost no adhesive restrictions. Therefore, LVL should be protected from the weather during job site storage and after installation [3, 37]. Wrapping the LVL materials for shipping to the job site is also critical for minimizing the moisture effect. End and edge sealing are the commonly used approach to avoid moisture penetration and protect LVL products in their services [37].

Both special cutting, notching, or drilling should be performed according to the manufacturer's instruction. LVL acts similarly to solid sawn timber or glue-laminated beams of equal height, which requires the same fastening and connection requirements as solid lumber [40]. The primary sources of knowledge for design, standard installation descriptions, and performance characteristics are provided in the manufacturer catalogs and inspection reports [37].

3.6 Applications

LVL is mainly used as structural framing in residential and industrial buildings. In the building industry, LVL is widely used for beams or headers over windows and doors on the edge, for hip and valley rafters, scaffold planking, and the flange material of I-joists [3, 10]. LVL may also be used as truck bed decking and road signposts. LVL is chiefly used as a structural component, most commonly

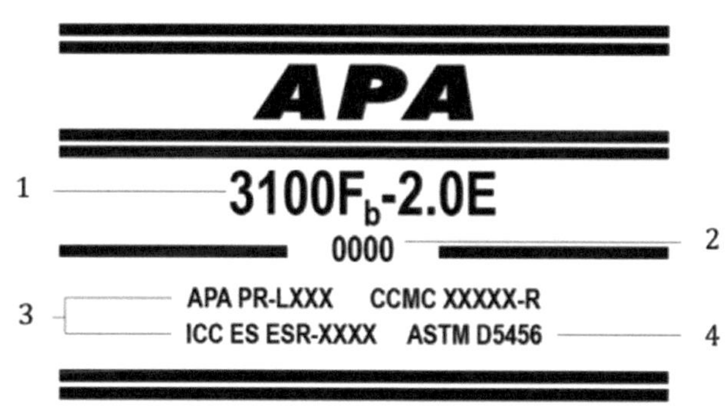

Figure 9.
LVL stamp: 1 – qualified LVL grade (usually represented by design values; 2 – APA mill number; 3 – product evaluation reports; 4 – standard specification for structural composite lumber (source: photo obtained from APA – the Engineered Wood Association).

Figure 10.
Curved LVL structures at Simon Fraser University (Canada) - Ripple Cone Canopy (source: photos obtained from StructureCraft [44]).

in hidden spaces where esthetics is not a concern. Certain manufacturers offer a finished or architectural grade look, but it typically comes at a cost. When using LVL in applications where esthetics is significant, standard wood finishing techniques may be used to accent the grain and preserve the surface layer. The finished wide outer layer of LVL looks like plywood [37]. **Figure 10** shows such a complex curved structure constructed Burnaby, British Columbia, Canada, which was designed and built with 53 parallel CNC-cut spruce LVL sections [44]. Each curved structure was panelized into six segments, shipped to the construction site, and assembled into one piece [44].

Veneer-based EWPs have also been used in the windmill industry, in which wood veneer sheets are used to make windmill blades [45]. Previously, the size of the wooden blade was constrained by the availability of large, consistent-quality tree trunks. Veneering, on the other hand, spreads out defects like knots, resulting in more substantial and more predictable stiffness properties. This makes it possible to make larger wooden blades. When compared to fiberglass, wood laminates provide substantial cost and reduced weight. There are examples of blades made primarily of LVL reinforced with carbon composite spars and coated with a fiberglass composite outer layer [46]. One of the largest windmill blades is 107 meters long (351 ft), which is longer than a football field, produced in Cherbourg, France. It was made from a high-tech sandwich structure consisting of thin layers of glass and carbon fibers and balsa wood veneer [47].

4. Parallel strand lumber (PSL)

4.1 Introduction

Parallel strand lumber (PSL) is known as a composite of veneer strands with wood fibers aligned primarily along the length of the member, i.e., the longitudinal direction of wood [3]. PSL is overall similar to laminated strand lumber (LSL) and oriented strand lumber (OSL) but is made up of veneer strands (sometimes called veneer strips). The length of veneer strands used in PSL is longer than the strands used in LSL and OSL, with a length-to-thickness ratio of around 300.

PSL was invented in 1975 by MacMillan Bloedel Ltd., in Vancouver, Canada, who set out to create a high-strength wood-based material [13]. The first PSL plant was opened in 1982, and its products were first commercially sold for Expo ‘86. MacMillan Bloedel, which is now called Weyerhaeuser, commercialized and patented its PSL products with the brand name Parallam®. The process has been improved over time to produce relatively giant and long beams, and the production and sales have steadily increased [48].

4.2 Manufacturing

The process of manufacturing PSL allows prominent members to be built from small trees, resulting in the more efficient use of forest resources [49]. The first stages in the production of PSL are similar to those used in the production of plywood or LVL. **Figure 4** illustrates the unique processes employed in the manufacturing of PSL, differentiating from those in plywood or LVL. To make veneer, logs are turned on a lathe [18]. The thickness of veneer is from 3 mm (1/8 in) to 6.4 mm (1/4 in) [3]. The veneer sheets are then dried to a moisture content of 2–3% before being sliced into long thin veneer strands parallel to one another [9]. After that, the veneer sheets are clipped into long, narrow veneer strands with a length of 2.4 m (8 feet), a width of 13 mm (1/2 in) [18], and a thickness from 2.54 mm (1/10 in) to 3.175 mm (1/8 in) [38].

The production process is designed to use materials from the log roundup and other less than full-width veneer in the veneer cutting stage. As a result, the process uses waste materials from a plywood or LVL operation [3]. The veneer strands are oriented to the length direction of a continuous billet using special equipment (**Figure 11**) and mixed with a waterproof exterior structural adhesive, such as phenol-formaldehyde, prior to hot-pressing.

The pressing process densifies the veneer strands to some degree, and the adhesive is cured with the aid of microwave technology [18, 49]. A continuous press is employed to produce PSL, which theoretically produces an unlimited length but is constrained only by transportation restrictions [3].

4.3 Typical species and sizes

Douglas fir is used to produce PSL in Canada, and southern yellow pines are employed in the USA. In addition to this, western hemlock and yellow poplar are also used [3, 49, 51]. In general, there are no restrictions on the use of other wood species.

The available stock sizes for PSL have to be compatible with existing wood framing materials and standard dimensions [18]. PSL beams are available in thicknesses

Figure 11.
Orientation of veneer strands in PSL (source: photos obtained from WorldPress.com [50]).

from 89 mm (3–1/2 in) to 178 mm (7 in), and in-depth from 235 mm (9–1/4 in) to 457 mm (18 in) [41]. PSL columns come in square and rectangular shapes of a dimension of 89 mm (3–1/2 in), 133 mm (5–1/4 in), or 178 mm (7 in) [41]. Smaller thicknesses can also be used, either individually as single plies or in combination for multi-ply applications [18, 49]. Steel connectors are usually required for larger dimensions [51]. PSL is available in lengths up to 20 meters (66 ft) [41].

The beam-like PSL products can be also ripped into thin boards, **Figure 12**, which opens a window for non-structural applications [18].

4.4 Grading

PSL is a proprietary product, the same as LVL. Therefore, specifications and dimensions are unique to each manufacturer. In North America, both PSL and LVL are treated as the same structural composite lumber [18]. The evaluation procedure and grade determination of PSL are the same as LVL (refer to Section 3.4). **Figure 13** presents a stamp of PSL (Parallam® Plus), including a description of the product and uses, the type of treatment, and the treatment facility. The treatment stamp can also reference the treating standards (such as AWPA U1/UC4A by the American Wood Protection Association) and third-party quality program monitor (SPIB - Southern Pine Inspection Bureau) [53].

4.5 Properties

Since natural defects such as knots, the slope of grain, and splits have been scattered across the material or eliminated during the manufacturing process. The combination of a structural adhesive used with dried wood veneer strands, heat, and pressure employed during pressing makes PSL less warping than solid timber. Therefore, PSL is a type of highly consistent, uniform EWPs [49], which exhibits much less variability and larger load-bearing capabilities than solid lumber [51]. The density of PSL is 720 kg/m^3 (45 lb./ft^3) [54], which is similar to that of the wood used. Other advantages of PSL are given to its high strength, stiffness, and dimensional flexibility [49]. PSL is less susceptible to shrinkage, warping, and splitting as it has a moisture content of 11% [49].

The texture of PSL is rich due to the grain of wood veneer strands and dark glue lines. PSL is a visually appealing construction material that fits well to the applications that require a high level of finished appearance [49]. The techniques applicable to sawn lumber can be used to machine, stain, and finish PSL. At the end of the manufacturing period, PSL is sanded to ensure exact dimensions and a high-quality appearance. Stain can be used to emphasize the warmth and texture of the wood [49]. It should be pointed out that the special cutting, notching, or drilling of PSL shall be performed in compliance with the manufacturer's instruction.

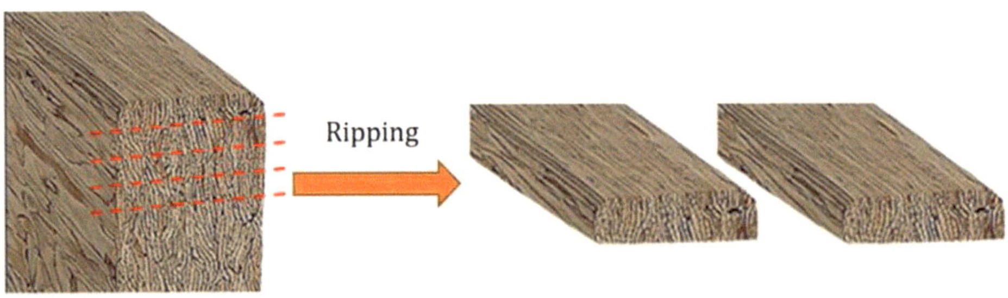

Figure 12.
Boards ripped from PSL (source: photos adopted from Goulddesigninc.com [52]).

Figure 13.
Stamp on PSL (Parallam® plus) (source: image obtained from Weyerhaeuser [53]).

Property	Value (MPa)
Modulus of elasticity, E	13,800
Allowable bending stress, f_b (300 mm depth)	37.0
Allowable shear stress, f_v (perpendicular to glue line or wide face of the strand)	3.7
Allowable bearing stress, f_{cp} (parallel to glue line or wide face of the strand)	9.4

Table 2.
Major strength properties of SCL (source: Canadian Wood Council [41]).

As mentioned above, both PSL and LVL are treated as the same SCL in Canadian practices; therefore, their design values are the same. **Table 2** lists the key strength properties of sample SCL. Some specific design values can be obtained from manufacturers [41].

4.6 Applications

PSL is mainly used in residential, commercial, and industrial construction as structural framing components, such as beams and columns in the post-and-beam construction, and headers, pillars, and lintels in the light-frame construction [18]. According to Part 3 of the National Building Code of Canada, CCMC has approved PSL for use as heavy timber construction [51]. Due to the excellent strength characteristics of PSL, it is possible to use it in the design of the roofs with a large span and rooms with open spaces. **Figure 14**(left) presents PSL beams that are used for an open floor plan at Vancouver Firehall No.15, Vancouver, Canada. Another example is presented in **Figure 14**(right), showing a portable Concord Pacific Display Pavilion at Vancouver, Canada, which uses a glass-enclosed superstructure made of exposed PSL columns. In addition, PSL is a visually attractive material; thus, it is well suited to applications where the finished look is essential. It can be also appropriate for hidden structural applications where appearance is unimportant [18].

Figure 14.
Floor and columns made of PSL in Vancouver Firehall No.15 (left) and Concord Pacific display pavilion (right), respectively (source: photos obtained from StructureCraft [55]).

5. Endnotes

EWPs are relatively recent structural members that have been widely incorporated in the building industry in North America and beyond. They have been invented and used for making timber buildings and furniture. The family of veneer-based EWPs mainly has three major members, i.e., plywood, LVL, and PSL. Because the majority of veneer-based EWPs are designed to handle relatively large loads, they must be manufactured in accordance with recognized standards or technical guides to ensure that the required engineering design values and applications are met. Veneer-based EWPs have been accepted and acknowledged in the building industry as premium structural materials. It is possible to make these products considerably large from small-diameter logs. The only restriction can be the length of LVL and PSL during transportation [38].

Non-traditional resources (such as under-utilized wood species) can be used to manufacture veneer-based EWPs of better physical and mechanical properties than other traditional structural products (such as solid timber products) [1]. Due to engineering design, removal of defects, drying of wood materials, application of adhesive, and layer-by-layer bonding, the veneer-based EWPs are stronger and more durable than solid wood of the same size. This is an outstanding advantage for constructing a building requiring high strength without a bulky appearance. Typically, LVL and PSL have about three times larger bending strength and 30% larger stiffness than the lumber products of comparable sizes [41].

Interest in veneer-based EWPs products will continue to grow for ecological reasons. For example, restrictions have been introduced in many countries on deforestation of large-diameter old-growth trees. Due to the needs in the construction market, alternative materials must be further developed. The rapid advancement in technology, along with the available raw materials, i.e., small-size fast-growth trees, would inevitably accelerate the development of EWPs [2]. Also, in recent years, the minimization of carbon footprints in construction has reached a consensus. Many architects and engineers have been designing and constructing buildings with 100% solid wood and EWPs. Meanwhile, research on the standardization of the veneer-based EWPs and expansion of their uses is no doubt required. Modernization of existing equipment and improvement of the gluing systems will allow the creation of innovative designs and special shapes that are currently not available for wood products. This will certainly expand the matrix of applications for veneer-based EWPs, as well as making them become more competitive in the market of building materials.

Acknowledgements

This piece of work was financially supported by the New Brunswick Innovation Research Chair Initiative Program by the New Brunswick Innovation Foundation (Canada) and the Collaborative Research and Development Grants by Natural Sciences and Engineering Research Council of Canada (CRDPJ 523922-18).

Conflict of interest

The authors declare no conflict of interest.

Author details

Elena Vladimirova* and Meng Gong
University of New Brunswick, Fredericton, Canada

*Address all correspondence to: evladimi@unb.ca

References

[1] Thelandersson S, Larsen HJ. Timber Engineering. New York, USA: Wiley; 2003

[2] Bodig J, Jayne BA. Mechanics of Wood and Wood Composites. Reprint edition. Malabar, FL: Krieger Pub Co; 1993

[3] Forest Products Laboratory (FPL). WoWood Handbook-Wood as an Engineering Material. General Technical Report FPL-GTR-190. Madison, WI: U.S. Department of Agriculture, Forest Service, Forest Products Laboratory; 2010

[4] Gong M. Lumber-Based Mass Timber Products in Construction. London, UK: IntechOpen; 2019. pp. 7-16. DOI: 10.5772/intechopen.85808

[5] Wooden Plywood Board. IndiamartCom 2021. https://www.indiamart.com/proddetail/wooden-plywood-board-14009756373.html [Accessed: June 16, 2021]

[6] ThinkWood CEU. Design and Construction of Taller Wood Buildings. 2021. https://www.thinkwood.com/wp-content/uploads/2019/09/ThinkWood-CEU-Design-and-Construction-of-Taller-Wood-Buildings.pdf [Accessed: June 3, 2021]

[7] Vogt F, Nauth M. Carpentry. Toronto, Canada: Nelson Education Limited; 2015

[8] Ansell MP. Wood Composites. Cambridge, United Kingdom: Elsevier Science & Technology; 2015

[9] Porteous J, Kermani A. Structural Timber Design to Eurocode 5. Hoboken, United Kingdom: John Wiley & Sons, Incorporated; 2013

[10] Canada NR. Plywood 2014. https://www.nrcan.gc.ca/our-natural-resources/forests-forestry/forest-industry-trade/forest-products-applications/taxonomy-wood-products/plywood/15855 [Accessed: March 24, 2021]

[11] Sellers T. Plywood and Adhesive Technology. Forest Products Utilization Laboratory, Mississippi State University, Mississippi, USA: CRC Press; 1985

[12] Filippovich A. Plywood history. From ancient Egypt to mid-twentieth century, part 1. Saint-Petersburg, Russia: Lesprominform. 2018;**3**:104-113

[13] Smulski S, editor. Engineered Wood Products: A Guide for Specifiers, Designers & Users. Madison, WI: Pfs Research Foundation; 1997

[14] Plywood - Sizes - Grades. The Canadian Wood Council - CWC 2021. https://cwc.ca/how-to-build-with-wood/wood-products/panel-products/plywood/. [Accessed: March 24, 2021]

[15] McIntyre RB. Design Features of the Mosquito. Aircraft Engineering and Aerospace Technology Emerald Insight, United Kingdom. 1944;**16**:64-67. DOI: 10.1108/eb031101

[16] Bowyer JL, Shmulsky R, Haygreen JG. Forest Products and Wood Science: An Introduction. 5th Ed. Oxford, UK: Wiley-Blackwell; 2007

[17] FAO. Yearbook of Forest Products 2000,2005,2010,2015,2019. Rome, Italy: FAO; 2021. DOI: 10.4060/cb3795m

[18] Canadian Wood Council. Wood Reference Handbook. Ottawa, Ontario, Canada: Canadian Wood Council; 1997

[19] Plywood. Wikipedia 2021. https://en.wikipedia.org/w/index.php?title=Plywood&oldid=1028259276 [Accessed: June 17, 2021]

[20] APA – The Engineered Wood Association. 2021. https://www.apawood.org/ [Accessed: April 14, 2021]

[21] certiWood™ Canply plywood. Plywood Design Fundamentals Canadian Plywood Association 2021. https://tolko.com/wp-content/uploads/2020/05/Plywood-Design-Fundamentals-Canadian-Plywood-Association.pdf [Accessed: May 18, 2021]

[22] CSA Group. CSA 086:19 Engineering Design in Wood. National Standard of Canada. Toronto, Canada: CSA Group; 2021

[23] CSA Group. CAN/CSA-O80 SERIES-15 (R2020) Wood Preservation. National Standard of Canada. Toronto, Canada: CSA Group; 2020

[24] ASTM International. D5516-18 Standard Test Method for Evaluating the Flexural Properties of Fire-Retardant Treated Softwood Plywood Exposed to Elevated Temperatures. West Conshohocken, PA: ASTM International; 2018. DOI: 10.1520/D5516-18

[25] Wood Design Manual 2017. Ottawa, Ontario, Canada: Canadian Wood Council Webstore; 2017

[26] APA. Design Capacities of APA Performance-Rated Structural-Use Panels Technical Note N375. Tacoma, Washington, USA: APA The Engineered Wood Association; 1995

[27] Wood Engineering Handbook. Second ed. Englewood Cliffs, NJ: Forest Products Laboratory; Prentice Hall; 1990 http://www.matweb.com/search/datasheet_print.aspx?matguid=bd6620450973496ea2578c283e9fb807 [Accessed: May 31, 2021]

[28] ASTM International. D3500-20 Standard Test Methods for Wood Structural Panels in Tension. West Conshohocken, PA: ASTM International; 2020. DOI: 10.1520/D3500-20

[29] ASTM International. D3043-17 Standard Test Methods for Structural Panels in Flexure. West Conshohocken, PA: ASTM International; 2017. DOI: 10.1520/D3043-17

[30] ASTM International. D3501-05a (2018) Standard Test Methods for Wood-Based Structural Panels in Compression. West Conshohocken, PA: ASTM International; 2018. DOI: 10.1520/D3501-05AR18

[31] ASTM International. D2718-18 Standard Test Method for Structural Panels in Planar Shear (Rolling Shear). West Conshohocken, PA: ASTM International; 2018. DOI: 10.1520/D2718-18

[32] ASTM International. D2719-19 Standard Test Methods for Structural Panels in Shear through-the-Thickness. West Conshohocken, PA: ASTM International; 2019. DOI: 10.1520/D2719-19

[33] BBC Research. MFG058A Plywood Manufacturing: Global Markets to 2022. Wellesley, Massachusetts, USA: BCC Publishing; 2018

[34] Georgia-Pacific Building Products. Plytanium Sturd-I-Floor Plywood Subfloor Panels. Atlanta, Georgia, USA: Georgia-Pacific Building Products; 2021. https://buildgp.com/product/plytanium-sturd-i-floor/ [Accessed: May 22, 2021]

[35] Georgia-Pacific Building Products. Plywood Roof Sheathing, OSB Roofing Boards. Atlanta, Georgia, USA: Georgia-Pacific Building Products; 2021. https://buildgp.com/wood/roof-sheathing/ [Accessed: May 22, 2021]

[36] American Wood Council. Visual Guide to Floor, Roof, and Wall Systems. Washington, DC, USA: American Wood Council; 2021

[37] Laminate Veneer Lumber. The Canadian Wood Council - CWC 2021. https://cwc.ca/how-to-build-with-wood/

wood-products/structural-composite/laminate-veneer-lumber/ [Accessed: April 14, 2021]

[38] Hsu WE. Structural Composite Lumber Manufacturing. Scotts Valley, California, USA: CreateSpace Independent Publishing Platform; 2016

[39] Gong M. Wood Technology, Lecture on Wood Products. Fredericton, Canada: University of New Brunswick; 2021

[40] Natural Resources Canada. Laminated veneer lumber. Ottawa, ON, Canada: Natural Resources Canada; 2014. https://www.nrcan.gc.ca/our-natural-resources/forests-forestry/forest-industry-trade/forest-products-applications/taxonomy-wood-products/laminated-veneer-lumber/15795 [Accessed: April 14, 2021]

[41] Introduction to Wood Design 2018. Canadian Wood Council; 2018

[42] Structural Grading. WoodSolutions; 2021 https://www.woodsolutions.com.au/articles/structural-grading [Accessed: April 14, 2021]

[43] Stora Enso Wood Products GmbH. LVL by Stora Enso Technical brochure. Helsinki. Finland: Stora Enso; 2021. https://www.storaenso.com/-/media/documents/download-center/documents/product-brochures/wood-products/lvl-by-stora-enso-technical-brochure-en.pdf [Accessed: June 4, 2021]

[44] Laminated Veneer Lumber | Engineered Wood | StructureCraft. StructureCraft Builders. 2021. https://structurecraft.com/materials/engineered-wood/laminated-veneer-lumber [Accessed: May 21, 2021]

[45] Moretti PM, Divone LV. Modern windmills. Scientific American. 1986;**254**:110-119

[46] Decker KD. Reorienting the economy to the rhythms of nature: Learning to live with intermittent energy supply. American Journal of Economics and Sociology. 2020;**79**:877-905. DOI: 10.1111/ajes.12333

[47] Extreme Measures: At 107 Meters, The World's Largest Wind Turbine Blade Is Longer Than A Football Field. Here's What It Looks Like | GE News. 2019. https://www.ge.com/news/reports/extreme-measures-107-meters-worlds-largest-wind-turbine-blade-longer-football-field-heres-looks-like [Accessed: May 30, 2021]

[48] Weyerhaeuser Uses Strong Scraps. Wood Business. 2014. https://www.woodbusiness.ca/weyerhaeuser-strong-scraps-1859/ [Accessed: April 24, 2021]

[49] Parallel Strand Lumber. The Canadian Wood Council – CWC. 2021. https://cwc.ca/how-to-build-with-wood/wood-products/structural-composite/parallel-strand-lumber/ [Accessed: April 16, 2021]

[50] Waste not Cambodia: Image. 2021. https://wastenotcambodia.files.wordpress.com/2012/05/parallam1.jpg [Accessed: June 16, 2021]

[51] Canada NR. Oriented Strand Lumber. 2014. https://www.nrcan.gc.ca/our-natural-resources/forests-forestry/forest-industry-trade/forest-products-applications/taxonomy-wood-products/oriented-strand-lumber/15827 [Accessed: April 24, 2021]

[52] Gould-Design-Inc. Parallel-strand-lumber. Gould Design, Inc. 2013. https://goulddesigninc.com/2013/11/14/the-abcs-of-ewp-engineered-wood-products/parallel-strand-lumber/ [Accessed: June 16, 2021]

[53] How can I be sure that I have genuine Parallam® Plus PSL? Trus Joist Technical Support. 2021. https://www.techsupport.weyerhaeuser.com/hc/

en-us/articles/205173710-How-can-I-be-sure-that-I-have-genuine-Parallam-Plus-PSL [Accessed: May 30, 2021]

[54] Parallam® PSL Deep Beam Specifier's Guide. 2015

[55] Parallel Strand Lumber. Engineered Wood. StructureCraft. StructureCraft Builders. 2021. https://structurecraft.com/materials/engineered-wood/parallel-strand-lumber [Accessed: May 21, 2021]

Chapter 4

Strand-Based Engineered Wood Products in Construction

Zizhen Gao and Meng Gong

Abstract

Strand-based engineered wood products (EWPs) have been widely employed in construction since their emergence in the 1970s. The use of strand-based EWPs can significantly increase the yield of forest resources by utilizing submarginal logs and branches. In this chapter, the strand-based EWPs, including oriented strand board (OSB), laminated strand lumber (LSL), and oriented strand lumber (OSL), are discussed in terms of their fabrication, properties, and uses in construction. Specifically, the manufacturing requirements for elements (i.e., strands), such as dimension, density, and moisture content, are introduced. The major manufacturing processes, such as selection of adhesives, pressing parameters, and thickness control, are also discussed. In addition, the engineering properties and uses of these EWPs are illustrated. Furthermore, some innovative applications of these products, such as hybrid cross-laminated timber, are presented in this chapter.

Keywords: strand-based engineered wood products, oriented strand board (OSB), laminated strand lumber (LSL), oriented strand lumber (OSL), manufacturing process, mechanical properties, structural applications

1. Introduction

Prior to availability of steel and concrete, wood has been widely used as the primary structural material in North America. In the recent past, the concept of green building has become a mainstream and people have been becoming aware of potential environmental benefits of wood and wood-based materials [1–3]. With the increase in population and wood buildings across the world, there is an observable increase in demand for structural wood products [4]. However, the production and properties of solid lumber are hard to meet the demand of construction since the logs become smaller and the wood quality becomes lower due to the change in raw materials that come from the faster-growing plantation species. As a result, engineered wood products (EWPs), such as glue-laminated timber (GLT), cross-laminated timber (CLT), laminated veneer lumber (LVL), and oriented strand board (OSB), were developed as the alternative, since they commonly provide better and more predictable physical and mechanical properties than solid wood lumbers [5].

Plywood used to be the most common panel material widely used as sheathing materials in wood buildings before the advent of strand-based wood products. Relatively, LVL, as a thicker structural form of veneer-based products, was developed for beams, studs, and other components in wood buildings as the substitution of solid lumber. However, small, twisted, and juvenile logs are not suitable for producing veneer. In addition, up to 60% of the log volume delivered

IntechOpen

to a plywood plant is treated as residues [6, 7]. The cost of manufacturing veneer-based EWPs keep increasing because the logs of satisfying quality, such as straight and large diameter stems, have been in short supply [6, 7]. Therefore, strand-based materials were invented as a substitute for plywood or other solid wood products to manufacture panel- and beam-like components in buildings. In the early 1970s, strand-based wood panels called waferboard, which consisted of randomly placed wood strands, were patented by Armin Elmendorf [8]. Waferboard was the parent product of what would evolve into modern strand-based products like OSB. When people realized that the properties along the board could be improved by controlling the direction of strand lay-up, OSB was established in its present form and was subsequently produced as a commercial wood panel in North American [8]. In the next decade after the 1970s, the OSB industry had developed rapidly and began to rival traditional plywood in production. By the mid-1990s, the production of OSB had increased to more than 50% of the plywood volume [8, 9]. Due to the advantages of OSB and its great success achieved in the wood panel market, laminated strand lumber (LSL) and oriented strand lumber (OSL), which are other two forms of strand-based products, were developed as a substitute for LVL and dimensional lumber in the 1990s. In the early 2000s, OSB surpassed plywood in production and sales, taking the largest market share of the wood panel products. Up to now, OSB is the leading wood panel product and occupies a 75% market share in North America [9]. Moreover, LSL and OSL have also gained some market share in the lumber market [8, 9]. **Figure 1** illustrates the production of OSB and plywood in North America over the last six decades [10]. Overall, the production of OSB exceeded that of plywood around 2000 and there was a decrease in the production of both OSB and plywood between 2005 and 2010 due to the decline in the housing market in the USA [10].

Strand-based EWPs have become important members of the family of EWPs with development and advance in technologies. Over veneer- or lumber-based EWPs, strand-based EWPs have the advantages such as low requirements for raw materials, high-dimensional flexibility, and stable physical and mechanical properties [1]. In this chapter, the strand-based EWPs, including OSB and strand-based SCL such as LSL and OSL, are discussed in terms of their fabrication, properties, and uses in construction.

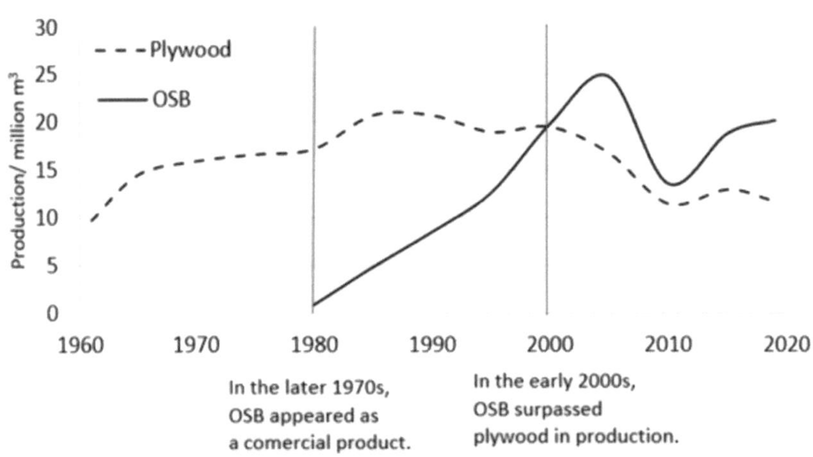

Figure 1.
Production of OSB and plywood in North America (source: FAOSTAT) [10].

2. Raw materials

2.1 Wood strands

One of the main advantages of strand-based EWPs compared to other types of EWPs is that a variety of wood sources are appropriate to be used in their manufacturing. In North America, aspen is one of the primary species for making OSB or other strand-based EWPs due to its low density, low cost, and rich abundance. In addition to aspen, southern pine, spruce, birch, yellow poplar, sweetgum, sassafras, and beech are also suitable raw materials for producing strands [1]. In reality, almost any low- to medium-density species (300–600 kg/m^3) can be utilized to make the elements, that is, strands also called wafers, chips or flakes, which have different sizes, in the fabrication of strand-based EWPs. In the real world, the strands made from high-density species, such as beech and birch, are often mixed with those made from low-density species to improve the mechanical properties of finished products [9]. In addition, wood with localized defects (such as knots), and wood-derived from small-diameter logs, forest residues (such as branches), or exotic and invasive species, can be also used in the manufacturing of strands. The yield of wood could reach 80–90% in the process of strand manufacturing since losses mainly coming from bark and broken strands during processing [1]. **Table 1** compares the quality requirements of raw materials between strand-based and other types of EWPs [11].

The manufacturing of strands contains the following key processes. Logs are first soaked in a hot water pond before debarking. This helps loosen the bark from logs and soften logs by increasing the moisture content of the logs to achieve a more desirable strand size during the stranding process. The fresh logs are then carried to a drum- or disc-type strander to make strands after debarking. Generally, strands produced are 75–300 mm in length, 15–25 mm in width, and 0.3–0.8 mm in thickness. Strand geometry has an essential influence on the performance of finished products [12, 13]. Therefore, strand grading using screens is critical to separate out strands that do not meet the dimensional requirement. The moisture content of strands has also a great influence on the bonding quality between strands. Typically, a cylindrical dryer is used to dry strands to a final moisture content of 2–6%, which is dependent on the type of adhesives [9, 11].

2.2 Adhesives

As the main raw material besides wood strands in the manufacturing of strand-based wood products, adhesives play a critical role in the determination of

Products	Generic raw material requirements	Preferred feedstock
Lumber-based EWPs such as GLT and CLT	Adequate strength, stiffness, and stability; small knots	Lumber from mature trees, but low-grade lumber for central layers
Veneer-based EWPs such as plywood and LVL	Large diameter logs; easy of being peeled	Mature trees with a large diameter
Strand-based EWPs such as OSB, LSL, and OSL	No specific requirements in mechanical properties and dimension of logs	Mature or juvenile logs, wood derived from small-diameter logs, forest residues, and wastes from the production of lumber or veneer

Table 1.
Quality requirements for raw materials of manufacturing EWPs.

the performance of finished products, in which phenol-formaldehyde (PF) and polymeric methylene di-isocyanate (pMDI) are the most commonly used structural adhesives [1, 6]. The adhesives to be used must meet the specific requirements for the mechanical properties and durability performance of the final products [1].

PF is one of the most common adhesives used in wood composites that are used as structural components for exterior uses due to its good mechanical properties and durability performance [1]. Therefore, PF resins are used in the manufacture of EWPs, which are exposed to weather or other moisture exposure situations during construction and use. Cured PF resins could remain stable chemical and physical properties in high temperature and humidity environments [14]. Therefore, PF has the ability to maintain the dimensional stability and mechanical properties of EWPs under wet conditions [1, 14]. The main disadvantage of PF is that its curing is a slow process compared with other structural adhesives, which requires a longer press time and higher press temperature during the manufacturing process [15]. As a result, phenol-resorcinol-formaldehyde (PRF) adhesives were developed as the substitution of PF [7]. As the name implies, it contains resorcinol in addition to phenol and formaldehyde. PRF has the advantage over PF of being curable at room temperature due to being much faster in reaction, and has similar good mechanical properties and durability performance to PF [7]. Some manufacturers have used PRF as the adhesives for strand-based wood products.

pMDI is usually used as an alternative to PF resin in strand-based EWPs. Although the costs of pMDI are higher than PF, it is taking the market share away from PF due to its rapid cure rate and higher tolerance of moisture content in the wood strands [1, 16]. Typically, pMDI is used in the core layer of strand-based EWPs since PF is used in the surface layers because pMDI provides good bonding strength, which makes the panel more waterproof and requires a lower curing temperature than PF. It is worth noting that mold release is needed if pMDI is used in the surface layer because it could bond strongly to the mental platens of the press [6, 16]. Special precautionary protective measures are required to use pMDI because the uncured resin can result in the chemical sensitization of persons exposed to it [1]. Fortunately, cured pMDI resin poses no recognized health concerns [1].

3. Oriented strand board (OSB)

3.1 Overview

OSB is multi-layered panels made from wood strands bonded together with adhesives under pressure and heat, which was first developed as the substitution of plywood in the 1970s [17]. Typically, OSB panels are made up of three orthogonal layers of strands, as shown in **Figure 2** [18]. The strands in surface layers are parallel to the long axis of the panel, whereas those in the core layer are perpendicular to the long axis of the panel or laid randomly. The performance of OSB panels can be engineering-designed by changing the strand size, orientation, and layered construction, allowing OSB to suit different uses [16]. OSB has been widely used for sheathing panels, web stocks of I-joists, packaging materials, and decorative and other purposes.

3.2 Manufacturing process

OSB can be produced using small-diameter logs from fast-growing trees, cores from veneering, and forest residues (such as branches and treetops). Only fresh

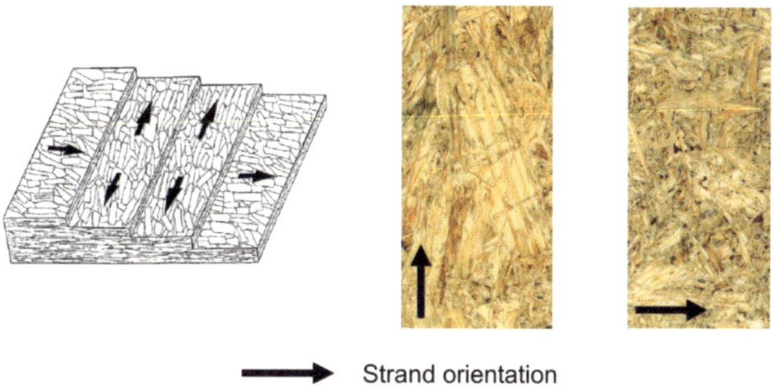

Figure 2.
Strand orientation of OSB panels: Schematic (left) (source: Picture obtained from FAPC) [18]; large-size strands in the surface layer (center) and small-size strands in the core layer (right).

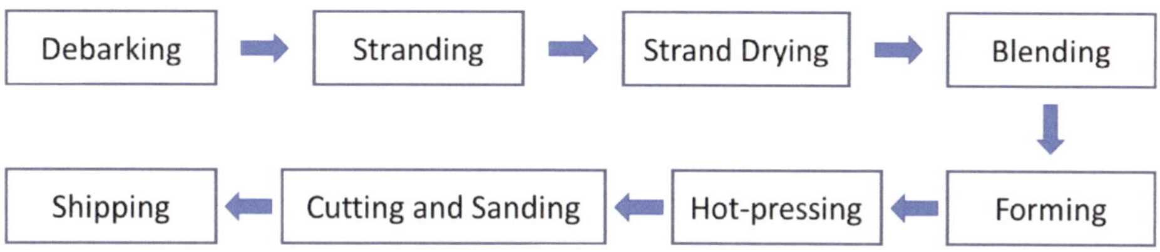

Figure 3.
The key manufacturing processes of OSB.

wood, rather than recycled wood, can be used for making stands and OSB [6]. A typical production process is given in **Figure 3**.

3.2.1 Stranding

The sizes of strands are given in Section 2.1, which results in a length-to-width ratio of about 3 [1, 19]. There are two types of stranders, drum and disc, producing strands. The disc stranders have advantages of simple structure, low price, and high production efficiency. The dimensions of the disc strands can be adjusted according to the requirements of different products when they are produced using a disc strander. However, it can only use large-size logs as raw materials. In contrast to disc stranders, the drum stranders have the adaptability to various shapes of raw materials, which means lower production costs for strand production. The larger-size strands are usually used in the surface layers, whereas the smaller-size strands in the core layer, constituting an optimized composite structure of better bending and shear properties.

3.2.2 Blending

Good adhesion between strands is a key factor to ensure the good performance of OSB panels. The process of adding a structural adhesive(s) and waxes to the strands is called blending. The adhesive content used in the surface layers is 3–6% while that in the core layer is 4–8%. Application of more adhesive in the core layer than the surface ones is a practical process in a production line. This is because the strands in the core layer have relatively smaller sizes than those in the surface ones, producing a larger surface area [6]. During blending, the wax with amount of 0.25–2% by weight is commonly added to improve the water resistance and reduce the thickness swelling of final OSB panels [4, 9]. Blending for surface and core layers of strands is usually done separately due to the difference in the spread amount of adhesive and wax, strand size, and type of adhesives between the surface and core layers. Therefore, at least two blenders are needed in a production line, one for surface layer strands and the other for core layer strands [9, 20].

3.2.3 Forming

The orientation and proportion of long strands in the surface layers significantly affect the bending properties of OSB. In general, about 80% of strands are oriented parallel to the long axis to meet the performance requirements of OSB [16]. Therefore, the process of depositing blended strands into the form of a mat is called forming. In addition to the orientation of the strands, the forming can also assist to produce a uniform weight across the area of a board, which can ensure the final OSB panels to have uniform density and mechanical properties. Although some new technologies, such as electrostatic strand orientation, have been developed, mechanical orientation is still the most reliable technology and is widely used in industrial manufacturing [6, 9]. Following the forming process, a heavy steel drum is often used to pre-press the mat before hot-pressing, albeit this step is not a must [16]. Pre-pressing can effectively reduce the risk of edge collapse during hot pressing. In addition, pre-pressing can also shorten press cycle time and reduce the risk of resin pre-curing, since pre-pressing squeezes much of air out of the mat [21].

3.2.4 Hot pressing

Hot pressing is the most important process, which cures the adhesives between strands under heat and pressure. This is the real step of turning the strands to a panel. A multi-opening press is widely used in the manufacturing of OSB in North America [9]. The size of press platens in a multi-opening press is generally 3.66 × 7.63 m (12 × 25 ft) but can be up to 3.66 × 12.2 m (12 × 40 ft). Large-press platens are favorable because they can greatly increase the production capacity of a production line. In addition, large single-opening presses and continuous presses are also used to manufacture OSB panels in some mills [9].

Heated press platens are enough to generate and transfer heat from the surfaces of a mat to the core to cure the adhesives since most of OSB panels are relatively thin with a thickness range of 7.9–28.6 mm (5/16 to 1–1/8 in.) [22]. The temperature during hot pressing ranges from 175 to 205°C (350–400°F) for PF adhesives. The pressing usually lasts for 3–6 min, which is depended on the type of adhesives, board density, and thickness [1, 9]. Theoretically, the relationship between the press time and the thickness of the OSB panel is quadratic when the densities are similar. For example, the time of hot pressing an 18-mm-thick OSB panel is about four times a 9-mm-thick one [9]. That means the time and cost of processing OSB panels greatly increase with increasing the thickness. To decrease the press time and reduce the cost, a high-frequency press or a steam injection press is adopted when making relatively thick OSB in some mills [1, 9]. Typically, the pressure used in the hot press is 4800–5500 kPa [9]. In short, the key to hot-press process is to reduce hot pressing time and improve production efficiency. Therefore, in industrial production, hot stacking of pressed materials is implemented shortly after emergence from the press to reduce the press time and energy consumption [6]. After the adhesive has completely cured, the OSB panels are cut to desired dimensions and sanded to nominal thickness. Finally, the edges of each panel are coated with zinc borate and oxine copper to improve its durability [18].

3.3 Properties and grading of OSB

As an alternative material of plywood, the nominal dimension of an OSB panel is typically 1.2 × 2.4 m (4 × 8 ft) in North America. Moreover, some OSB manufacturers also provide oversized panels with a dimension of 2.4 × 7.2 m (8 × 24 ft) for panelized roof and wall systems, facings for structural insulated panels (SIPs), and modular floors [22]. The physical and mechanical properties of OSB are directly related to wood species, strand size, adhesive type, and processing parameters. The density of OSB panels is usually 500–800 kg/m^3 [1]. Because the layers of OSB are oriented, its mechanical properties are different in the parallel-to- and perpendicular-to-orientation directions. Since both plywood and OSB are used as wood structural panels in wood buildings, the structural properties of OSB are similar to plywood. **Table 2** presents the major properties of OSB and plywood [1, 7, 18].

As a type of construction materials, the fire performance of OSB is critical. The flame spread index (FSI) is one of important indexes, which is commonly evaluated in conformance with ASTM E84"Standard test method for surface burning characteristics of building materials" [1, 23]. According to the International Building Code (IBC), the FSI is classified into Class A (0–25), Class B (26–75), and Class C (76–200) [1, 24]. The OSB panels without fire-treatment typically fall into Class C, while that of plywood panels are Class B or Class C [25]. The FSI values of OSB and plywood are also given in **Table 2**.

In North America, OSB panels are manufactured in conformance with Voluntary Product Standard PS 2 "Performance Standard for Wood Structural Panels" or Canadian Standard CSA O325 "Construction sheathing" [22]. According to PS2 and CSA O325, OSB panels are large, based on their intended use in construction, classified into three grades, namely Sheathing, Structural I sheathing, and Single floor [26, 27]. The minimum requirements of the typical properties and their testing methods of OSB panels with different grades are specified by PS2 and CSA O325, respectively. For

Property	Direction	OSB	Plywood
Density		500–800 kg/m^3	400–600 kg/m^3
Modulus of rupture	Parallel	21–48 MPa	33–48 MPa
	Perpendicular	8–30 MPa	—
Modulus of elasticity	Parallel	4.8–8.3 GPa	7–11 GPa
	Perpendicular	1.9–3.2 GPa	—
Tensile strength	Parallel	6.9–10.3 MPa	10.3–27.6 MPa
Compressive strength	Parallel	10–17 MPa	20.7–34.5 MPa
Shear strength	—	6.9–10.3 MPa	4.1–6.9 MPa
Shear modulus	—	1.2–2.0 GPa	—
Linear hygroscopic expansion in thickness (30–90% RH)	—	0.15%	—
Internal bond strength	—	0.28–0.57 MPa	—
Flame spread index	—	100–172	35–180
Flame spread class	—	C	B or C

Source: Data from Handbook of Wood Chemistry and Wood Composite, American Wood Council and Wood Handbook [1, 7, 25].

Table 2.
Typical properties of OSB and plywood.

Span rating or performance category		Stiffness, EI × 10^3		Maximum moment	
		Major	Minor	Major	Minor
		N·mm^2	N·mm^2	N·mm	N·mm
Sheathing	Roof-24	292	85	330	130
	Roof-24/subfloor-16	395	94	390	140
	Roof-32/subfloor-16	490	113	460	190
	Roof-40/subfloor-20	1240	358	810	360
	Roof-48/subfloor-24	1790	763	920	510
Structural I sheathing	3/8	292	85	330	130
	7/16	395	141	390	220
	15/32	490	245	460	320
	1/2	490	273	460	330
	19/32 & 5/8	1240	471	810	500
	23/32 & 3/4	1790	763	920	650
Single floor	Single floor-16	876	198	650	230
	Single floor-20	1110	264	710	240
	Single floor-24	1600	546	910	320
	Single floor-32	4170	1270	1570	600
	Single floor-48	8660	2110	2080	820

Source: Data from the standard PS 2–18 [26, 28].
Note: Major and minor represent the stress applied parallel- and perpendicular-to-strength axis, usually the length direction of a panel.

Table 3.
Quality assurance minimum reference values for the bending stiffness and strength of dry small specimens.

example, the bending properties of OSB shall be evaluated in conformance with ASTM D3043 "Standard Test Methods for Structural Panels in Flexure," and the minimum reference values for its quality assurance are given in **Table 3** [26, 28]. In addition, PS2 and CSA 0325 specify that the bonding performance of OSB for construction shall meet the Exposure 1 bond classification, which means that the OSB panels are suitable for uses subjected to non-permanent exposure to the weather [26].

According to PS 2, the grade and end-use of an OSB panel shall be evaluated and classified by a third-party inspection agency, such as APA-The Engineered Wood Association and TECO-Sun Prairie [26]. The third-party agency has to visit the mills on a regular unannounced basis and confirm the performance of the OSB panels still meets the requirements stipulated in the standard [1, 26]. OSB panels conforming to product performance standards are marked with grade stamps by the third-party agency [1]. The sample grade stamps are shown in **Figure 4** [28, 29].

3.4 Applications

3.4.1 Sheathing

One of the major uses of OSB is for sheathing such as roof sheathing, subflooring, and wall sheathing in light-frame construction (**Figure 5**). Typically, the OSB used as sheathing materials is the sheathing grade or structural-I sheathing grade [30, 31]. The common nominal thickness of OSB for sheathing is 7.9 mm (5/16 in.), 9.5 mm (3/8 in.), 11.1 mm (7/16 in.), 11.9 mm (15/32 in.), 12.7 mm (1/2 in.), 15.1 mm

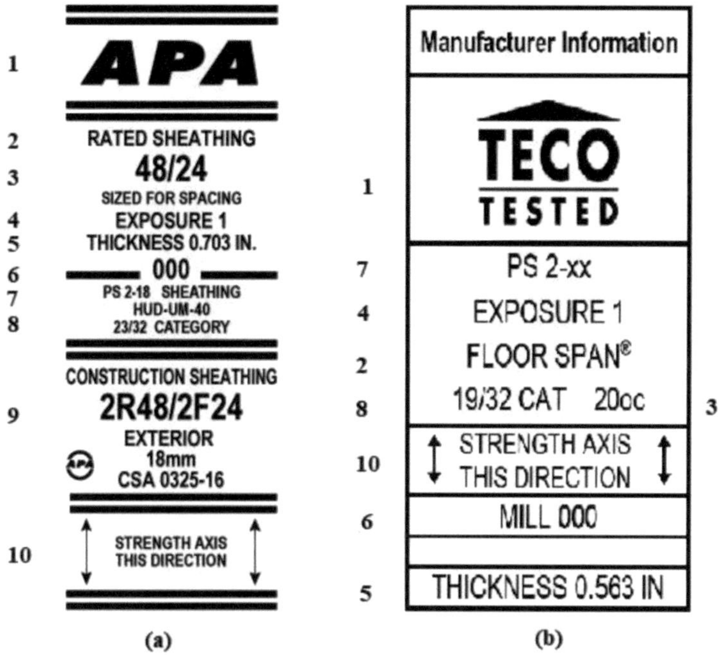

Figure 4.
Typical grade stamps for OSB: (a) APA grade stamp; and (b) TECO grade stamp. (1. Third-party inspection agency; 2. Panel grade; 3. Span rating; 4. Bond classification; 5. Thickness; 6. Mill number; 7. Product standards; 8. Performance category; 9. Canadian construction sheathing standard; 10. Panel face orientation indicator. Source: Pictures obtained from APA at https://www.apawood.org/apa-trademark and TECO at https://www.pfsteco.com).

(19/32 in.), 15.9 mm (5/8 in.), 18.2 mm (23/32 in.), and 19.0 mm (3/4 in.) [22]. When the OSB with the same performance category (same grade and same thickness) is used for different sheathings, the allowable maximum spacing of supports is different [30, 31]. Generally, the span rating of OSB, which donates the maximum spacing of supports for different sheathing, is marked on the grade stamps [28, 29]. For example, the span rating in **Figure 4(a)** is 48/24, which means the maximum spacing of supports is 48 in. when the panel is used for roof sheathing, and the maximum on center spacing of supports is 24 in when it is used for subfloors. According to PS 2 and CSA O325, OSB with roof span ratings of 24 or larger can be used as wall sheathing when the distance of studs is 24 in., whereas the panels with a roof span rating of less than 24 can only be used for wall sheathing with a distance of studs of 16 in. [26, 27, 30]. In addition, the minimum fastening schedule (such as nail spacing and nail size) and allowable live load of sheathing are also given in PS 2 and CSA O325.

3.4.2 Floor

The single-floor grade OSB is intended for flooring panels under carpet or pad (**Figure 6**). The common nominal thickness of OSB for flooring panels are 15.1 mm (19/32 in.), 15.9 mm (5/8 in.), 18.2 mm (23/32 in.), 19.0 mm (3/4 in.), 22.2 mm (7/8 in.), 25.4 mm (1 in.), and 28.6 mm (1–1/8 in.) [22]. Similar to the sheathing, the span rating of flooring panels is also given on the grade stamps; for example, the spacing rating in **Figure 4(b)** is 20 °C, which indicates that the maximum span of the flooring panel is 20 in. on center [28, 29]. Typically, the span ratings of single-floor OSB are 16, 20, 24, 32, and 48 oc [26, 27]. Allowable uniformly

Figure 5.
OSB sheathing used on floors (top), roofs (center), and walls (bottom) (source: Pictures obtained from APA at https://www.apawood.org/photography).

distributed live load at maximum span for flooring panels is 100 psf. The allowable uniformly distributed live load can be increased by reducing the spacing of supports of a floor [22]. The relationship between allowable uniformly distributed live load and span is given in **Table 4** [30].

3.4.3 Other applications

In addition to being used as panels, I-joints web stock is another major use of OSB, **Figure 7**, in wood buildings [9]. I-joints are manufactured using lumber or SCL flanges and OSB webs. Typically, flanges and webs are bonded using exterior-type adhesives such as PF and pMDI. The shear properties of I-joints are provided by OSB webs, while the flanges provide the bending properties of I-joints since OSB has higher shear properties than plywood (**Table 2**). OSB is also used in industrial applications such as shelving and packaging in commercial and industrial structures

Figure 6.
OSB used for floor sheathing and I-joist webs (source: Picture obtained from APA at https://www.apawood.org).

Span rating	Maximum span (in.)	Allowable live load (psf)						
		Span of supports						
		12	16	20	24	32	40	48
16 oc	16	185	100	—	—	—	—	—
20 oc	20	270	150	100	—	—	—	—
24 oc	24	430	240	160	100	—	—	—
32 oc	32	—	405	295	185	100	—	—
48 oc	48	—	—	425	290	160	100	55

Table 4.
Allowable live load of flooring panels with the different spans of supports (source: Data from APA at engineered wood construction guide) [30].

(**Figure 8**). In addition, some researchers tried to develop new applications of OSB panels. For example, Ma et al. developed a new type of lumber-like wood products made of OSB panels, namely laminated OSB, using a cold-cured structural adhesive, considering the cost and difficulty of manufacturing thick OSB [32].

4. Laminated strand lumber (LSL) and oriented strand lumber (OSL)

4.1 Overview

LSL and OSL are derivatives of OSB since they are all made from flaked wood strands (**Figure 9**). These two strand-based EWPs have more similarities than differences. The main difference is that the strands used in LSL are longer than OSL and OSB [1]. The length-to-thickness ratio of the strands used in LSL is approximately 150, while that of the strands for OSL is 75 [33]. As the two typical SCL, LSL and OSL were developed in response to the increasing demand for dimensional lumber [1]. LSL and OSL have become substitutes to solid lumber,

Figure 7.
I-joist made of OSB web and LVL flanges (source: Picture obtained from APA at https://www.apawood.org).

Figure 8.
OSB deck board of a pallet (source: Picture obtained from APA at https://www.apawood.org).

LVL, and PSL in many applications, such as studs, beams, and columns. The outstanding advantages of LSL and OSL are, in comparison with LVL and PSL, their lower material requirements, high yield of wood fibers, and stable properties [1]. LSL and OSL are, in terms of their manufacturing processes and product properties, thought to be exchangeable in the industry [34]. Therefore, the following discussion is mainly given to LSL, which can be used as references when readers are interested in OSL.

4.2 Manufacturing process

The manufacturing processes of LSL/OSL are similar to that of OSB [34], which can be seen in **Figure 3**. The low-density hardwoods, such as aspen and yellow poplar, are commonly used species for making LSL/OSL [34]. As mentioned in Section 4.1, the main difference between LSL and OSL is the dimensions of the strands. The dimensions of the strands used in LSL are 230–356 mm in length and

Figure 9.
LSL products (source: Pictures obtained from Weyerhaeuser at https://www.weyerhaeuser.com).

15–25 mm in width, while those of the strands for OSL are generally 100–150 mm in length and 15–25 mm in width [4, 9]. In addition, the thickness of LSL is larger than OSB; thus, the PF or PRF is not suitable adhesives for making LSL. Instead, the diphenylmethane diisocyanate (MDI) adhesive is used, which can significantly reduce the press time as compared with PF [9]. The blending and forming processes of LSL manufacturing are also similar to those of OSB manufacturing. The big difference between OSB and LSL goes to the orientation of strands in a mat/panel. The strands in the surface layers of OSB are mainly oriented parallel to the length direction, and those in core layers are perpendicular to the length direction or randomly. However, the strands in LSL are always aligned in its panel length direction through its thickness [4, 34]. This can help provide relatively high mechanical properties of the final products in the axis direction. The mat of LSL can be 406 mm (16 in.) in thickness and 2440 mm (8 ft) in width. The single opening is exclusively used in the manufacturing of LSL because it has the advantages of shortening press cycle time and being easy to make a uniform vertical density profile [4]. Sometimes, the steam injection is used during hot pressing to quickly increase the temperature of the core layers, reduce the press time, keep a uniform density profile through the thickness, and create a great resistance to warping [4]. It is worth noting, in some cases, specialized curved press platens could be used in the hot-pressing process of OSL to produce the products with some architectural shapes [9]. The LSL panels are cut to size and tested before receiving a protective end or edge seal. The moisture content of LSL products ranges from 7 to 10%.

4.3 Properties and grading of LSL and OSL

In North America, the nominal dimensions of LSL are generally similar to those of dimensional lumber on the market. Some manufacturers also provide special size products for special applications; for example, the size of TimberStrand®, which is the trade name of LSL manufactured by Weyerhaeuser Company, is up to 14.6 m in length, 1.2 m in width, and 140 mm in thickness [4, 35].

The density of LSL is about 15% higher than that of OSB, ranging from 640 to 670 kg/m^3 (40–42 lbs/ft^3) [34]. The structural properties of LSL are evaluated in conformance with the standards ASTM D 5456 "Standard Specification for

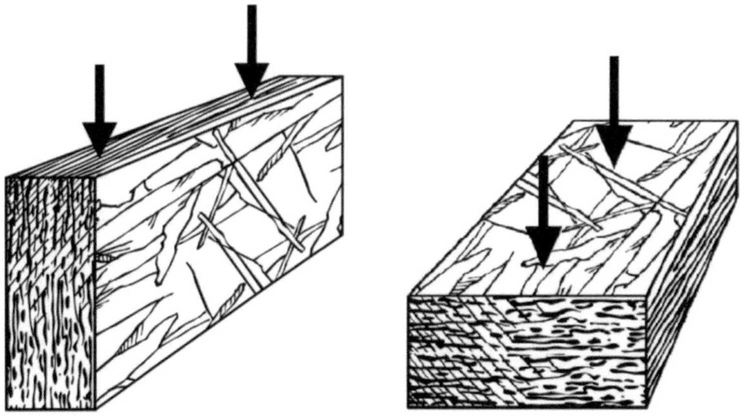

Figure 10.
Different directions of LSL: Beam direction (left) and plank direction (right) (source: Pictures obtained from Weyerhaeuser at https://www.weyerhaeuser.com).

Evaluation of Structural Composite Lumber Products" [36]. The stiffness of LSL is larger than that of LVL, and the tensile strength of LSL is similar to structure dimension lumber [34]. In general, the allowable modulus of elasticity (MOE) is used as the basis of LSL grading; for example, 1.5E grade indicates that the allowable MOE is 1.5×10^6 psi (8960 MPa) [4]. The allowable values of flexural strength, tensile strength, compression strength, and other structural properties of LSL produced by different manufacturers are slightly different [37, 38]. In addition, the allowable stress of LSL in the beam direction differs from that in the plank direction (**Figure 10**) [38, 39]. As an example, the allowable mechanical properties of LSL manufactured by Norbord Inc., Canada, are shown in **Table 5** [38].

Nail withdrawal and lateral resistance connection properties are also important for LSL when used in wood buildings. The equivalent specific gravity, which means the nail withdrawal and lateral resistance connection properties are deemed to be equivalent to that of the solid lumber with a specific gravity, is always used to estimate the connection properties of LSL [4, 39]. For example, the equivalent specific gravity of TimberStrand® LSL with 1.3E and 1.5E grade is 0.5, which means the standard bolted connection design values could be obtained in the adopt code for Douglas fir with a specific gravity of 0.5 [39, 40]. The grade of LSL/OSL

Grade	MOE	Axial		Beam direction			Plank direction		
		F_t	F_c	F_b	F_v	$F_{c\perp}$	F_b	F_v	$F_{c\perp}$
	MPa	MPa	MPa	MPa	MPa	MPa	MPa	MPa	MPa
1.7E LSL	11,720	14.13	14.13	14.82	2.76	8.27	19.31	1.03	2.24
1.55E LSL	10,690	13.34	15.00	16.27	3.62	7.93	18.06	1.07	3.76
1.35E LSL	9310	9.86	13.13	12.76	3.07	7.93	14.20	1.03	3.90
1.5E OSL	10,345	12.24	12.24	12.07	2.76	7.93	17.58	0.90	2.24
1.3E OSL	8960	8.96	8.96	11.20	2.41	7.93	13.79	0.79	1.97
0.8E OSL	5515	4.69	7.58	7.79	2.45	9.76	—	—	—

Source: Data from APA at https://www.apawood.org.
Note: MOE is Modulus of Elasticity; F_t is tensile strength; F_c is compression strength; F_b is bending strength and F_v is shear strength; $Fc_\perp$ is compression strength perpendicular to the axial direction.

Table 5.
Allowable mechanical properties of LSL and OSL with different grades.

Figure 11.
A grade stamp for SCL: 1. Qualified SCL grade; 2. APA mill number; 3. Product evaluation reports (APA and ICC-ES in the United States and CCMC in Canada); 4. Standard specification for SCL (source: Pictures obtained from APA at https://www.apawood.org).

is evaluated and classified by APA-The Engineered Wood Association in North America. The structural design values for LSL/OSL are published on a proprietary basis by manufacturers, which are certified by APA. It is worth noting that this classification is for all SCL products including LSL and OSL. A sample grade stamp on SCL products is shown in **Figure 11** [33].

4.4 Applications of LSL and OSL

LSL/OSL has lower requirements for raw materials than veneer-based SCL and similar structural properties to dimension lumber, which has become a good alternative to solid timber. Presently, LSL products with different dimensions and grades have been used as headers, studs, and rim boards in residential wood frame construction (**Figure 12**) [4, 39]. Generally, the specific design values of LSL, such as allowable stresses, the distance between connections and allowable live load, can be found in the user's guides provided by the manufacturers [37–39]. LSL products are also used as nonstructural rated materials in the non-construction field such as molding, millwork, and furniture [9]. Moreover, some researchers have identified and developed the new applications of LSL. For example, the University of New Brunswick and the University of Alberta, Canada, have developed a hybrid cross-laminated timber (HCLT) panel product made of LSL/OSL and dimension lumber and explored its bending and rolling shear properties [41]. The racking performance of shear walls made of HCLT panels has been also examined under both monotonic and cyclic loadings. Another example is given by Wang et al. [42], who investigated the feasibility of manufacturing HCLT made of LSL (used as the surface layers or core layers) and Lodgepole pine lumber. Their results indicated that the bending properties and rolling shear properties of HCLT were significantly higher than those of generic CLT made of Lodgepole pine lumber only.

In comparison with LSL, OSL is a relatively newer strand-based SCL product, and its market is still under development. The OSL products have similar applications to LSL, which are usually used for band joists in floor construction and as substitutes for rafters in roof construction [1]. The feasibility of using OSL as a transverse layer(s) for manufacturing CLT has been proven [5, 41]. In addition, because OSL can be manufactured to an architectural shape, it also can be used in architectural windows and doors, furniture parts, and other specialty applications [9].

Figure 12.
Applications of LSL in residential frame construction. Top: LSL header; center: LSL studs; bottom: LSL rim board (source: Pictures obtained from Weyerhaeuser at https://www.weyerhaeuser.com/woodproducts/).

5. Endnotes

The strand-based EWPs, such as OSB, LSL, and OSL, has all the environmental advantages of other EWPs, such as low carbon emissions. However, the strand-based EWPs have, due to their relatively complex manufacturing process by adding more fossil-based adhesives, higher environmental impacts than lumber-based and veneer-based EWPs. A study by the University of British Columbia, Canada, indicated that the veneer-based EWPs were more sustainable products than the strand-based wood products, which showed lower energy utilized and fewer emissions of CO_2, SO_x, NO_x, and other volatile organic compounds (VOCs) and particulate matters generated during the manufacturing [20]. Therefore, the manufacturing process of strand-based EWPs requires further research and development. For example, use of bio-based adhesives to produce strand-based EWPs can assist to reduce their environmental footprint.

In summary, the strand-based EWPs have become important members of the family of EWPs, providing a way of producing relatively large sizes of man-made lumber with a high-yield wood fibers from small-diameter and low-quality logs. With the anticipated increase in demand for EWPs in construction and change in accessibility and quality of mature forest resources, the development and application of strand-based EWPs keep increasing in the future. With the rise of tall wood buildings, strand-based EWPs will have great market potential and will be a new research hotspot.

Acknowledgements

This piece of work was financially supported by the New Brunswick Innovation Research Chair Initiative Program by the New Brunswick Innovation Foundation (Canada) and the Discovery Grant by Natural Sciences and Engineering Research Council of Canada (RGPIN 2017-03994).

Conflict of interest

I confirm there are no conflicts of interest.

Author details

Zizhen Gao* and Meng Gong
University of New Brunswick, Fredericton, NB, Canada

*Address all correspondence to: zz.gao@unb.ca

References

[1] Ross RJ. Wood handbook: Wood as an engineering material. USDA Forest Service, Forest Products Laboratory, General Technical Report FPL-GTR-190. 2010. 509 p

[2] Pajchrowski G, Noskowiak A, Lewandowska A, Strykowski W. Wood as a building material in the light of environmental assessment of full life cycle of four buildings. Construction and Building Materials. 2014;**52**:428-436. DOI: 10.1016/j.conbuildmat.2013.11.066

[3] Gustavsson L, Pingoud K, Sathre R. Carbon dioxide balance of wood substitution: Comparing concrete-and wood-framed buildings. Mitigation and Adaptation Strategies for Global Change. 2006;**11**(3):667-691. DOI: 10.1007/s11027-006-7207-1

[4] Thelandersson S, Larsen HJ, editors. Timber Engineering. Chichester: John Wiley & Sons; 2003. 446 p

[5] Gong M. Lumber-based mass timber products in construction. In: Concu G, editor. Timber Buildings and Sustainability. Rijeka: IntechOpen; 2019. pp. 7-16. DOI: 10.5772/intechopen.85808

[6] Nishimura T. Chipboard, oriented strand board (OSB) and structural composite lumber. In: Ansell M, editor. Wood Composites. Cambridge: Woodhead Publishing; 2015. pp. 103-121. DOI: 10.1016/B978-1-78242-454-3.00006-8

[7] Rowell RM, editor. Handbook of Wood Chemistry and Wood Composites. Boca Raton, Florida, USA: CRC Press; 2012. 473 p

[8] Zerbe JI, Cai Z, Harpole GB. An evolutionary history of oriented strand board (OSB). USDA Forest Service, Forest Products Laboratory, General Technical Report, FPL-GTR-236. 2015. 10 p

[9] Shmulsky R, Jones PD. Forest Products and Wood Science: An Introduction. 6th ed. Chichester: Wiley Blackwell; 2019. 477 p

[10] Food and Agriculture Organization of the United Nations (FAO). 2021. Available from: http://www.fao.org/faostat/en/#data/FO [Accessed: 20 March 2021]

[11] Moore JR, Cown DJ. Processing of wood for wood composites. In: Ansell M, editor. Wood Composites, Cambridge: Woodhead Publishing; 2015. pp. 27-45. DOI: 10.1016/B978-1-78242-454-3.00002-0

[12] Kruse K, Dai C, Pielasch A. An analysis of strand and horizontal density distributions in oriented strand board (OSB). European Journal of Wood and Wood Products. 2000;**58**(4):270-277

[13] Beck K, Cloutier A, Salenikovich A, Beauregard R. Effect of strand geometry and wood species on strandboard mechanical properties. Wood and Fiber Science. 2009;**41**(3):267-278

[14] Pizzi A, Mittal KL. Handbook of Adhesive Technology. Boca Raton, Florida, USA: CRC Press; 2017. 643 p

[15] Walker JC. Primary Wood Processing: Principles and Practice. Dordrecht, NED: Springer Science & Business Media; 2006. 576 p

[16] Cai Z. Wood-based composite board. In: Wiley Encyclopedia of Composites. Chichester: John Wiley & Sons; 2012. DOI: 10.1002/9781118097298.weoc262

[17] Thoemen H. Wood-based Panels: An Introduction for Specialists. London: Brunel University Press; 2010. 283 p

[18] Hiziroglu, S. Oriented strand board as a building material. Stillwater, Oklahoma, USA: Oklahoma State university Extension Fact Sheet, FAPC-146, Food and Agricultural Products Research Center; 2006. 4p.

[19] Lam F, Craig BA. Shear strength in structural composite lumber. Journal of Materials in Civil Engineering. 2000;**12**(3):196-204

[20] Chan G. Sustainability Assessment of OSB and Softwood Plywood Manufacturing in North America. Vancouver, BC, Canada: University of British Columbia; 2012. 71 p

[21] Hsu WE. Oriented Strand Board Manufacturing. Scotts Valley, California, USA: CreateSpace Independent Publishing Platform; 2013. 316 p

[22] The Engineered Wood Association (APA). Product Guide: Oriented Strand board. 2021. Available from: https://www.apawood.org [Accessed: 25 March 2021]

[23] American Society for Testing and Materials. ASTM D E84-21a. Standard Test Method for Surface Burning Characteristics of Building Materials. 2021

[24] International Code Council. International Building Code. Country Club Hills, Ill: ICC; 2018

[25] American Wood Council. Flame spread performance of wood products. Design for code acceptance No. 1. 2019. Available from: https://www.awc.org/codes-standards/publications/dca1 [Accessed: 02 September 2021]

[26] PS VPS. Performance Standard for Wood Structural Panels. Washington, D.C., USA: National Institute of Standards and Technology (Department of Commerce); 2019

[27] CSA O325, Construction Sheathing, CSA O325. Toronto, ON, Canada: Canadian Standards Association; 2003

[28] The Engineered Wood Association (APA). APA Panel Trademark. 2021. Available from: https://www.apawood.org [Accessed: 25 March 2021]

[29] PFS-TECO. PS 1 and PS 2 Gradestamp Explanation. Sun Prairie, Wisconsin: TECO; 2021. Available from: https://www.pfsteco.com/library. [Accessed: 30 March 2021]

[30] The Engineered Wood Association (APA). Engineered Wood Construction Guide. 2021. Available from: https://www.apawood.org [Accessed: 10 April 2021]

[31] PFS-TECO. OSB design and Application Guide. Sun Prairie, Wisconsin: TECO; 2021. Available from: https://www.pfsteco.com/library [Accessed: 10 April 2021]

[32] Ma, ZY. Effect of surface treatments on the bond quality of laminated OSB products [thesis]. Fredericton, NB, Canada: University of New Brunswick; 2019

[33] The Engineered Wood Association (APA). Structural Composite Lumber Selection and Specification. 2021. Available from: https://www.apawood.org [Accessed: 25 March 2021]

[34] Blau K. Personal Communications. Product Development and Quality. Vernon, BC, Canada: Tolko Industries Ltd. Canada; 2021

[35] Weyerhaeuser Company. 2021. Available from: https://www.weyerhaeuser.com/ [Accessed: 25 March 2021]

[36] American Society for Testing and Materials. ASTM D 5456-19. Standard Specification for Evaluation of Structural Composite Lumber Products. 2019

[37] The Engineered Wood Association (APA). Joint Evaluation Report ESR-2725, 2021. Available from: https://www.apawood.org [Accessed: March 25, 2021]

[38] The Engineered Wood Association (APA). Joint Evaluation Report ESR-1053. 2019. Available from: https://www.apawood.org [Accessed: 25 March 2021]

[39] Weyerhaeuser. Specifier's Guide for Beams Headers and Column. 2021. Available from: https://www.weyerhaeuser.com/ [Accessed: 25 March 2021]

[40] Canadian Standards Association (CSA). Engineering Design in Wood. Mississauga, ON, Canada: CSA; 2014. 280 p.

[41] Chui YH, Gong M, Niederwestberg J. Development of a lumber-SCL massive timber panel product. Final Report #: WSTC2013-015. Fredericton, NB, Canada: Wood Science and Technology Centre, the University of New Brunswick; 2015. p. 55

[42] Wang Z, Gong M, Chui YH. Mechanical properties of laminated strand lumber and hybrid cross-laminated timber. Construction and Building Materials. 2015;**101**:622-627. DOI: 10.1016/j.conbuildmat.2015.10.035

Chapter 5

Perceptions, Attitudes, and Interests of Architects in the Use of Engineered Wood Products for Construction: A Review

Hüseyin Emre Ilgın and Markku Karjalainen

Abstract

Increased use of engineered wood products (EWPs) and thus decreasing share of non-biobased materials such as concrete reduces the impact of buildings on the climate by mitigating the primary energy use and greenhouse gas emissions in construction. A construction project includes many parameters, where the selection of construction material is one of the crucial decisions with its numerous criteria e.g. cost, strength, environmental impact. Furthermore, this complicated process includes different parties such as architects, engineers, contractors. Architects are among the key decision-makers in material selection, and their perceptions influence what they propose and hence an increase in wood construction. In literature, many studies have been conducted on the technological, ecological, economic aspects of EWPs, while limited studies are focusing on EWPs for construction from stakeholders' perspective. In this chapter, architects' attitudes towards the use of EWPs in buildings were scrutinized.

Keywords: Engineered wood products, construction, architects, attitude

1. Introduction

By its very nature, one cubic meter of wood stores almost a ton of CO_2, so wood reduces the carbon footprint of the construction industry while evaluating the entire life cycle from raw material to production, use, and recycling, and timber buildings play an important role in supporting to the sustainable bioeconomy [1, 2].

Studies indicate that increasing the use of EWPs in the construction sector has environmental benefits, as wood is a renewable and lightweight material [3, 4], where utilization of EWPs instead of conventional building materials e.g. concrete or steel, the total fossil fuel footprint of building construction can be diminished to a considerable extent through environmental-friendly-material replacement [5, 6]. Also, based on some estimates, substituting concrete with wood could lessen the energy used by construction processes by 40%, while greenhouse gas emissions could be reduced by 35% [7, 8]. Extensive use of EWPs may have helped make the transition towards more carbon-free production of building materials [9, 10].

A construction project involves a large number of participants with various roles, goals, and concerns [11] as in the case of high-rise timber buildings e.g. Mjøstårnet (Norway, 2019) (**Figure 1**), HoHo (Austria, 2020) (**Figure 2**). Besides, the building material selection process includes many parameters, e.g. cost, structural performance, environmental friendliness, fire performance, availability, and speed of construction [12, 13]. Moreover, the material selection procedure is a complex non-linear process involving various actors such as contractors, structural designers, developers, and architects [14–16].

Among these parties involved, architects play a critical role in material selection [17, 18], their perceptions influence their choice of material in the structural frame [13, 15, 19, 20]. Perhaps more importantly, architects' perceptions could lead to an increase in the use of EWP for construction [19, 21].

In the literature, many studies have been carried out on the ecological, technological, economic, and social aspects of engineered wood products and various technical solutions in buildings [22]; while there is a relatively limited number of studies concentrating on the EWPs for construction from the stakeholders' perspective (e.g. [11, 16, 20–25]).

This chapter presented a comprehensive overview of the perceptions, attitudes, and interests of architects in the use of EWPs together with their perceived benefits and barriers for construction. This study gathered, mapped out, and systematized scattered and multifaceted knowledge on architects' attitudes towards EWPs employment in buildings, and chronologically presented them in an accessible and manageable discourse. Notably, the chapter also revealed how this perception has changed over time.

Figure 1.
Mjøstårnet (Norway, 2019) (Source: Wikipedia).

Figure 2.
HoHo (Austria, 2020) (Source: Wikipedia).

An increasing emphasis on the climate impact of building materials in the construction sector should therefore increase the likelihood that future EWPs will be favored to a greater extent. In this sense, attempts to increase the awareness of architects about EWPs will have a positive effect, e.g. regarding the economic aspects. In this study, wood or timber refers to engineered timber products such as cross-laminated timber, laminated veneer lumber, glue-laminated timber/glulam.

2. Studies on architects' attitudes towards the use of EWPs for construction and discussion

In the literature, several studies from different countries such as Australia, America, Sweden are concentrating on wood as a structural material in the buildings from architects' perspective through questionnaires and/or interviews (e.g. [13, 20, 23, 24, 26, 27]).

Among the studies, Truskett undertook surveys and interviews with architects in Victoria (Australia) to explore factors influencing the specification of wood products [28]. The findings showed that while the vast majority (90%) of those surveyed 'always or mostly' used structural timber for residential purposes, only 20% frequently used structural timber for non-residential applications. According to the architects, timber as a structural and finishing material has strong aesthetic appeal, but the factors such as maintenance and durability, professional networks, industry practice, information, and environmental issues hamper its general use.

Kozak and Cohen focused on the construction material selection of architects (n = 594 out of 3,986) and structural engineers (n = 384 out of 1,822) for non-residential buildings through an online survey in the United States and Canada [26]. The results showed that steel and concrete continue to be the most common material in non-residential applications, whereas wood governs the construction market for elderly housing and is also frequently employed in religious buildings, restaurants, and commercial/residential combinations. It was also pointed out that as the building height increases, the use of wood decreases. However, the attitudes of architects were encouraging towards the use of timber-frame if they had previously specified it.

In 2004, Wagner and Hansen scrutinized material preferences among architects and engineers in America and Chile by a cross-cultural comparison to establish a procedure for choosing a customer group of a company and then classifying its demands and needs [27]. They stated that American architects did not give much weight to issues concerning cost and ecological properties of the building material when deciding on wood construction, whereas other considerations e.g. dimensional stability were thought to have more potential for development when compared with competing materials like steel. Similarly, surveyed architects from Chile were not interested in the environmental features of the wood. Besides this, both sample groups of architects had a positive attitude towards the aesthetical properties of wood, and they perceived uniform quality as an essential asset. It was also noticed that fire-related issues received less concern among the architects than the relevant literature suggests [26].

O'Connor et al. studied architects' and engineers' perceptions regarding the use of wood in the North American non-residential construction sector through an extensive mail survey applied to a series of specifier's focus groups [15]. This research identified several perceived challenges for wood utilization in construction: building code concerning fire safety; cost-competitiveness with steel especially in terms of complicated structures or structures with a longer span; design performance related to strength, durability, stiffness, and lack of established practices. They also proposed short- and long-term recommendations for addressing these obstacles.

The research of Bayne and Taylor also examined the barriers to the use of wood in Australian non-residential buildings to understand the reasons behind the common use of non-wood products e.g., concrete and steel [29]. They aimed to identify the reasons for the lack of confidence in specifying wood as a structural material in these buildings. As a result of interviews with 34 architects, engineers, building designers, and project managers, a range of strengths and weaknesses regarding the structural application of wood to non-residential use were underlined. Findings suggested that aesthetics, easy construction, and adaptability of design and fire performance were evaluated as advantages, while cost and speed of erection were taken as the most common obstacles by the specifiers. It was also concluded that the use of wood was found more suitable and promising by the architects for smaller building types such as housing for the elderly, schools, public buildings, and churches.

In 2008, Roos et al. presented perceptions of 23 Swedish architects and structural engineers about the material selection process, which also included a comparison of wood with other materials, the effects of main stakeholders, and the relation of wood construction with professional roles and knowledge [30]. Both architects and engineers were interested in using wood but perceived it as complicated. This study as a prospect highlighted the issues such as clear demonstration of business-sound wood compared with concrete and steel; functional information flows from the construction industry to wood sector; expression of smart solutions enabling flexibility and appropriate span lengths from an architectural standpoint; putting emphasis on aesthetic and visual aspects of timber-frame, and importance of providing more information about environmental benefits of wood by suppliers.

Bysheim and Nyrud investigated Norwegian architects' perceptions regarding the use of wood as a structural material in urban construction via a questionnaire to measure attitudes towards the physical, mechanical, and fire-related properties of wood [31]. They found that many architects show a tendency to use structural timber, but few are planning to do so. They were also positive; fire and aesthetically related properties, costs of using structural timber compared to substitutes, and the energy-related properties of the material, the physical and mechanical properties of timber-frame. They felt qualified to specify timber-frame in buildings but did not perceive the choice of using structural timber as being entirely up to them. The architects had positive attitudes towards timber-frame in residential buildings but were negative towards use in other building types. Additionally, they were positive to utilize in buildings up to 3-story. The architects had little experience with timber-frame in other building types other than residential ones. Most of the respondents strongly agreed that other architects would have a positive tendency to them employing timber-frame, whereas contractors and real estate developers were perceived as negative towards the use of timber-frame.

Robichaud et al. explored the challenges to the use of wood in the North American non-residential construction sector [32]. This study aimed to examine the possible role of communication in this emerging market among architects through a survey (n = 165 out of 5,000). The results showed that generally, the architects assessed wood to be a 'sincere' but 'unexciting' structural material. When compared to concrete and steel, wood was perceived as the most environmentally friendly material. However, wood was rated with the lowest score in terms of durability, fire-resistant, structural performance, and contribution to the high building value. In the recommendation part of the study, the issue of better communication on the part of wood producers and product information was highlighted.

The research of Roos et al. analyzed architects' and structural engineers' attitudes and perceived factors that hinder or facilitate the specification of wood construction in Sweden [20]. The main finding of this research was that the material preference of architects and structural engineers is affected by attitudes concerning the properties of wood and beliefs about the control and ease of building in wood. Wood was generally perceived as a suitable building material. Issues about decay, instability, and sound transmission were assessed as negative aspects, while the features of strength, environmental friendliness, simple handling, and suitability for use along with industrial methods were taken as advantages of wood. Besides this, developers and contractors were perceived as the most influential parties by both Swedish architects and structural engineers in the material selection process.

The result suggests that if the following measures are taken into consideration, perceived obstacles could be lessened: (i) developing clearer business concepts for timber-based transparent and affordable construction approaches that decrease the uncertainty, (ii) creating prefabrication methods for wood to reduce the risk factor in the construction, (iii) developing education and training in building design and construction in wood, (iv) providing information about the environmental performance of wood as a building material, (v) improving the 'professional status' of wood via interesting design, (vi) supporting architects and engineers in pursuing wood construction and developing a dialogue among all the related professions.

Hemström et al. assessed Swedish architects' perceptions, attitudes, and interest towards steel-, concrete-, and timber-frames in multi-story buildings through a web-based questionnaire (n = 412 out of 3,600) [11]. The results indicated that concrete was found the most favorable frame material for multi-story construction mainly because of its performance of engineering-based issues e.g., stability and fire safety that was considered critical for the selection of frame material. The general attitude towards, and interest in, timber-frame utilization was positive and related to its perceived environmental features. Differing from findings in North America [15], this study showed that costs and time to construct a building are not perceived as major barriers to the use of wood among architects in Sweden. Contradictory to the perception of wood being a less suitable multi-story frame material than concrete and steel, the interest in the use of wood frames was large. Contrary to Norwegian findings [16], the overall attitude towards the use of wood frames in residential buildings presented here was not different from the attitude towards the wood in non-residential buildings. Besides this, the results showed that contractors, structural engineers, and building commissioners have a great influence on the choice of frame material.

In 2014, Xia investigated the reasons as perceived obstacles for comparatively limited use of the timber-frame in multi-story non-residential buildings (compared to low-rise housing) among industry professionals - also including architects - in Australia by a questionnaire survey (n = 74 out of 176) [13]. The results indicated five main groups of identified obstacles: (1) lack of support in official regulations, (2) lack of interest in the industry, (3) lack of experience in professionals, (4) perception of drawbacks, and inadequate knowledge about merits of timber-frame utilization. This study also made several recommendations concerning more supportive legislation by the governmental side to stimulate the use of wood in multi-story building construction, industry training to raise the awareness and knowledge of the technological improvements regarding EWPs, the attitudes of developers and investors as the most influential decision-makers towards increasing the awareness of timber-frame advantages.

Viluma and Bratuškins conducted research among architects and other stakeholders in Latvia to find out the main barriers to using wood for buildings through 38 interviews and questionnaires [33]. There were 73 answers from 85 registered persons of which 36 were architects, 25 were students and lecturers, as well as representatives from timber production and media. In this study, the main motivating factors and seven main barriers to the selection of wood constructions were identified. Research results showed that architects' attitude towards timber-frame, in general, is positive, but they thought that due to the Latvian Fire Safety Regulation, it is not easy to find solutions for wood construction. Additionally, the architects emphasized the *stereotypes, legislation,* and *the specialist's qualification* as the main barriers, while the architecture students opted for the lack of knowledge, lack of experience, and lack of information as the main obstacles out of seven given possibilities. Two of the seven barriers formulated - *stereotypes* and *legislation* - were not stated in other research, however, these are the most mentioned barriers in the Latvian case.

Conroy et al. investigated familiarity, use, and perceptions of EWPs among the AIA-certified architects across Washington, Oregon, and California through an online questionnaire (n = 533 out of 3,469) [18]. The results indicated that durability, fire resistance, and strength were assessed as weaknesses of engineered wood products, unlike other studies such as [16, 29] that found architects saw wood fire performance as a strength. The architects from Washington and Oregon projected the use of wood in the construction industry to develop more in the next five years compared to steel and concrete. To boost the use of wood as a construction material for the structure and building enclosure in non-residential buildings, it was recommended that the forest products industry enhances its internet presence, developing interdisciplinary communication strategies.

As one of the most recent studies, Kuzman et al. attempted to better understand the specification process of EWPs and to provide an updated overview of the perceived identity of these materials among architects in Slovenia, Austria, France, Sweden, Croatia, and Bosnia, and Herzegovina [34]. The results indicated that generally, participating architects have a positive attitude towards wood utilization in all countries. Thermally modified wood was perceived as positive, whereas the architects were unfamiliar with more recently introduced wood modification methods e.g., acetylation, furfurylation (which are not well known). Their findings suggested that the opportunities for wood to gain a greater market share will grow.

Markström et al. probed Swedish architects' perception of the use of EWPs in buildings and the parameters which positively influence the preference of these products via a survey questionnaire [24]. Findings highlighted that in general, and as per the more recent study by [34], the perception of EWPs is positive among Swedish architects, and most of them think that their use will increase in the future. They also added that other decision-makers with greater influence over the material selection, such as contractors, developers choose other materials. A lack of knowledge, as well as uncertainties about the quality over time, were other common reasons for not preferring EWPs. The results also indicated that environmental concerns and aesthetic appearance are the main reasons to select these materials for the architects involved in building projects. It was stated that knowledge, familiarity, and architects' attitude play a role in increasing the use of EWPs.

Therefore, it can be said that perceived positive aspects of EWPs have markedly changed from earlier studies in 1997. By that year, in one of the studies by [28] entitled *'Factors influencing architects in their specification of timber and timber products'*, architects perceived EWPs positively, mostly due to a strong image of their aesthetic appeal. However, later studies showed that according to the architects, EWPs had numerous perceived positive aspects besides their aesthetic quality as seen in **Table 1**. Among these positive aspects, *environmental performance* (e.g. [11, 20, 30, 32]), *energy efficiency* (e.g. [11, 16, 29]) and *speed of erection* (e.g. [30]) were the highlights. On the other hand, *durability, fire resistance,* and *strength* were assessed as the most critical barriers to EWPs in the last two decades of research among architects (e.g. [15, 18, 20 26, 32]). In general, in the light of the above-mentioned studies, it can be stated that architects adopted positive attitudes towards the use of EWPs in construction.

In terms of the major benefits and barriers to using EWPS for construction, similar findings were identified in different studies such as [11, 24, 32]. Architects mostly had a positive attitude towards aesthetic quality (e.g. [24, 31]) and environmental performance (e.g. [18, 26]) of EWPs.

Also, in Bayne and Taylor's study on the barriers to the use of EWPs, it was found that the use of timber is more suitable for smaller buildings such as housing development [29]. Similarly, Bysheim and Nyrud found that architects took a positive attitude towards the use of timber as frame material in three-story houses [31].

Some other studies among architects such as [11, 18, 24, 32] highlighted *ecological characteristics* of EWPs such as low climate impact and environmental friendliness.

In studies conducted among architects in the USA [18, 32] and those in Sweden [30], *fire safety issues* were considered as one of the most important obstacles to the widespread use of timber frame structures. On the other hand, exceptionally, fire performance was assessed as a positive consideration for the structural application of timber for non-residential purposes by [29].

Study by (chronologically)	Target groups	Main perceived benefits of EWPs	Main perceived barriers to EWPs
[24]	architects in Sweden	• aesthetics • low impact on the environment • speed of construction	• lack of knowledge
[18]	architects in the US West Coast	• ease of use • aesthetics • cost • environmental friendliness	• durability • fire resistance • structural performance
[33]	stakeholders (including architects) in Latvia		• stereotypes • legislation • lack of knowledge
[13]	industry professionals (including architects) in Australia		• maintenance costs • fire resistance • limited awareness of emerging technologies
[11]	architects in Sweden	• sustainability • ease of renovation and demolition • ease of recycling	
[20]	architects and structural engineers in Sweden	• strength • environmental friendliness • simple handling • compatibility with industrial methods	• durability • instability • sound insulation
[32]	architects in North America	• environmental friendliness	• durability • fire resistance • structural performance
[31]	architects in Norway	• aesthetics • fire resistance • cost • physical and mechanical properties	
[30]	architects and building engineers in Sweden	• lightweight • aesthetics • good indoor climate • environmental friendliness	• fire resistance • sound insulation • form stability • insecure supply • unsuitability for large-span-structure
[29]	architects, engineers, building designers, and project managers in Australia	• aesthetics • ease of use • adaptability of design • fire resistance	• structural performance • cost

Study by (chronologically)	Target groups	Main perceived benefits of EWPs	Main perceived barriers to EWPs
[27]	architects and engineers in the US and Chile	• warmness • consistent material with a unique texture	
[26]	architects and structural engineers in the US and Canada	• warmness • functionality • cost • environmental friendliness	• fire resistance • durability • inconsistent price • insecure supply

Table 1.
Studies on architects' attitudes towards the use of EWPs for construction.

As many studies (e.g. [13, 33]) reported, *more experience with concrete construction* compared to timber was recognized as the main disadvantage of EWPs for construction. Additionally, *knowledge gaps or lack of expertise* in timber construction were underlined as weaknesses in the studies by [30, 35]. Although *lack of cost competitiveness* was generally seen as one of the major barriers to the use of EWPs (e.g. [15, 29]), Roos et al. pointed out that while some architects surveyed claim that EWPs are cost-effective when applied correctly, others fear high costs due to perceived risk factors in timber construction [30].

Besides, according to the participants in the study by Xia et al., *lack of developer interest* was the strongest obstacle to the use of EWPs [13]. Regarding this, the surveyed architects may not be able to perceive their impact on the choice of frame material as strongly as building contractors or clients. This issue may support the relatively low perceived influence of Swedish architects [11, 30], while it may differ from the findings by O'Connor et al. that architects played the most critical role in the multidisciplinary material selection process [15].

Overall, architects' perceptions of EWPs' engineering performance can deter them from employing EWPs for construction. Such a change can be driven by an increase in examples of promising timber building applications. e.g. high-rise buildings (over 8-story), and so the general attitude of architects towards EWPs will be more positive in terms of engineering-based features such as sound insulation, fire safety, durability, and structural performance.

Future scenarios for wooden buildings could improve if there is a new trend towards greater importance of environmental factors in the choice of structural material facilitated by policies, which can make a difference in the demands of customers and the tendencies of contractors.

Moreover, architects can play an important role as prime marketers in increasing EWPs for construction, but it seems that more initiatives are required to enhance their familiarity since the lack of experience and level of knowledge may prevent architects from proposing timber in their projects.

3. Conclusions and recommendations

The aim of this chapter was to understand the architects' perceptions, attitudes, and interests in the use of EWPs for construction. In doing so, this research attempted to identify perceived major benefits and barriers to EWPs utilization.

Overall, architects mostly had a positive attitude towards the use of EWPs. Among EWPs' positive aspects, aesthetic quality, environmental performance, energy efficiency, and speed of erection were the highlights. Durability, fire resistance, and strength were assessed as the most critical barriers to the common use of EWPs. These were followed by a lack of cost competitiveness, knowledge gaps or lack of expertise, and lack of developer interest.

In this sense, the following recommendations may help to overcome identified barriers by improving overall attitudes towards EWPs for construction:

(1) providing architecture students with more education and inspiration at university, more information about wood-based products, better design aids, and more design examples (2) supporting architects in timber construction by creating a sharing environment with members of these professions (3) developing more active participation in EWP-based problem solving and better interdisciplinary communication strategies among timber suppliers, the timber construction industry and the architectural community (4) developing business-oriented approaches for timber compared to traditional materials e.g. concrete in construction (5) developing effective timber prefabrication methods to reduce the risk factor in construction (6) Enhance the collaboration of different stakeholders such as government, client, designer, contractor, and supplier by issuing more supportive regulations and guidelines to increase the use of EWPs for construction.

It is believed that this chapter will help to deepen the understanding of various considerations shaping the decision-making process in the use of EWPs for construction.

Author details

Hüseyin Emre Ilgın* and Markku Karjalainen
Tampere University, Tampere, Finland

*Address all correspondence to: emre.ilgin@tuni.fi

References

[1] Tolppanen J, Karjalainen M, Lahtela T, Viljakainen M. Suomalainen puukerrostalo: rakenteet, suunnittelu ja rakentaminen. Helsinki: Opetushallitus: Puuinfo. 2013.

[2] Aaltonen JA. Viewpoints of Residential Multi-Story Construction Industry's Productivity, Environmental Regulations and Future in Finland, MSc Thesis, Forest Economics and Marketing Faculty of Agriculture and Forestry, University of Helsinki, May 2019.

[3] Gustavsson L, Joelsson A, Sathre R. Life cycle primary energy use and carbon emission of an eight-storey wood-framed apartment building. Energy and Buildings. 2010;42:230-242.

[4] Høibø O, Hansen E, Nybakk E. Building material preferences with a focus on wood in urban housing: Durability and environmental impacts. Canadian Journal of Forest Research. 2015;45:1617-1627.

[5] Milaj K, Sinha A, Miller TH, Tokarczyk J.A. Environmental utility of wood substitution in commercial buildings using life-cycle analysis. Wood and Fiber Science. 2017;49: 338-358.

[6] Hurmekoski E, Myllyviita T, Seppälä J, Heinonen T, Kilpeläinen A, Pukkala T, Mattila T, Hetemäki L. Impact of structural changes in wood-using industries on net carbon emissions in Finland. Journal of Industrial Ecology. 2020;24:899-912.

[7] Herczeg M, McKinnon D, Milios L, Bakas I, Klaassens E, Svatikova K, Widerberg O. Resource Efficiency in the Building Sector. Final Report, Prepared for European Commission by ECORYS and Copenhagen Resource Institute, Rotterdam, the Netherlands. 2014.

[8] Kylkilahti E, Berghäll S, Autio M, Nurminen J, Toivonen R, Lähtinen K, Vihemäki H, Franzini F, Toppinen A. A consumer-driven bioeconomy in housing? Combining consumption style with students' perceptions of the use of wood in multi-storey buildings, Ambio. 2020;49:1943-1957.

[9] Hildebrandt J, Hagemann N, Thrän D. The contribution of wood-based construction materials for leveraging a low carbon building sector in Europe. Sustainable Cities and Society. 2017;34:405-418.

[10] Viholainen N, Kylkilahti E, Autio M, Toppinen A. A home made of wood: Consumer experiences of wooden building materials. International Journal of Consumer Studies. 2020;00:1-10.

[11] Hemström K, Mahapatra K, Gustavsson L. Perceptions, Attitudes and Interest of Swedish Architects towards the Use of Wood Frames in Multi storey Buildings. Resources, Conservation and Recycling. 2011;55:1013-1021.

[12] Castro-Lacouture D, Sefair JA, Flórez L, Medaglia AL. Optimization Model for the Selection of Materials Using a LEED-Based Green Building Rating System in Colombia. Building and Environment. 2009;44:6:1162-1170.

[13] Xia B, O'Neill T, Zuo J, Skitmore M, Chen Q, Perceived Obstacles to Multi-storey Timber Frame Construction an Australian Study, Architectural Science Review.2014;57:3:169-176, 2014.

[14] Emmitt S. Architectural Technology. London: Blackwell Science. 2002.

[15] O'Connor J, Kozak R, Gaston C, Fell D. Timber Use in Non-residential Buildings: Opportunities and Barriers. Forest Products Journal. 2004;5:3:19-28.

[16] Bysheim K, Nyrud, AQ. Using A Predictive Model to Analyse Architects'

Intention of Using Wood in Urban Construction. Forest Products Journal. 2009;59:7/8:65-74.

[17] Gosselin A, Blanchet P, Lehoux N, Cimon Y. Main Motivations and Barriers for Using Wood in Multi-Story and Non-Residential Construction Projects, “Wood for large buildings,” BioResources. 2017;12:1:546-570.

[18] Conroy K, Riggio M, Knowles C. Familiarity, Use, and Perceptions of Wood Building Products: A Survey Among Architects on the United States West Coast, BioProducts Business. 2018.

[19] Mahapatra K, Gustavsson L. Multi-storey Timber Buildings: Breaking Industry Path Dependency. Building Research and Information. 2008;36:6: 638-648.

[20] Roos A. Woxblom, L. and McCluskey, D. The Influence of Architects and Structural Engineers on Timber in Construction - Perceptions and Roles. Silva Fennica. 2010;44:5:871-884.

[21] Bregulla JR, Grantham R, Johansson HE, Enjily V. Barriers to the enhanced use of wood in Europe: particular attention to the regulatory barriers. Report prepared by Building Research Establishment (BRE) as a part of the “Roadmap 2010” programme of the European Confederation of Woodworking Industries, CEI-Bois, Brussels. 2003.

[22] Toivonen R, Lähtinen K. Sustainability - A Literature Review on Concealed Opportunities for Global Market Diffusion for the Cross-laminated Timber (CLT) in the Urbanizing Society. The manuscript for Bioproducts Business. 2019.

[23] Mallo MFL, Espinoza O. Awareness, Perceptions and Willingness to Adopt Cross-laminated Timber by the Architecture Community in the United States. Journal of Cleaner Production. 201;94:198-210.

[24] Markström E, Kuzman MK, Bystedt A, Sandberg D, Fredriksson M. Swedish Architects View of Engineered Wood Products in Buildings. Journal of Cleaner Production. 2018;181:33-41.

[25] Markström E, Kuzman MK, Bystedt A, Sandberg D. Use of Wood Products in Multi-storey Residential Buildings: Views of Swedish Actors and Suggested Measures for an Increased Use, Wood Material Science & Engineering. 2019;14:6:404-419.

[26] Kozak RA, Cohen DH. Architects and Structural Engineers: An Examination of Wood Design and Use in Non-residential Construction. Forest Products Journal. 1999;49:4.

[27] Wagner ER, Hansen EN. A method for Identifying and Assessing Key Customer Group Needs. Industrial Marketing Management. 2004;33:643-655.

[28] Truskett B. Factors Influencing Architects in their Specification of Timber and Timber Products. Market research report no. 1. University of Tasmania. March 1997.

[29] Bayne K, Taylor S. Attitudes to the Use of Wood as a Structural Material in Non-residential Building Applications: Opportunities for Growth. Australian Government: Forest and Wood Products Research and Development Corporation. 2006.

[30] Roos A, Woxblom L, McCluskey D. Architects', and building engineers', and stakeholders' perceptions to wood in construction: results from a qualitative study. In: Delbeck G EB, Hoen H-F, editors. Proc of the biennal meeting of the Scandinavian society of forest economics. Norway: Scandinavian Forest Economics. 2008;184-94.

[31] Bysheim K, Nyrud AQ. Architects' Perceptions of Structural Timber in Urban Construction, Conference COST E53, 29-30 October 2008, Delft, The Netherlands. 2008.

[32] Robichaud F, Kozak R, Richelieu A. Wood Use in Nonresidential Construction: A Case for Communication with Architects. Forest Products Journal. 2009;59:1-2:57-65.

[33] Viluma A, Bratuškins U. Barriers for Use of Wood in Architecture: The Latvian Case. Architecture and Urban Planning. 2107;13:43-47.

[34] Kuzman MK, Sandberg D, Haviarova E. Architects´ Perception of EWPs and Modified Wood in Contemporary Timber Architecture, 2018 World Conference on Timber Engineering (WTCE 2018), August 20-23, Seoul, Republic of Korea. 2018.

[35] Bengtson A. Framing Technological Development in a Concrete context - The Use of Timber in the Swedish Construction Industry. Ph.D. Dissertation, Department of Business Studies, Uppsala University, Sweden. 2003.

Chapter 6

A Study on Contractors' Perception of Using Wood for Construction

Hüseyin Emre Ilgın, Markku Karjalainen, Olli-Paavo Koponen and Anu Soikkeli

Abstract

Construction work is very resource-intensive, and construction projects contain many parameters, in which the choice of building material is one of the critical decisions with numerous criteria, e.g., cost, durability, and environmental impact. Moreover, this complex process includes different parties such as contractors, architects, engineers, where contractors are the most influential decision-makers in material selection. Increasing the use of renewable materials such as wood, which is a technically, economically, and environmentally viable alternative in buildings, can make construction more sustainable. The perceptions of the contractors influence what they propose and therefore the increase in wood construction. With the increasing resource efficiency and the need to adapt to climate change in the construction industry, there is need for contractors to implement sustainable practices. In this chapter, contractors' perceptions of the use of wood in buildings were examined. The results are expected to contribute to environmental remediation by developing strategies to counter perceived barriers and providing insight into new solutions to a conservative space and expanding the use of wood to achieve a more sustainable construction industry. In addition, recommendations for future research, e.g., adhesive- and metal-fastener-free dovetail wood board elements as sustainable material alternatives were presented.

Keywords: engineered wood products, construction, contractors, perception, carbon footprint

1. Introduction

The last two decades have witnessed a dramatic increase in environmental awareness and concern about the impacts of business activities on climate and natural resources on a global scale, and environmental degradation is often addressed as a worldwide problem [1–6]. In this sense, the building construction industry can contribute significantly to the reduction of CO_2 emissions, high energy consumption, excessive waste, and the development of a more resource-efficient and sustainable building environment [7–10]. More specifically, construction activities involve the use of a wide variety of materials, such as concrete, steel, timber where the choice of these materials has significant impacts on the environment [11, 12] as in the cases of tall timber buildings such as the 25-story and 87 m high Ascent

(Milwaukee, structurally topped out) (**Figure 1**), the 22-story and 73 m high HAUT (Amsterdam, under construction) (**Figure 2**), and the 18-story and 58 m high Brock Commons Tallwood House (Vancouver, 2017) (**Figure 3**) [13].

There are several criteria to consider in the selection of building materials, including stability, durability, environmental impact, speed of assembly, cost, and availability [14, 15]. Although design professionals are often involved in this process [16], contractors have the most influence in material selection decisions and therefore play an important role in supporting sustainable development in the context of the construction industry [8, 17, 18].

With the increasing resource efficiency and the need to adapt to climate change in the construction industry, contractors need to execute sustainable practices. However, contractors' decision-making and perceptions of structural frameworks remain largely unexplored, and there are few studies on the selection of structural frameworks involving contractors' perspectives (e.g., [7, 8, 19]).

Wood is the primary building material used by mankind throughout history, a sustainable and renewable building material [20, 21]. The use of wood in construction can affect carbon balance by reducing fossil fuel consumption in manufacturing compared with alternative materials, preventing emissions from cement processing, and storing carbon in wood products and forests. Thus, increasing the use of wood in construction and other long-lasting uses will help achieve sustainable development goals, where timber is recognized as a sustainable material in all major green building rating tools, e.g., Leadership in Energy and Environmental Design (USA) and the BRE Environmental Assessment Method (UK).

Wooden buildings are characterized by a lower carbon construction concept than non-wood buildings [22–26], and timber construction represents a lower embodied energy consumption compared with steel and concrete production [7]. Wooden structures provide significant advantages in combating climate change, because wood can be used as an alternative to other materials to reduce greenhouse gas emissions, but also has unique features such as storing large amounts of carbon in the structure [27, 28]. For example, estimated environmental impact of wood use in Brock Commons Tallwood House (**Figure 3**) was calculated as 1753 metric tons of CO_2 in terms of carbon stored in the wood [29]. Besides being used as a building material during the construction, wood can be reused as a raw material for other structures after the building's service life or, as a last resort, burned

Figure 1.
Ascent (Image courtsey of Jason Korb/Korb + Associates Architects).

Figure 2.
HAUT (Photo courtsey of Jannes Linders).

Figure 3.
Brock Commons Tallwood House (Photo by Michael Elkan and courtsey of Acton Ostry Architects).

instead of fossil fuels [30–32]. The use of wood has many other advantages such as esthetic value, better design adaptability, ease of construction, living comfort, and indoor quality [15, 33, 34].

A better understanding of the perspectives of key actors, such as contractors, in the selection of structural frameworks, can improve insights into concrete path dependency. Because of the CO_2 effects of structural frameworks, such knowledge contributes to understanding the key factors for the development of a more sustainable built environment.

Overall, this chapter examines the perceptions of the contractors regarding the use of wood in buildings. It is believed that the results will contribute to environmental remediation by developing strategies against perceived barriers and providing insight into new solutions to a conservative space and expanding the use of wood to achieve a more sustainable construction industry.

In this chapter, wood or timber refers to engineered wood products (EWPs) [35, 36] such as CLT (a prefabricated multilayer EWP, manufactured from at least three layers of boards by gluing their surfaces together with an adhesive under pressure), laminated veneer lumber (LVL; made by bonding together thin vertical softwood veneers with their grains parallel to the longitudinal axis of the section, under heat and pressure), and glue-laminated timber (glulam) (abbreviated as GL; made by gluing together several graded timber laminations with their grains parallel to the longitudinal axis of the section).

2. Studies on contractors' perceptions of the use of wood for construction

In the literature, many studies focus on the technological aspects of EWPs, their use in construction, and different building solutions [37–50]. Several studies address wood as a structural material in buildings from the perspectives of key professionals (e.g., [8, 17, 51–58]) and consumers or residents (e.g., [59–64]); while there is a very limited number of works focusing on EWPs from the contractors' perspective in the literature. They are from different countries such as Sweden, Finland, Australia focusing on the use of wood as a structural material through surveys and/or interviews.

Among the studies, Hemström et al. [8] conducted interview-based research among contract managers working in contracting companies about their role in the sociotechnical regime, the choice of structural framework, and their perceptions of different alternatives. The results showed that because of their critical position in the firm, managers greatly influence the choice of the structural framework for multifamily buildings managed by the established concrete-based sociotechnical regime. The results also indicated that, due to cognitive rule-based decision-making processes, when assessing the cost of different structural frameworks, they applied their previous experience with concrete solutions as a structural material rather than making deep cost assessments. This approach has prevented timber-framed multifamily buildings from entering common use. While strong incentives for the use of concrete have made it difficult for timber frames to become more common, initiatives promoting wood could contribute to increased awareness and perceptions of wood construction and expectations for future developments in Sweden.

Riala and lola [19] conducted a study through 18 interviews with representatives from the entire value chain to identify the barriers to the adoption of multistory timber construction in Finland, the ways that wood competes with established solutions, and the possibilities for partially integrating construction into the bioeconomy. The results showed that although barriers to its adoption still exist, multistory timber construction can offer competitive solutions for more sustainable construction. A noteworthy finding of concern for the wood products industry was that interviewees with the most experience in multistory timber construction were more critical than those with less experience. This showed that building more wood and gaining more experience is not enough to increase the popularity of wood construction. Additionally, limited possibilities were found to relate the construction industry to bioeconomy. The best way to ensure greater use of renewable materials in multistory construction would be to focus on increasing the competitiveness of multistory timber construction. For this, it was necessary to take advantage of the strengths of the wooden structure such as lightness and prefabrication possibilities and focus on improvement.

Through the qualitative analysis of the data from 36 interviews, Wang et al. [7] examined the perceptions and insights of British construction experts (e.g., industry interest group, timber manufacturer, construction material merchant) to increase understanding of Green Building and the potential of using wood for the UK construction industry. The results confirmed the important role played by the British government in the creation, promotion, and development of Green Building and showed a positive increase in the use of wood in the UK construction industry, supporting the idea that the environmental performance of wood was the main factor in wood adoption. Experts with sound knowledge of wood as a building material were also shown to agree on wood's superior environmental properties; however, end-users who do not know about wood products often have a strong prejudice against their use. Finally, it was shown that legislation, environmental awareness, attitudes and traditions, market and competition, publicity and communication, and technology and know-how are among the main drivers promoting wood as a sustainable solution for Green Building in the UK construction industry. Additionally, most respondents rated the lack of education as one of the most prominent challenges in the current construction industry, hindering the expansion of potential applications of wood products. Following the discussion on wood providing the optimal solution for Green Building (e.g., [21, 65]), the study found a generally positive attitude toward the use of wood in the UK construction industry and found support for the idea that environmental performance is the main driver for wood adoption in the Green Building concept.

Through a survey of 74 experienced construction industry participants (e.g., architects, contractors, developers, and government officials), Xia et al. [11] explored the main barriers to the use of timber framing in multistory construction in Australia. According to the results, the barriers identified can be broadly divided into five groups: lack of legal support, lack of industry interest, lack of experienced professionals, perception of wood framing disadvantages, and limited awareness of wood framing advantages. The survey confirmed the limited awareness of the new wood technologies available, as well as the biggest barriers to the perceived increase in maintenance costs and fire risk. The results are expected to benefit the government and the timber industry by contributing to environmental remediation by developing strategies to increase the use of multistory timber technologies by countering perceived barriers in the Australian context. One approach to overcoming these barriers might be the collaboration of various stakeholders such as governments, customers, designers, and contractors. It was suggested that the government may introduce more supportive legislation and regulations to encourage the use of wood for structural purposes. Industry training (e.g., workshops and seminars) and education in timber structures might have contributed to increasing awareness and knowledge of technological innovations in wood products in the construction industry.

Tan et al. [66] extensively reviewed studies on the relationship between sustainability performance and contractors' competitiveness. The results indicated that there was no unique relationship between the two variables. Therefore, a framework for the implementation of sustainable construction practices was developed to increase competitiveness to help contractors develop their sustainability policies, strategies, and practices to meet the growing need for sustainable development in the construction industry.

Qi et al. [4] aimed to identify the factors affecting contractors' adoption of green construction practices through a survey. The results showed that managerial concern is the most important driver in the adoption of green practices. Significant relationships were also found between government regulation and enterprise size, along with the adoption of green construction practices, while there was no substantial evidence of the relationship between the adoption of green construction practices and perceived stakeholder pressures. This study aimed to contribute to better decision-making regarding the implementation of green construction practices.

3. Conclusions and recommendations

This study aimed to understand the contractors' perceptions of the use of wood for construction. In doing so, this chapter attempted to identify perceived major barriers to timber utilization. However, as the focus of conducted research among contractors differed as noted above, it was not possible to compare these studies with each other, but still, some conclusions were reached as follows: (a) their previous experience with concrete solutions, especially as a structural material, could have prevented them from conducting in-depth analysis for new materials; (b) building more wood and gaining more experience may not be enough to increase the popularity of wood construction; (c) lack of legal support, lack of industry interest, lack of experienced professionals, lack of education, perception of wood framing disadvantages, and limited awareness of wood framing advantages as well as the perceived

Figure 4.
Adhesive- and metal fastener-free dovetail wood board element.

increase in maintenance costs and fire risk were cited as barriers to wood use; (d) managerial concern was among the most important drivers for the adoption of green practices related to wood use.

In this context, the following recommendations can help address identified issues among contractors by improving general attitudes toward the use of wood: (1) providing initiatives that promote wood for increasing awareness and perception of wood construction and prospects for future developments; (2) increasing competitiveness by highlighting the strengths of wooden structure such as lightness and prefabrication possibilities; (3) establishing collaboration of critical stakeholders such as governments, customers, designers, and contractors;

(4) introducing more supportive legislation and regulations at the government level to encourage the use of wood for structural purposes; (5) organizing industry training (e.g., workshops and seminars) and education in wooden structures to increase awareness and knowledge of technological innovations in wood products in the construction industry; (6) conducting more research projects (e.g., the DoMWoB project/Dovetailed Massive Wood Board Elements for Multi-Story Buildings – *see Acknowledgments and Funding*) [67] and developing more innovative and environmentalist EWPs (e.g. adhesive- and metal-fastener-free dovetail wood board elements) (**Figure 4**) [68] to demonstrate the potential of wood for use in construction.

It is believed that this chapter will help deepen the understanding of the various aspects that shape the decision-making process particularly among contractors in the use of EWPs for construction.

Acknowledgements

This project has received funding from the European Union's Horizon 2020 research and innovation program under the Marie Skłodowska-Curie grant agreement No [101024593].

Funding

This project has also received funding (60 000 EUR) from the Marjatta and Eino.

Kolli Foundation for funding the technical performance tests including fire safety, structural, moisture transfer resistance and air-tightness, and sound insulation.

Author details

Hüseyin Emre Ilgın[1*], Markku Karjalainen[1], Olli-Paavo Koponen[1] and Anu Soikkeli[2]

1 Tampere University, Tampere, Finland

2 University of Oulu, Oulu, Finland

*Address all correspondence to: emre.ilgin@tuni.fi

References

[1] Tse RYC. The implementation of EMS in construction firms: Case study in Hong Kong. Journal of Environmental Assessment Policy and Management. 2001;**3**(2):177-194

[2] Tam WYV, Tam CM, Shen LY, Zeng SX, Ho CM. Environmental performance assessment: Perceptions of project managers on the relationship between operational and environmental performance indicators. Construction Management and Economics. 2006;**24**(3):287-299

[3] Shen LY, Yao H. Improving environmental performance by means of empowerment of contractors. Management of Environmental Quality. 2006;**17**(3):242-254

[4] Qi GY, Shen LY, Zeng SX, Jorge OJ. The drivers for contractors' green innovation: An industry perspective. Journal of Cleaner Production. 2010;**18**(14):1358-1365

[5] Shrinkhal R. Economics, technology, and environmental protection: A critical analysis of phytomanagement. In: Pandey VC, Bauddh K, editors. Phytomanagement of Polluted Sites, Market Opportunities in Sustainable Phytoremediation. Amsterdam: Elsevier; 2019. pp. 569-580

[6] Maurya PK, Ali SA, Ahmad A, Zhou Q, Castro JS, Khan E, et al. An introduction to environmental degradation: Causes, consequence and mitigation. In: Environmental Degradation: Causes and Remediation Strategies. Uttarakhand, India: Agro Environ Media; 2020. DOI: 10.26832/aesa-2020-edcrs-01

[7] Wang L, Toppinen A, Juslin H. Use of wood in green building: A study of expert perspectives from the UK. Journal of Cleaner Production. 2014;**65**:350-361

[8] Hemström K, Gustavsson L, Mahapatra K. The sociotechnical regime and Swedish contractor perceptions of structural frames. Construction Management and Economics. 2017;**35**(4):184-195

[9] Kong A, Kang H, He S, Li N, Wang W. Study on the Carbon Emissions in the Whole Construction Process of Prefabricated Floor Slab. Applied Sciences. 2020;**10**:2326

[10] Sijakovic M, Peric A. Sustainable architectural design: Towards climate change mitigation. Archnet-IJAR: International Journal of Architectural Research. 2021;**15**(2):385-400

[11] Xia B, O'Neill T, Zuo J, Skitmore M, Chen Q. Perceived obstacles to multi-storey timber-frame construction: An Australian study. Architectural Science Review. 2014;**57**(3):169-176

[12] Khoshnava SM, Rostami R, Zin RM, Štreimikienė D, Mardani A, Ismail M. The role of green building materials in reducing environmental and human health impacts. International Journal of Environmental Research and Public Health. 2020;**17**(7):2589

[13] CTBUH. Council on Tall Buildings and Urban Habitat. Illinois Institute of Technology, S.R. Crown Hall, 3360 South State Street, Chicago, Illinois, USA. Available from: www.ctbuh.org [Accessed: February 10, 2022]

[14] Dodoo A, Gustavsson L, Sathre R. Effect of thermal mass on life cycle primary energy balances of a concrete- and a wood-frame building. Applied Energy. 2012;**92**:462-472

[15] Sandak A, Sandak J, Brzezicki M, Kutnar A. Biomaterials for Building Skins. In: Bio-based Building Skin. Environmental Footprints and Eco-design of Products and Processes.

Singapore: Springer; 2019. p. 16. DOI: 10.1007/978-981-13-3747-5_2

[16] Bayne K, Taylor S. Attitudes to the use of Wood as a Structural Material in Non-residential Building Applications: Opportunities for Growth. PN05.1020. Forest and Wood Products Australia: Melbourne, VIC; 2006

[17] Roos A, Woxblom L, McCluskey D. The influence of architects and structural engineers on timber in construction - Perceptions and roles. Silva Fennica. 2010;**44**(5):871-884

[18] Hemström K, Mahapatra K, Gustavsson L. Perceptions, attitudes and interest of Swedish architects towards the use of wood frames in multi-storey buildings. Resources, Conservation and Recycling. 2011;**55**:1013-1021

[19] Riala M, Ilola L. Multi-storey timber construction and bioeconomy: Barriers and opportunities. Scandinavian Journal of Forest Research. 2014;**29**(4):367-377

[20] Herbert S. The oldest wood construction of the world, (Der aelteste Holzbau der Welt). Artikel aus der Zeitschrift: GWF Wasser Abwasser. 1993;**134**(12):711-714

[21] Sathre R, Gustavsson L. Using wood products to mitigate climate change: External costs and structural change. Applied Energy. 2009;**86**(2):251-257

[22] Ritter M, Skog K, Bergman R. Science Supporting the Economic and Environmental Benefits of Using Wood and Wood Products in Green Building Construction; General Technical Report FPL-GTR-206. Madison, WI, USA: U.S. Department of Agriculture, Forest Service, Forest Products Laboratory; 2011. pp. 1-9

[23] Pierobon F, Huang M, Simonen K, Ganguly I. Environmental benefits of using hybrid CLT structure in midrise non-residential construction: An LCA based comparative case study in the U.S. Pacific Northwest. Journal of Building Engineering. 2019;**26**:100862

[24] Dong Y, Qin T, Zhou S, Huang L, Bo R, Guo H, et al. Comparative whole building life cycle assessment of energy saving and carbon reduction performance of reinforced concrete and timber stadiums—A case study in China. Sustainability. 2020;**12**:1566

[25] Lolli N, Fufa MS, Wiik MK. An assessment of greenhouse gas emissions from CLT and glulam in two residential nearly zero energy buildings. Wood Material Science and Engineering. 2019;**14**:342-354

[26] Lukić I, Premrov M, Leskovar ŽV, Passer A. Assessment of the environmental impact of timber and its potential to mitigate embodied GHG emissions. IOP Conference Series: Earth and Environmental Science. 2020;**588**:1.01-1.05

[27] Kazulis V, Muizniece I, Zihare L, Blumberga D. Carbon storage in wood products. Energy Procedia. 2017;**128**:558-563

[28] Green MC, Karsh JE. The Case for Tall Wood Buildings—How Mass Timber Offers a Safe, Economical, and Environmentally Friendly Alternative for Tall Building Structures; Mgb ARCHITECTURE + DESIGN, Equilibrium Consulting, LMDG Ltd., Eds. Vancouver, BC, Canada: BTY Group; 2012

[29] The Canadian Wood Council – CWC. The Advent of Tall Wood Structures in Canada, a Case Study. Ontario: Brock Commons Tallwood House, University of British Columbia Vancouver Campus; 2018. Available from: https://cwc.ca/wp-content/uploads/2019/03/CS-BrockCommon.Study_.23.pdf Accessed: February 10, 2022

[30] Werner F, Taverna R, Hofer P, Richter K. Carbon pool and substitution effects of an increased use of wood in buildings in Switzerland: First estimates. Annals of Forest Science. 2005;**62**:889-902

[31] Kutnar A, Hill C. Life cycle assessment—Opportunities for forest products sector. BioProducts Business. 2017;**2**:52-64

[32] Bergman RD, Falk RH, Gu H, Napier TR, Meil J. Life-Cycle Energy and GHG Emissions for New and Recovered Softwood Framing Lumber and Hardwood Flooring Considering End-of-Life Scenarios; Res. Pap. FPL-RP-672. Madison, WI, USA: U.S. Department of Agriculture, Forest Service, Forest Products Laboratory; 2013. p. 35

[33] Gold S, Rubik F. Consumer attitudes towards timber as a construction material and towards timber frame houses – Selected findings of a representative survey among the German population. Journal of Cleaner Production. 2009;**17**(2):303-309

[34] Nolan G. Timber in Multi-Residential, Commercial and Industrial Building: Recognising Opportunities and Constraints. PNA140-0809. Melbourne, VIC: Forest and Wood Products Australia, 2011

[35] Wang Z, Yin T. Cross-Laminated Timber: A Review on Its Characteristics and an Introduction to Chinese Practices. IntechOpen: London, UK; 2021

[36] Rahman T, Ashraf M, Ghabraie K, Subhani M. Evaluating Timoshenko method for analyzing CLT under out-of-plane loading. Buildings. 2020;**10**:184

[37] Zwerger K. Recognizing the similar and thus accepting the other: The European and Japanese traditions of building with wood. Journal of Traditional Building, Architecture and Urbanism. 2021;**2**:305-317

[38] Van Damme B, Schoenwald S, Zemp A. Modeling the bending vibration of cross-laminated timber beams. European Journal of Wood and Wood Products. 2017;**75**:985-994

[39] Chiniforush AA, Akbarnezhad A, Valipour H, Xiao J. Energy implications of using steel-timber composite (STC) elements in buildings. Energy and Buildings. 2018;**176**:203-215

[40] Toivonen R, Lähtinen K. Sustainability—A Literature Review on Concealed Opportunities for Global Market Diffusion for the Cross-Laminated Timber (CLT) in the Urbanizing Society; The Manuscript for Bioproducts Business. Curitiba, Brazil: IUFRO; 2019

[41] Yusof NM, Tahir PM, Lee SH, Khan MA, James RMS. Mechanical and physical properties of cross-laminated timber made from Acacia mangium wood as function of adhesive types. Journal of Wood Science. 2019;**65**:20

[42] Karjalainen M, Ilgın HE. A statistical study on multi-story timber residential buildings (1995-2020) in Finland. In: Proceedings of the LIVENARCH VII Livable Environments & Architecture 7th International Congress OTHER ARCHITECT/URE(S), Trabzon, Turkey; 28-30 September 2021. Vol. I. Trabzon, Turkey: KTU Printing Center; 2021. pp. 82-94

[43] Li M, Zhang S, Gong Y, Tian Z, Ren H. Gluing techniques on bond performance and mechanical properties of cross-laminated timber (CLT) made from Larix kaempferi. Polymers. 2021;**13**:733

[44] Bahrami A, Nexén O, Jonsson J. Comparing performance of cross-laminated timber and reinforced concrete walls. International Journal of

Applied Mechanics and Engineering. 2021;**26**:28-43

[45] Sun Z, Chang Z, Bai Y, Gao Z. Effects of working time on properties of a soybean meal-based adhesive for engineered wood flooring. The Journal of Adhesion. 2021:1-20 (ahead-of-print)

[46] Ilgın HE, Karjalainen M, Koponen O. Review of the Current State-of-the-Art of Dovetail Massive Wood Elements. IntechOpen: London, UK; 2021

[47] Ilgın HE, Karjalainen M, Koponen O. Dovetailed massive wood board elements for multi-story buildings. In: Proceedings of the LIVENARCH VII Livable Environments & Architecture 7th International Congress OTHER ARCHITECT/URE(S), Trabzon, Turkey; 28-30 September 2021. Vol. I. Trabzon, Turkey: KTU Printing Center; 2021. pp. 47-60

[48] Karjalainen M, Ilgın HE, Yli-Äyhö M, Soikkeli A. Complementary Building Concept: Wooden Apartment Building: The Noppa toward Zero Energy Building Approach. IntechOpen: London, UK; 2021

[49] Tulonen L, Karjalainen M, Ilgın HE. Tall Wooden Residential Buildings in Finland: What Are the Key Factors for Design and Implementation? IntechOpen: London, UK; 2021

[50] Rinne R, Ilgın HE, Karjalainen M. Comparative study on life-cycle assessment and carbon footprint of hybrid, concrete and timber apartment buildings in Finland. International Journal of Environmental Research and Public Health. 2022;**19**:774

[51] Ilgın HE, Karjalainen M. Perceptions, Attitudes, and Interest of Architects in the Use of Engineered Wood Products for Construction: A Review. IntechOpen: London, UK; 2021

[52] Karjalainen M, Ilgın HE, Tulonen L. Main design considerations and prospects of contemporary tall timber apartment buildings: Views of key professionals from Finland. Sustainability. 2021;**13**:6593

[53] Aaltonen A, Hurmekoski E, Korhonen J. What about wood?—"Nonwood" construction experts' perceptions of environmental regulation, business environment, and future trends in residential multistory building in Finland. Forest Products Journal. 2021;**71**:342-351

[54] Ilgın HE, Karjalainen M, Pelsmakers S. Finnish architects' attitudes towards multi-storey timber-residential buildings. International Journal of Building Pathology and Adaptation. 2021;ahead-of-print

[55] Markström E, Kuzman MK, Bystedt A, Sandberg D, Fredriksson M. Swedish architects view of engineered wood products in buildings. Journal of Cleaner Production. 2018;**181**:33-41

[56] Karjalainen M, Ilgın HE, Somelar D. Wooden additional floors in old apartment buildings: Perspectives of housing and real estate companies from Finland. Buildings. 2021;**11**:316

[57] Karjalainen M, Ilgın HE, Metsäranta L, Norvasuo M. Wooden Facade Renovation and Additional Floor Construction for Suburban Development in Finland. IntechOpen: London, UK; 2022

[58] Karjalainen M, Ilgın HE, Metsäranta L, Norvasuo M. Suburban residents' preferences for livable residential area in Finland. Sustainability. 2021;**13**:11841

[59] Karjalainen M, Ilgın HE, Somelar D. Wooden Extra Stories in Concrete Block of Flats in Finland as an Ecologically Sensitive Engineering Solution, Ecological Engineering—Addressing

Climate Challenges and Risks. IntechOpen: London, UK; 2021

[60] Karjalainen M. The Finnish Multi-Story Timber Apartment Building as a Pioneer in the Development of Timber Construction. Oulu, Finland: University of Oulu; 2002 (In Finnish)

[61] Karjalainen M, Ilgın HE. The change over time in Finnish Residents' attitudes towards multi-story timber apartment buildings. Sustainability. 2021;**13**:5501

[62] Lähtinen K, Harju C, Toppinen A. Consumers' perceptions on the properties of wood affecting their willingness to live in and prejudices against houses made of timber. Wood Material Science and Engineering. 2019;**14**:325-331

[63] Kylkilahti E, Berghäll S, Autio M, Nurminen J, Toivonen R, Lähtinen K, et al. A consumer-driven bioeconomy in housing? Combining consumption style with students' perceptions of the use of wood in multi-story buildings. Ambio. 2020;**49**:1943-1957

[64] Karjalainen M, Ilgın HE, Metsäranta L, Norvasuo M. Residents' attitudes towards wooden facade renovation and additional floor construction in Finland. The International Journal of Environmental Research and Public Health. 2021;**18**:12316

[65] Gustavsson L, Joelsson A, Sathre R. Life cycle primary energy use and carbon emission of an eight-storey wood-framed apartment building. Energy and Buildings. 2010;**42**(2): 230-242

[66] Tan YT, Shen LY, Yao H. Sustainable construction practice and contractors' competitiveness: A preliminary study. Habitat International. 2011;**35**:225-230

[67] Ilgın HE, Karjalainen M. Preliminary design proposals for dovetail wood board elements in multi-story building construction. Architecture. 2021;**1**:56-68

[68] Ilgın HE, Karjalainen M, Koponen O. Various Geometric Configuration Proposals for Dovetail Wooden Horizontal Structural Members in Multistory Building Construction. IntechOpen: London, UK; 2022

Chapter 7

Tall Wooden Residential Buildings in Finland: What Are the Key Factors for Design and Implementation?

Lassi Tulonen, Markku Karjalainen and Hüseyin Emre Ilgın

Abstract

This chapter examines tall residential buildings using engineered wood products (EWPs) in Finland. As specified in the National Building Code of Finland, 'tall wooden building' is defined as a structure with more than 8-story. Currently, there are two wooden residential buildings, 14-story Lighthouse Joensuu (2019) with laminated veneer lumber (LVL) structure and 13-story HOAS Tuuliniitty (under construction) with cross laminated timber (CLT) structure, that fit the definition above. This study analyses the phenomena associated with the design and implementation processes of these remarkable buildings, the starting points for the projects as well as the prospects of tall wooden housing in Finland through the case study method and interviews with the key actors in the projects. These cases are mapped with extremely detailed information, comprising a valuable source both for designers, engineers as well as developers. As a result, the current state-of-the-art and the critical factors influencing the design and implementation of these challenging sustainable projects in Finland have been identified. It is believed that this chapter will aid and direct key stakeholders in the construction industry in the sound planning and development of tall wooden residential projects in Finland.

Keywords: Engineered wood products (EWPs), construction, tall residential building, Finland, architectural and structural design considerations

1. Introduction

Timber has typically been used for centuries in single and multi-residential construction projects around the world. Timber in the form of EWPs, such as CLT (manufactured from at least three layers of boards by gluing their surfaces together with an adhesive under pressure), has been available on small scale projects in Europe, particularly in Germany and Austria, since its invention in the early 1990s. Moreover, recent advances in EWP technology have overcome major challenges in material strength and fire safety, making these products competitive against concrete and steel for tall building design.

Increasing wood construction has been a governmental policy in Finland for nearly three decades, and wooden buildings are being promoted by public authorities through various supporting wood programs and legislative changes. The key to encouraging wood construction has been to increase the degree of processing of the wood industry, improve exports and enhance ecology locally [1].

In the 2010s, climate targets became the main driver of wooden construction. Technical solutions in this promising area have so far been perceived to develop mainly as state-supported housing so that public solutions and operating models developed through assisted residential construction gradually become established in self-sufficient housing [2].

Apartment building construction has been seen as an opportunity to boost wood utilization [3]. Approximately 70% of residential construction completed in Finland in the last five years has consisted of apartment buildings, many of which are still built with reinforced concrete. Although structural systems for industrial timber construction have been known in Finland for nearly thirty years, wooden apartments have gradually become widespread.

Fire regulations played a central role in spreading the factor contributing to the wooden structure. In 1997, the revised fire code version resulted in permission being granted for the construction of residential and office buildings with wooden frames and façades up to 4-story high, due to which momentarily growth started in the construction of wooden apartment buildings in Finland. In 2011, the building regulation was revised to allow the construction of wooden apartment buildings up to a maximum of 8-story under certain conditions [4].

Especially in recent years, the number of wooden flats has started to increase significantly, and wooden structures have also attracted a lot of attention. Multi-story wooden apartments, e.g. Lighthouse Joensuu also gained great international visibility. Especially in the media, there is a lot of focus on the height of wooden buildings, and the development and innovation of wooden structures are widely presented through tall buildings.

However, very few wooden apartment buildings have been implemented so far, and their share in all Finnish apartment production is still very small. In 2019, less than 32,000 apartment buildings were completed in Finland, of which less than 2% were in wooden apartment buildings [5]. However, as industrial manufacturing and design and construction know-how improves and costs drop, larger-scale wooden apartment construction is becoming increasingly common. In particular, the construction of low-rise wooden apartment buildings is beginning to settle down so well that industrial wood construction, especially for them, can no longer be considered purely experimental production [6, 7].

As apartment building with EWPs has begun to take root, critical views on new pilot projects have also emerged. For example, more than 8-story wooden apartment buildings can be considered as such, which require functional fire safety analysis and a special procedure for construction projects. Instead, less than 10-story wooden buildings and increased competition in established markets are more important for the construction sector [8].

Most residential apartment buildings in Finland have a maximum of 9-story. At the end of 2018, there were a bit more than 61,000 residential apartment buildings in Finland, of which only 236 were 10-story or higher [9]. Thus, of all apartment buildings in Finland, only 0.4% were relatively tall construction, which contributes to the question of whether there is a need or justification for the development of tall wooden apartment buildings, especially as subsidized housing production. On the other hand, although there are so far few apartment buildings in Finland, tall construction has also increased significantly in recent years [10].

The main purpose of this study was to reveal the current state of tall wooden apartment construction with EWPs in Finland. The research examined the architectural and structural considerations of tall wooden residential buildings in the Finnish context. To achieve this goal, this chapter scrutinized the design and implementation process as well as the architectural and structural features of the two wooden residential buildings (over 8-story), Lighthouse Joensuu (2019), and HOAS Tuuliniitty (under construction), in Finland.

The scope of the chapter was limited by using five main points to identify key factors for design and implementation in Finnish tall wooden residential construction: Literature survey, structural considerations of wooden apartment buildings, special features of tall building construction, case studies including general information together with architectural and structural design considerations, and finally, interviews.

By revealing the up-to-date status with key factors for the design and implementation of contemporary Finnish tall wooden residential practices, this research provides insights into the making of more technically and economically sound planning decisions for future tall wooden residential building designs. In this study, wood or timber refers to EWPs including CLT, laminated veneer lumber (LVL) (made by bonding together thin vertical softwood veneers with their grain parallel to the longitudinal axis of the section, under heat and pressure), and glue-laminated timber/glulam (GL) (made by gluing together several graded timber laminations with their grain parallel to the longitudinal axis of the section).

In this chapter, the three main research questions were: (i) What are the key factors for the design and implementation of tall wooden apartment buildings in Finland? (ii) What are the special features of tall residential buildings in Finland? (iii) What are the prospects for tall wooden apartment construction in Finland?

2. Literature survey

While wood construction research in Finland has increased in recent years, the use of EWPs in the construction industry has become progressively more common. However, particularly in the Finnish context, the peculiarities of the tall wooden apartment building have not been analyzed very extensively and studies have focused mainly on low-rise wooden construction. Previous research on tall wooden construction in Finland has primarily scrutinized individual wooden apartment buildings and so, no comparative study has been conducted between projects.

Among the above-mentioned studies, the research conducted at the Karelia University of Applied Sciences in Joensuu was particularly on the construction of tall wooden apartment buildings. This is mostly because currently, the only wooden apartment building with more than 8-story in Finland is Lighthouse Joensuu, which is also widely used as a case study. Several theses were published related to Lighthouse Joensuu construction process [11], the use of wood in tall construction in general [12], and more specifically individual technical and structural solutions. In addition to these studies, a collection of articles on Lighthouse Joensuu in tall urban construction was published [13].

On the other hand, the factors affecting the overall development of wooden apartment buildings were examined in several Finnish studies (e.g [14–18]).

2.1 Structural considerations of wooden apartment buildings

In Finland, wooden apartment buildings are currently implemented with three different structural systems: a volumetric modular system, a load-bearing large element system, and a post-beam system. The case studies examined in this chapter and the other tallest wooden apartments in Finland were applied with either large element or volumetric modular system solutions.

Wooden post-beam structures were considered a competitive structural system in Finland, especially in low-rise wooden apartment buildings with a maximum of 5-story

[19]. The structural system chosen affects the architectural design of the building, through the location of the load-bearing lines and the possible spans of the subfloors [20]. Steel fasteners or screw fasteners are usually used for joints in wooden structures. The method of fastening is influenced by the selection of the structural system.

Compared to reinforced concrete construction, the maximum spans of the subfloors of wooden apartment buildings are typically somewhat shorter and, depending on the construction method, spans of about 4-7meters can be cost-effectively achieved with wooden subfloors. The load-bearing structures and partitioning walls between apartments are also usually thicker than concrete structures, which is particularly due to fire and sound insulation requirements [21].

In addition to the above-mentioned structural systems, it is possible to use various combinations of structural systems or wood/timber-concrete hybrid structures. For example, the 14-story and 49 m high Treet (2015) wooden apartment building in Bergen (Norway) consists of a GL frame and volumetric modular system. The load-bearing structure of the building is an adhesive lattice resembling a bridge structure and two concrete subfloors on the fifth and tenth floors. Volumetric CLT modules make up four-story-high entities that are separate from the load-bearing frame and do not contribute to the total frame stiffening [22].

Currently, the 18-story and 85 m high Mjøstårnet (2019) in Brumunddal (Norway) (**Figure 1**) is the tallest wooden apartment building in the world, and the 24-story and 84 m high HoHo (2020) in Vienna (Austria) (**Figure 2**) is the

Figure 1.
Mjøstårnet (Norway, 2019) (source: Wikipedia).

Figure 2.
HoHo (Austria, 2020) (source: Wikipedia).

second tallest [23]. Mjøstårnet's structural system is like a massive post-beam grid as in the case of Treet, but the apartments were built with prefabricated walls and floor elements rather than a volumetric modular system. The partition walls are implemented as CLT structures and the facades as sandwich panels attached to the GL frame. The wall structures do not contribute to the structural stiffness of the building. The elevator shafts and staircases are implemented as a CLT structure. Subfloors are made of LVL up until the tenth floor. The upper floors have concrete subfloors to counter the swaying of the building caused by lateral forces [24].

HoHo was designed as a hybrid structure. In the middle of the building was a reinforced concrete stiffening staircase, the surrounding spaces of which were constituted by load-bearing CLT walls. The subfloors are wood-concrete joint-structures that are supported by the concrete stairwell and the exterior walls. In total, about 75% of the building was made of wood [25].

Other possible joint solutions are, for example, the steel corner joint used in the 10-story and 34 m high Dalston Works wooden block of flats in London (UK). The steel parts used to connect walls to the CLT-subfloors above and below the floor slab, are also connected through the subfloor. This solution transfers the upthrust caused by wind forces to the lower floors [26]. At the 8-story Bridport House in London, CLT wall and floor elements are connected through a comb joint that reduces compression of the floor slab [12].

In volumetric modular system construction, the building is assembled from prefabricated modules in the factory, which in apartment building construction form either individual apartments or parts of an apartment. These elements are connected at the construction site, for example with steel brackets, and they usually form the load-bearing and stiffening frame of the building as such. In Finland, most of the volumetric modular system construction is currently carried out with solid wood CLT elements, but the elements can also be implemented as a rigid structure.

In a solid wood volumetric modular system, the walls and the roofs are typically formed of CLT panels. The floor of this element, i.e. the load-bearing subfloor structure, can also be implemented with a CLT board, but recently the beam subfloor has been a more common solution, especially due to better sound insulation. For example, Puukuokka 1 in Jyväskylä was completed with CLT slab floors, but buildings 2 and 3 were made with beam floors. In volumetric modular system construction, the walls and mezzanines between the apartments consist of two adjacent or overlapping elements, which allows the wooden surface to be left visible inside the dwellings in some cases if sound insulation is implemented between the double structure. The CLT solid wood panel acts as load-bearing and bracing structure in the walls at the same time, but due to the way the volumetric modular system is installed, only one of the adjacent walls can be fixed so that it acts as a bracing wall [27]. Solid wood also performs as a vapor barrier, and in part also as warm and sound insulation.

Volumetric modular system construction imposes quite a few boundary conditions on architectural design. The element has certain maximum dimensions, which depend, among other things, on the manufacturer, transport, and lifting possibilities. The narrow rectangular shape of these elements limits the types of housing that can consist of the elements. On the other hand, the volumetric modular system technology allows for a more cost-effective implementation of recessed balconies than concrete construction, for example. The basic solutions in architectural design that affect the cost-effectiveness of such constructions are the size and shape of the element and their total number and frequency.

2.2 Special features of tall residential construction

In addition to wooden construction in Finland, interest in tall buildings has also increased over the past decade. This is largely related to the urbanization trend and the rapid population growth in the major cities [28]. In a Finnish urban structure, buildings that are already quite low can be considered tall construction. Tall construction definitions can vary by city in Finland, depending on how tall the current building stock is. The definition of tall buildings starting from 16-story in Helsinki is accepted as 12 in Tampere. In many smaller cities of Finland, construction of more than 8-story already stands out as a taller construction than the rest of the building stock. In the view of the authors of this study, a tall wooden building is assumed to be a building of more than 8-story height based on the National Building Code of Finland.

There are a total of about 50 buildings over 50 m in height in Finland [29]. At present, the tallest residential building in Finland is the 35-story and 132 m high Majakka in Helsinki (also known as Lighthouse, Redi Kalasatama 1) [23]. Other Finnish tallest apartment projects include the 26-story and 86 m high Cirrus (2006) in Helsinki, and the 24-story and 78 m high Niittyhuippu (2017) in Espoo. The tallest building project in Finland is the Trigoni Tower 1 with 51-story planned for Pasila, which will be 180 m high. In the 2010s, some rather tall student apartments with reinforced concrete have also been implemented, which were perhaps the best benchmark for the current tall wooden apartment building construction.

3. Research methods

The study was conducted through literature surveys, the case study method, and interviews with the key actors in the projects.

3.1 Case studies

In this section, two wooden residential buildings (over 8-story), Lighthouse Joensuu (**Figure 3**) and HOAS Tuuliniitty (**Figure 4**) were analyzed with exceptionally detailed information in terms of architectural and structural design considerations. Also, general information e.g. number of stories, floor area, frame system, fire consultant about case studies were provided in **Table 1** below.

In Lighthouse Joensuu project, in particular, the start-up of the entire project and wooden construction was strongly the goals set by the city's corporate management, according to which the builder acted, while HOAS Tuuliniitty was the builders' first wooden apartment building projects.

Lighthouse Joensuu and HOAS Tuuliniitty have were designed based on the existing town plan. Moreover, the Tuuliniitty is a special wood construction area,

Figure 3.
Lighthouse Joensuu (Joensuu, 2019) (drawn by Emre ILGIN).

Figure 4.
HOAS Tuuliniitty (Tuuliniitty, under construction) (drawn by Emre ILGIN).

	Lighthouse Joensuu	HOAS Tuuliniitty
# of stories & height	14 & 48 m	13 & 42 m
Completion date	2019	2021
Floor area	4787 m^2	4787 m^2
Gross area	5934 m^2	7584 m^2
Residential area	3764 m^2	4433 m^2
Space efficiency	63%	58%
# of apartment & area (av.)	117 & 32 m^2	165 & 27 m^2
Duration of the project	38-month	
Duration of the construction	20-month	
Frame system	LVL/CLT large elements	CLT volumetric elements
Builder/client	Joensuun Elli opiskelija-asunnot Oy	Helsingin seudun opiskelija-asuntosäätiö HOAS
Architect	Arcadia Oy	Ark.tsto. Jukka Turtiainen (nowadays Arkkitehtipalvelu)
Structural designer	Joensuun Juva (nowadays A-Insinöörit)	A-Insinöörit
Wood supplier	Stora Enso	Elementti-Sampo
Contractor	Eero Reijonen	JVR Rakenne
Fire consultant	Markku Kauriala Oy	KK-Palokonsultti Oy

Table 1.
General information about Lighthouse Joensuu and HOAS Tuuliniitty.

the town plan of which defines wood as the mainframe and façade material for all new buildings.

3.1.1 Architectural design considerations

The architectural plan of the residential floors is largely based on the structural system selected (**Figure 5**). In HOAS Tuuliniity, most of the apartments are studios and all the space elements are located transversely to the building frame, which means that the frame depth of the buildings becomes quite large and almost all the apartments open in only one direction. The building has two staircases, one of which is an exit staircase outside the building envelope. The first floor is made of concrete construction and accommodates storage and common areas. Floors 2–13 are similar residential floors. Floor 13 is an attic floor not included in the building height, which houses the residents' club room and the sauna facility. The staircase and elevator shaft are implemented as separate space elements, which are located against the outer wall along the middle corridor. On the facades of the Tuuliniity, horizontal wood panel cladding is employed.

In Lighthouse Joensuu, the apartments are grouped around the staircase in the middle of the building (**Figure 6**). There are two compartmentalized staircases and an elevator that opens in both directions in the middle of the frame. All residential units are located on the periphery of the building, and the staircase is completely opaque. Most of the apartments are placed parallel to the staircase and the depths of the apartments are very narrow. Especially the studios consist of different models from the student units in Lighthouse Joensuu. Also, all interior surfaces are upholstered, and the wooden surface is invisible. The main reason for this is that the LVL panels used as the frame material are not esthetic surface material like CLT. The facades of the building are covered with fiber cement boards in three different shades of white and gray.

3.1.2 Structural design considerations

Lighthouse Joensuu is implemented with a large-element system, where the load-bearing walls are LVL-structured and floors are CLT-structured. In the foundation system of the building, steel pipe piles are used by anchoring to the rock. The lower

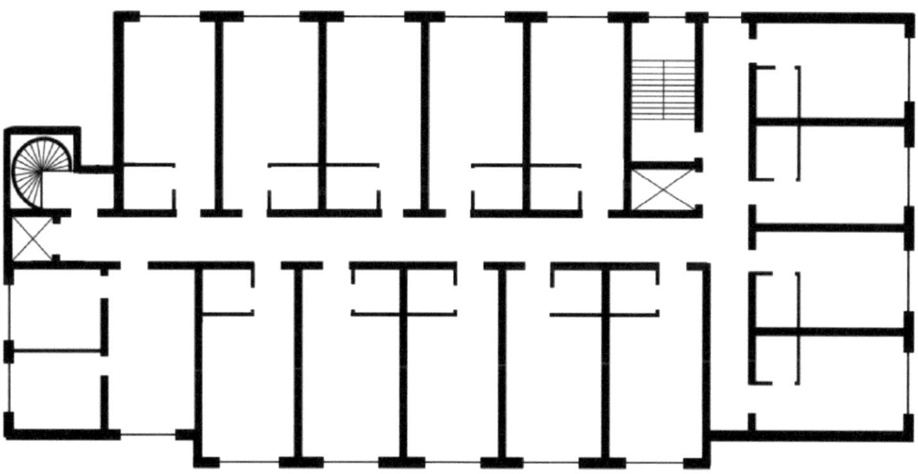

Figure 5.
HOAS Tuuliniity typical floor plan (drawn by Emre ILGIN).

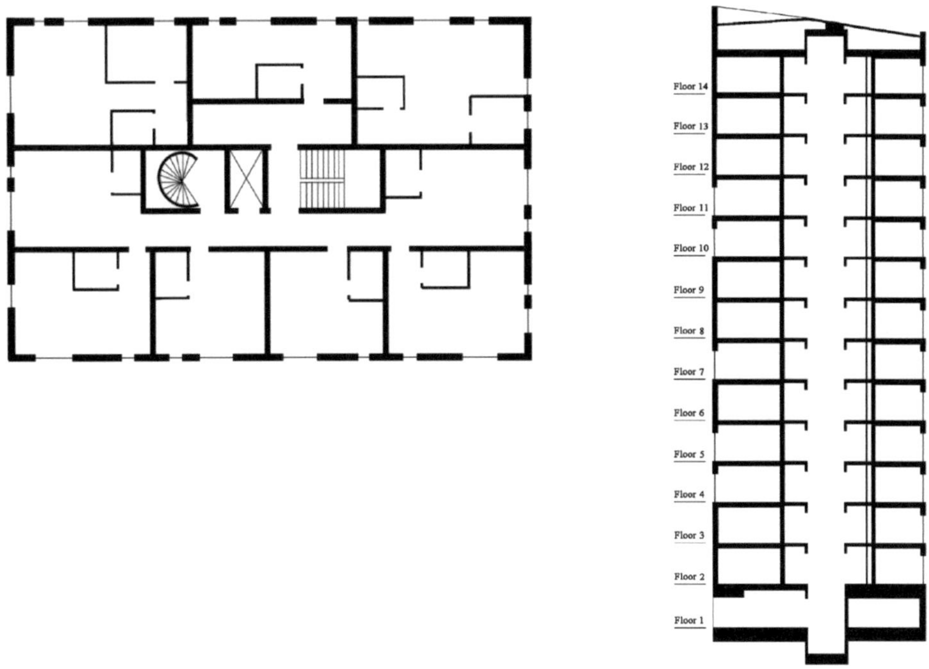

Figure 6.
Lighthouse Joensuu typical floor plan and section (drawn by Emre ILGIN).

floors have a concrete structure with 30 cm thick load-bearing reinforced concrete. Similarly, 20-30 cm thick concrete and 50 mm thick concrete subfloor are utilized on the first floor. A thicker intermediate floor slab increases the mass of the structure, which helps to control the lateral forces on a light wooden apartment building.

The thickness of the load-bearing walls decreases in the upper floors as the load decreases. In the three lowest wood-structured floors, the thickness of the load-bearing LVL board is 162 mm, the next seven floors, and the three top floors, 144 mm and 126 mm, respectively. The fire and sound insulation inside the apartments are made of 50 mm stone wool with double plasterboard on top.

The intermediate base subfloors of the staircases are made of solid wood elements connected by butt or half-point joints. The longer corridor has an LVL structure, and the shorter corridor and staircases have a CLT structure. The stairs are prefabricated elements made of CLT panels with the steps milled into the panel. Under the intermediate floor slab of the staircase, there is an LVL beam supported by the walls, to which the stair element is attached with screw connections.

The wet rooms are built on-site. The implementation of large-element construction did not allow the use of wet room elements, as the sound insulation on the lower surface of the subfloor structure had to be implemented before the construction of wet rooms.

HOAS Tuuliniitty is built from modular CLT elements. The load-bearing walls are constructed of CLT panels and the intermediate floor structure is built with LVL beam structure. Overlapping space elements are connected with dowel pins solder casting from inside the modular element. The partitions between the modular elements are load-bearing and the outer walls are mainly non-structural. Longitudinal bracing is mostly carried out by the walls of the central corridor and transverse bracing by the walls between apartments. The Tuuliniity is located near the sea, which causes the building to belong in terrain class 0. This affects the wind loads significantly compared to, for example, the Joensuu Lighthouse which belongs in terrain class 2.

The walls of the first floor are reinforced concrete elements and a cast 25 cm reinforced concrete slab. Similarly, on wooden floors, the intermediate floors of the staircase are made of reinforced concrete slabs. Reinforced concrete piles and a load-bearing lower base are employed in the foundation system. Wet rooms are implemented as pre-installed elements.

The load-bearing walls between the apartments consist of two 120 mm CLT panels with 30 mm of mineral wool as sound insulation between them. The apartments have stone wool insulation with a double gypsum board. There are sound insulation pads between the CLT frame and the studs of the insulation frame. The walls between the apartments and the staircase an otherwise similar structure, but there is only one CLT panel in the wall.

The ceilings of the apartment modules are 80 mm thick CLT panels that are kept exposed as a visible ceiling surface. The load-bearing floors have a 220 mm LVL-beam structure and a 100 mm layer of mineral wool as sound insulation. Beneath the beam structure, there is a double fire gypsum board and a 30 mm layer of mineral wool. On top of the beams, there is an 18 mm veneer, sound insulation layer, and a cast floor. In a tall building, a flexible frame structure would cause excessive swaying of the building, so instead of separating apartment modules with flexible insulation pads, the sound insulation is implemented as an acoustic inner shell inside the apartments. The intermediate floors of the central corridors are made of 250 mm thick reinforced concrete slabs that are connected to the surrounding CLT walls with steel.

3.2 Interviews

The semi-structured interview study (see Appendix) dealt with the specific issues through a realized case study and the general issues about tall wooden apartment building construction in Finland based on the views of different stakeholders as project partners on the success of the overall process and the roles of different actors in these challenging projects. Also, Lighthouse Joensuu and HOAS Tuuliniity, as well as other important apartments such as PuuMera, Puukuokka, DAS Kelo, were included in the survey. The 21 interviews highlighted the following key findings:

- The builder, structural designer, and architect/chief designer were deemed to have the most significant influence on the outcome of projects, and construction supervision municipal building control was the least important.

- According to interviews at HOAS Tuuliniity, the design process was architecturally oriented although the project was implemented with the same volume modular element supplier's system. This was partly because the architectural office was responsible for the project's BIM coordination. Tuuliniity did not happen either.

- Overall, it was estimated that the most important factor affecting the general result of the projects was the cost of wood construction. The projects tried to compensate for the construction cost, for example by simplifying the floor plan and the shape of the building. In volumetric modular system construction, in particular, costs are largely determined by the reproducibility of the total number of modules, so the cost savings have a significant impact on the overall architecture of the building.

- Architects estimated that wood construction, compared to the more conventional concrete building, has had the most significant impact on the shape and main dimensions of the building mass, as well as on the amount and schedule of design work. In particular, the design accuracy of wood construction and the frontloading of design work were emphasized. The architects interviewed also underlined the understanding of the structural system used and the boundary conditions it set right at the beginning of the design.

- The influence of wood construction on the facade and surface materials has been assessed by the architects more broadly concerning what type of building is considered a wooden apartment building. Non-architects' view, too, was quite often expressed that a wooden facade is not automatically related to wooden construction, but wood is seen mainly as a frame material. The use of wood as a visible surface material was also generally considered more essential indoors than in the facade of a building.

- No major problems were identified in the implementation of wooden apartment buildings with up to 8-story and they were believed to become more common as experience and cost competitiveness improve. However, according to interviews, it seems unlikely that tall wooden apartment buildings will proliferate much further unless the tall wooden building is strongly supported by public authorities.

- The builders of HOAS Tuuliniity and Lighthouse Joensuu stated that the solutions for tall timber construction in the projects were well suited for the construction of student housing. One of the major problems in the development of tall wooden structures was that the projects were one-off and the development of experience and building concepts through repetition has not materialized.

- Overall, the interviewees stated that there is little need for tall buildings in Finland, regardless of whether the buildings are made of wood or any other material. However, the construction of the first Finnish tall wooden apartments was considered an important factor in the credibility of wooden construction.

- While the focus of construction was found to be on buildings up to 8-story, the construction of taller buildings was seen as a tool to improve technical solutions and develop design expertise. The interviews also highlighted opportunities for Finnish wooden apartment buildings and relevant knowledge in foreign markets, so it might make sense to develop new tall wooden apartments even if there is no real demand for them in Finland.

4. Conclusions

The actual competitiveness factors of wood construction, such as the speed of construction or the possibilities of wood architecture, have not been very significant reasons for wood construction, but the projects have been chosen to be wood-based, mainly due to other objectives such as ecology or sustainable image of wood construction.

The implementation process for wooden apartment buildings is currently considerably different between buildings with a maximum of 8-story and taller.

8-story wooden apartment buildings are comparable to the low-rise wooden construction and similar structural solutions can be employed for low-rise construction.

The architectural design of a tall wooden building requires close interaction between the architect and the structural designer, where the structural designer also plays a major role in the architectural design in practice. With industrial prefabrication, the design process for tall wooden blocks of flats apartment buildings is more forward-looking frontloaded than conventional construction, which requires an open flow of information from all related parties and the ability to make solutions decisions early in the project. Additionally, the special technical design should not be done based on ready-made architectural plans, but solutions should be developed from the beginning in cooperation with all design fields.

Especially for projects up to 8-story high, it seems that different networks of developers, contractors, and wood suppliers are evolving, that develop their building concept and are widely responsible for the entire implementation chain of the building project.

Appendix: sample questions used in interviews

1. What would you give to the completed project and the implementation process?

   Finished building (1–5).

   construction/project management (1–5).

   construction (1–5).

   architectural/main design (1–5).

   permitting process and official activities (1–5).

   building components manufacturing process (1–5).

   Do you want to justify your answer or make suggestions for improvement?

2. How much influence do you think the following parties had on the outcome of the project?

   builder (1–5).

   principal designer/architect (1–5).

   structural designer (1–5).

   contractor (1–5).

   wood supplier (1–5).

   city planner (1–5).

   construction supervisor (1–5).

   Do you want to justify your answer or make suggestions for improvement?

3. How much impact do you think the following parties had on the cost of the project?

   builder (1–5).

   principal designer/architect (1–5).

   structural designer (1–5).

   contractor (1–5).

   wood supplier (1–5).

city planner (1–5).

construction supervisor (1–5).

Do you want to justify your answer or make suggestions for improvement?

4. How much influence do you think the following parties had on the project schedule?

builder (1–5).

principal designer/architect (1–5).

structural designer (1–5).

contractor (1–5).

wood supplier (1–5).

city planner (1–5).

construction supervisor (1–5).

Do you want to justify your answer or make suggestions for improvement?

(1) very low (2) quite low (3) neutral (4) quite significant (5) very significant.

Author details

Lassi Tulonen, Markku Karjalainen and Hüseyin Emre Ilgın*
Tampere University, Tampere, Finland

*Address all correspondence to: emre.ilgin@tuni.fi

References

[1] Karjalainen M. The Finnish multi-story timber apartment building as a pioneer in the development of timber construction (in Finnish). PhD Dissertation. Oulu, Finland: University of Oulu. 2002.

[2] Ijäs V. Puukerrostalojen rakentamisen esteet ja mahdollisuudet: Keskeisten suomalaisten rakentamis- ja Kiinteistöalan sidosryhmien vertaileva asennemittaus. Tampereen Teknillinen Yliopisto. 2013.

[3] Kaihlanen J. Puun Käyttö Rakentamisessa Aiotaan Kaksinkertaistaa Seuraavan Neljän Vuoden Aikana – Eniten Kasvumahdollisuuksia Nähdään Puukerrostaloissa. Maaseudun Tulevaisuus. 2019.

[4] The National Building Code of Finland - Structural Fire Safety, Decree of the Ministry of the Environment 2017;Section 8:5.

[5] Number of buildings, dwellings and persons by type of building and number of storeys 31 Dec. 2018 in Finland. Available from http://www.stat.fi/til/rakke/2018/rakke_2018_2019-05-21_tau_001_en.html. [Accessed: 2021-03-30]

[6] Responsible promoter of the Finnish wood. Puuinfo. Available from https://puuinfo.fi/sites/default/files/Puutuoteteollisuuden%20ja%20puurakentamisen%20kilpailukyvyn%20varmistaminen%20koulutuksen%20kehitt%C3%A4misen%20avulla.pdf [Accessed: 2021-03-30]

[7] Häkkinen A. "Jyväskylän Puukuokka-kortteli ei ole enää mikään yksittäinen pilotti." Rakennuslehti, 29.1.2018. Available from https://www.rakennuslehti.fi/2018/01/jyvaskylan-puukuokka-kortteli-nayttaa-tieta-puukerrostalon-asuntosuunnittelulle/. [Accessed: 2021-03-30]

[8] Hurmekoski E, Korhonen J. Pitkän aikavälin katsaus puurakentamisen markkinoihin. Metsätieteen aikakauskirja 2017-7757. Tieteen tori. 2017.

[9] Tables. Tilastokeskus. Suomen virallinen tilasto (SVT): Rakennus- ja asuntotuotanto [verkkojulkaisu] Available from http://www.stat.fi/til/ras/tau_en.html. [Accessed: 2021-03-30].

[10] Aatsalo J. "Tornitalot Leviävät Helsingistä maakuntiin, Kuopio rakentaa niitä peräti neljä – Taustalla uudet rakentajia helpottavat rakennusmääräykset." Available from Helsingin Sanomat, 11.2.2018. https://www.hs.fi/koti/art-2000005562311.html. [Accessed: 2021-03-30].

[11] Repo J, Luttinen E. Lighthouse Joensuun Rakennuttamisprosessi. Karelia-ammattikorkeakoulu, Thesis, Rakennustekniikan koulutus, Finland, 2019.

[12] Ikonen V, Moilanen M. Use of Wood in High-Rise Residential Buildings. Karelia-ammattikorkeakoulu, Thesis, Rakennustekniikan koulutus, Finland, 2019.

[13] Mertanen V. Puu korkeassa kaupunkirakentamisessa artikkelikokoelma. Thesis, Karelia-ammattikorkeakoulu. 2019.

[14] Määttä K, Hietala J, Jutila K. Puurakentaminen: Sääntelyn kapeikot ja kehittämisvaihtoehdot. Pellervon taloustutkimus PTT, 2016.

[15] Hurmekoski E. Long-Term outlook for wood construction in Europe: Itä-Suomen yliopisto. 2016.

[16] Karjalainen M. Puukerrostalojen asukas- ja rakennuttajakysely 2017 Loppuraportti. Ympäristöministeriö. 2017.

[17] Franzini F. Wooden Multistory Construction in Finland: Perceptions of municipality civil servants. MSc Thesis. University of Helsinki. Finland. 2018.

[18] Palokangas T. Julkisen puurakentamisen edistäminen kunnallisessa päätöksenteossa. MSc Thesis. University of Helsinki. Finland. 2019.

[19] Puurakentamisen asema ja mahdollisuudet Suomessa. Puuinfo. Available from https://www.puuinfo.fi/puutieto/puurakentaminen/puurakentamisen-asema-ja-mahdollisuudet-suomessa. [Accessed: 2021-03-30].

[20] Runkojärjestelmän vaikutukset puukerrostalon arkkitehtisuunnitteluun. Puuinfo. Available from https://www.puuinfo.fi/suunnitteluohjeet/runkoj%C3%A4rjestelm%C3%A4n-vaikutukset-puukerrostalon-arkkitehtisuunnitteluun. [Accessed: 2021-03-30].

[21] Yleisimmät rakennejärjestelmät. Puuinfo. Available from https://www.puuinfo.fi/puutieto/puusta-rakentaminen/yleisimm%C3%A4t-rakennej%C3%A4rjestelm%C3%A4t. [Accessed: 2021-03-30].

[22] Malo KA, Abrahamsen RB, Bjertnæs MA. Some structural design issues of the 14-Storey timber framed building Treet in Norway. European Journal of Wood and Wood Products. 2016;74:3:407-424.

[23] CTBUH. Council on Tall Buildings and Urban Habitat. Illinois Institute of Technology, S.R. Crown Hall, 3360 South State Street, Chicago, Illinois, USA. Available from http://www.ctbuh.org [Accessed: 2021-03-30].

[24] Abrahamsen, Rune. Mjøstårnet - Construction of an 81 m Tall timber building: Internationales Holzbau-Forum IHF. 2017.

[25] HoHo. Available from http://www.hoho-wien.at/ [Accessed: 2021-03-30]

[26] Technical case studies Dalston Lane:Tall Timber. Available from https://www.building.co.uk/technical-case-studies/dalston-lane-tall-timber/5079749. [Accessed: 2021-03-30].

[27] Sorri S. Design instructions for CLT multi-storey modular element buildings, MSc Thesis, Tampere University of Technology, Finland. 2017.

[28] Koste O. Aleksi Neuvonen, and Kaisa Schmidt-Thomé. Kaupungistumisen käännekohdat – Skenaarioita Suomen kaupungistumisen tulevaisuudesta 2039. Demos Helsinki. 2018.

[29] Mäenpää M. Rationalities of tall building in Finnish cities, MSc Thesis, University of Turku, Finland, 2019.

Chapter 8

Carbon Impacts of Engineered Wood Products in Construction

Hongmei Gu, Prakash Nepal, Matthew Arvanitis and Delton Alderman

Abstract

Buildings and the construction sector together account for about 39% of the global energy-related CO_2 emissions. Recent building designs are introducing promising new mass timber products that have the capacity to partially replace concrete and steel in traditional buildings. The inherently lower environmental impacts of engineered wood products for construction are seen as one of the key strategies to mitigate climate change through their increased use in the construction sector. This chapter synthesizes the estimated carbon benefits of using engineered wood products and mass timber in the construction sector based on insights obtained from recent Life Cycle Assessment studies in the topic area of reduced carbon emissions and carbon sequestration/storage.

Keywords: life cycle assessment, mass timber products, forest carbon, wood products carbon, carbon sequestration/storage, avoided emissions

1. Introduction

Wood utilization in construction is a practice as old as human civilization. Value-added wood products harvested from forests have been used as building materials for millennia. With global forest resources strained, research in efficient use of harvested wood materials with engineered conceptions has increased significantly in recent years. Traditional wood building products, such as plywood, oriented strandboard (OSB) and I-Joists, are now complemented by emerging mass timber products, such as Cross Laminated Timber (CLT) [1], Glue Laminated Timber (GLULAM), Nail Laminated Timber (NLT), Dowel Laminated Timber (DLT), and Mass Ply Panels (MPP). These products are all engineered for widespread and efficient use in construction.

Mass timber is a term used to describe innovative wood product systems that utilize large, solid wood panels for wall, floor, and roof construction. These panels are six feet or more in width and length, and are manufactured with resin, nails, or by the use of dowels. Each layer of boards is oriented perpendicular to the adjacent layer and dowelled, glued, or nailed on the wide face of each board, in a symmetric manner in order that the outer layers have the same orientation. Panels can be used in CLT, DLT, GLULAM, NLT, and MPP systems. Mass timber products can be used to build traditional houses, office buildings, and high-rise structures.

Mass timber use is increasing around the world not only because of the desirable properties of engineered designs, but also due to their low-carbon footprint and carbon storage benefits. In addition, mass timber products are renewable materials; carbon loss in forests due to increased harvests suggest the need for the manufacture of mass timber products, which can offset carbon loss over time as forest biomass regrowth occurs on forestland. Until recently the use of wood in high-rise buildings had been limited due to building code restrictions. A recent change in the International Building Code now allows for mass timber use in buildings up to 18 stories. Sustainability of the mass timber buildings has been studied since the first USDA-FS Tall Wood Building competition winner "Framework" (**Figure 1**) building was designed, and progresses with the current "Ascent" building, to-be-the-tallest mass timber building (**Figure 2**) in the North America. Research and code development have paved the way for architects and building developers who are increasingly turning to mass timber building designs for both new construction and renovations. One of their aspirations is to help achieve the UN Climate Change Paris Agreement [2] goal of limiting global warming to no more than 1.5°C temperature rise from pre-industrial times. Given that the building sector accounts for 39% [3] of total global warming potential (GWP), researchers have conducted systematic assessment on whole-building environmental impacts for potential GWP reductions, using an internationally accepted method called Life Cycle Assessment (LCA) [4].

LCA is a holistic and scientific method for assessing environmental impacts from all life cycle stages of a product, process, service, or even a whole-building system. International standards ISO 14040 [5] and ISO 14044 [6] provide guidelines, principles, and framework to conduct an LCA. Standards such as ISO 21930 (2017) [7] and EN 15978 (2011) [8] provide guidelines for assessing whole-building system environmental performance over the entire building life cycle.

Environmental impacts of engineered wood products and mass timber use in construction can be evaluated within two LCA frameworks. An attributional LCA evaluates environmental impacts of manufacturing, installation, use, and disposal of a product (9–10). In contrast, a consequential LCA evaluates change in environmental impacts due to a change in product output or a change in a service

Figure 1.
12-story framework building with mass timber from ground floor (not built) designed for Portland Oregon US (Credit: Lever Architecture).

Figure 2.
25-story Ascent building built with mass timber products in Milwaukee Wisconsin US (Credit: Thornton Tomasetti and Korb & associates).

or a system [9–11]. In the case of wood products LCA, an attributional LCA, for example, provides information about the amount of GHG released during manufacturing of a unit (m^3) of lumber or CLT. In contrast, a consequential LCA evaluates overall greenhouse gas (GHG) emission, of increased demand for lumber or CLT use in buildings, including both direct (within system boundary) and indirect (e.g., change in forest land use) effects resulting from such increased demand.

Engineered wood products (EWP) used in buildings have a long history. Glulam timber was used to construct an auditorium in Basel Switzerland [12], and plywood became a popular engineered wood building material in the early 1900's, followed by OSB and Laminated Veneer Lumber (LVL) invented in the mid 1960's. In the mid-1970's, CLT was first used as a building material in a roof system. Now, with architects and developers aiming to achieve more sustainable designs, mass timber products such as NLT, DLT, and MPPs are becoming more and more popular.

With the Tall Wood building movement in North America ramping up [13–15], the environmental benefits of using mass timber and other engineered wood products have been scientifically examined in multiple studies around the world, and consider global forest resource availability and depletion, and the impact on world forest regeneration and protection. This chapter highlights carbon reduction and the storage benefits of engineered and mass timber products utilizing insights obtained from recent LCA studies.

2. Forest carbon, harvested wood products carbon, and avoided carbon emissions of substituting wood for non-wood materials in the construction sector

Increased demand for wood used in the construction sector may lead to changes in timber growth, growing stock inventory, and harvest that affect the storage of forest carbon. It may also alter the quantity of wood products manufactured and used to displace non-wood materials affecting carbon stored in wood products and avoided manufacturing emissions. Potential increased timber prices resulting from increased demand may encourage increased investment and forest management activities contributing to higher forest growth, growing stock, and reforestation (and forest carbon). Price increases might also lead to unsustainable harvesting in a region that supplies wood to wood-consuming industries. All of these potential effects should be considered when evaluating the overall carbon benefit of substituting wood for non-wood materials in construction. A consequential LCA provides a framework to consider all potential changes over a given time frame; the projected environmental effects for increased wood demand scenarios are compared with the effects projected for the business-as-usual reference case. The difference between the two scenarios constitutes the net emissions effects of increased wood use in the construction sector.

Using a consequential LCA, Nepal et al. [16] evaluated a scenario of increased softwood lumber and structural panel use in nonresidential construction in the United States, compared to a reference scenario without such an increase in wood consumption. They found that for each ton of CO_2e (tCO_2e) in additional wood consumption that replaced non-wood material in nonresidential construction in the United States, there was a net savings of 2.33 tCO_2e emissions over a period of 50 years. This estimate considered changes in carbon storage in forests (including logging residues) due to biological regrowth and market induced investment in forest management, the carbon stored in harvested wood products, and manufacturing emissions. In another study, Hildebrandt et al. [17] evaluated various scenarios of increases in annual demand for CLT-based solid wood structures and the Glulam-based frame structures in Europe by 2030. Their results indicated that these structures would result in lower annual GHG emissions and higher carbon stored in wood products (used in buildings), with an estimated combined annual carbon benefit of about 29.6 to 60.5 million tCO_2e per year during the 2015–2030 time-frame. The results were dependent upon the assumed future increases in the growth rates of different wood products used for construction. The derived combined carbon benefit per unit of CO_2 in additional wood consumed, given their projected changes in carbon stored in the wood products used in housing stock during the same period (11.8 to 15.4 million tCO_2e per year), was 2.50 to 3.93 tCO_2e/tCO_2e. The changes in forest biomass carbon that would result from increased harvests was not considered in these estimates, which may increase or decrease the derived avoided emissions depending on how forest management would change in response to increased timber harvests in Europe. Similarly, Smyth et al. [18] estimated the avoided emissions benefits of wood substitution in a built environment in Canada that included single-family and multi-family housing, six-story multiuse buildings, residential flooring, furniture, and decking reported an average avoided emissions benefit of 0.54 tCO_2e/tCO_2e for sawnwood and 0.45 tCO_2e/tCO_2e for plywood. Their estimate is based on a potential increase in wood use in those structures and excludes changes in forest ecosystem carbon and carbon stored in harvested wood products.

Findings from these studies indicate that there are considerable variabilities and uncertainty in the estimated GHG reduction benefits of wood products substitution in buildings. This is mainly due to the differences in the system boundary (e.g., consideration of direct or market induced effects on forest carbon), non-wood materials substituted (steel vs. concrete), energy mix assumed to be used in manufacturing wood products (e.g., wood energy and fossil energy), and types of buildings analyzed (low-rise vs. high-rise, single-family vs. multi-family, residential vs. nonresidential, etc.), and whether carbon storage in wood products is considered. This suggests that the carbon benefits of wood substitution in the construction sector should be evaluated holistically, considering all potential direct and indirect effects (e.g., land use change) caused by increased wood consumption in the building sector.

3. Carbon emission reductions with engineered wood products and mass timber use in the building sector

A whole building LCA (WBLCA) is typically performed to evaluate the total environmental performance of a building from materials utilized and operation during their entire life cycle. Consistent with International Standards [7–8], the system boundary of a WBLCA (**Figure 3**) starts with the product manufacturing stage (stage A1-A3), followed by construction (A4-A5), use (B1-B7) and the end of life (C1-C4) process. Environmental benefits or burdens after the end of life of a building are typically considered beyond the system boundary and depend on whether building materials, after demolition, are reused, recycled to produce new products, burned with or without energy capture, or dumped in landfills.

Buildings constructed with a large quantity of EWPs are usually compared to traditional concrete and steel buildings in order to examine the materials impact on the whole-building life cycle environmental performance. **Table 1** summarizes the embodied carbon of select building materials as a comparison. Concrete and steel, the main building materials used in traditional buildings, are carbon intensive materials. In contrast, EWPs are low-carbon impact materials. Therefore, CLT and other mass timber products use are emerging into the building sector

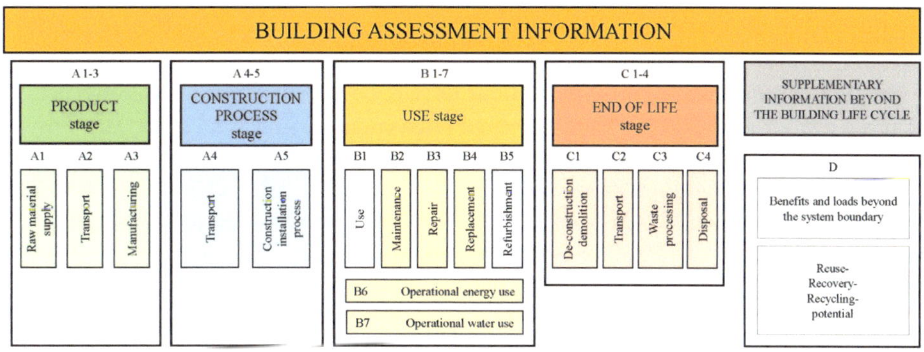

Figure 3.
Building life cycle stages and modules for building life cycle assessment.

Building Material	GWP[1]
1 m^3	kg CO_2 eq
Concrete[2]	225~550
Galvanized steel sheet	2929
Cross Laminated Timber (North America)	110~158
Glue Laminated Timber (North America)	81~515
Laminated Veneer Lumber	423
Plywood	368
Oriented Strand Board	361
Wood I-Joist[3]	17

[1]*Global warming potential from US LCI database or North American EPD.*
[2]*Concrete mix between 2500 psi~5000 psi, and values are from National Ready Mixed Concrete Association report.*
[3]*From American Wood Council EPD for 10 meter long I-Joist.*

Table 1.
Common building materials carbon emissions from product stage (A1-A3 in **Figure 1***).*

to replace some portion of concrete and steel use in mid-to-high rise residential buildings [14–15]. I-joists, used as EWPs in construction, are typically used in floor and roof framing, due to their extremely high-strength relative to their mass (or volume), thus they are assessed in LCA with length as their functional unit rather than volume.

Comparative building LCA is often applied to assess the total carbon emission difference between traditional concrete buildings and buildings with mass timber or other engineered wood products under the assumption of functional equivalency. A significant reduction in carbon emissions or global warming potential from mass timber buildings, has been discovered as a result of several whole-building LCA comparative studies [19–23]. Embodied carbon (EC) refers to the building's total upfront carbon emissions from the manufacture of all materials, transportation, and installation of construction materials. EC does not include the carbon stored in materials, or of the impact from building operational energy. EC, from building life cycle Stage A (module A1 to A5), was reported to drop by between 18% and 50% for mid-to-high rise mass timber buildings as compared to traditionally-built concrete and steel counterparts, depending on the amount of EWPs used in the buildings. If 384 $kgCO_{2e}/m^2$ of floor area is assumed as a median value for EC (for traditional concrete and steel buildings), as suggested in Simone et al. [23], then with an 18–50% reduction from mass timber used to replace concrete and steel, the mass timber buildings EC would be between 157 to 315 $kgCO_{2e}/m^2$ of floor area. Globally 230 billion m^2 [24] of total floor area is projected to be built by 2060 in order to meet projected world urban population growth demand, a reduction of about 16 to 44 billion tonnes CO_2e emissions may be achieved with the construction of mass timber EWP buildings rather than with concrete and steel materials. For perspective, CO_2 emissions reduction of this magnitude would be equivalent to the removal of 3.5 to 9.6 billion passenger vehicles from the road in one year (*4.6 metric tons CO_{2eq}/vehicle/year quoted from EPA's GHG equivalencies Calculator* [25]), or reducing electricity use by 22 to 62 trillion kWh (*7.09×10^{-4} metric tons CO_2/kWh* [25]) which equivalent to eliminating the carbon footprint of the world's electricity consumption over a year.

4. Forest carbon sequestration and harvested wood products carbon storage benefits from mass timber buildings

Trees sequester carbon from the atmosphere during their growth. Harvested trees manufactured into wood products continue to store carbon during the products' lifespans. EWPs used in construction store carbon for extended periods as buildings usually are in service for more than 50-years. With mass timber buildings emerging into the global building sector, even longer lifetimes are expected [26] for buildings that utilize large quantities of mass timber products. Thus, with such buildings, greater service duration and increased carbon storage benefits will potentially be realized.

Carbon stored in buildings with EWPs can be calculated simply using a static method as described in Eq. (1) below:

$$\begin{aligned} CO_2 \text{ stored (kg)} &= \text{Volume of EWP}\left(m^3\right) \text{x Density}\left(kg/m^3\right) \text{x}\left(1-MC\right) \\ &\quad \text{x}\left[\text{carbon in wood}\right] \text{x}\left[\text{Molar mass of } CO_2 / \text{Molar mass of C}\right] \quad (1) \end{aligned}$$

- MC – moisture content (%) of EWP
- The carbon in wood is about 50%
- the molar mass of CO_2 is 44; and the molar mass of C is 12.

From Eq. (1), it is clear that carbon stored in EWP depends on the wood species. Common species used to produce mass timber products include fir (*Abies* spp.), Spruce (*Picea* spp.), Douglas fir (*Pseudotsuga menziesii*), larch (*Larix* spp. *Nutt.*), pines (*Pinus* spp.), Western hemlock (*Tsuga heterophylla*), and yellow poplar (*Liriodendron tulipifera Linnaeus*). Their densities (at 12% moisture content) range from 470 kg/m^3 to 627 kg/m^3. Using Eq. (1), it was estimated that carbon storage per m^3 of wood used in mass timber buildings can range from 758 kg CO_2 to 1,012 kg CO_2. Data collected from a few mass timber constructions built in North America, Europe, and Australia indicated that the mass timber volume used in construction ranged from a minimum of 0.08 m^3/m^2 of floor area to a maximum of 0.62 m^3/m^2, with an average of 0.29 m^3/m^2 of floor area. This indicates that an average of 220~293 kg CO_2 can be stored for every m^2 of floor area in a mass timber building.

The Softwood Lumber Board [27] projects that 3.82 billion board feet (bf) (equivalent to 9 million m^3) of softwood lumber will be consumed in mass timber products per year in the U.S. residential and non-residential construction sectors by 2035. Under this assumption, an estimated 6.8 to 9.1 metric tons of CO_2 per year will be stored in those buildings during the buildings' lifespans. In this way, timber buildings fabricated in urban areas may be considered as transferring carbon storage from the forest to the city.

A mass timber building can serve as a carbon reservoir over the building lifespan, which can potentially assist in the mitigation of climate change. Mass timber buildings not only store carbon in the structure, but they can also help sequester more carbon in forests via two mechanisms. First, in a sustainably managed forest, harvesting mature stands and replanting lead to an increased carbon sequestration rate compared to merely maintaining mature stands. Second, increased demand for wood products use in buildings may lead to timber price increases, which may provide economic incentives to invest in intensified forest management activities and/or plantation, which can further lead to increased forest growth and carbon sequestration [16]. When building with wood products, carbon sequestered by trees is transferred into urban buildings and stored for extended periods of time before being released back into the atmosphere upon the building's decommissioning or renovation, through landfill emissions or burning for energy, as depicted in **Figure 4**. Such an extended time delay of carbon emissions has been recognized as an effective way to mitigate climate change.

Figure 4.
Carbon life cycle of the EWPs in construction.

5. Potential environmental impacts from end-of-life design for engineered wood products in buildings

When buildings reach the end of their service life, the materials, after demolition, will either go to landfill or are recycled for reuse and/or substituted for fossil energy. Traditional wood materials used in buildings have a recycle rate of only 27% with the remaining 73% of wood waste from demolition estimated to go to landfills according to the EPA's 2018 C&D management report [28]. Among 27% that are recycled, 22% are reused as compost and mulch, 11% are used in manufacturing wood products, and 67% are burned to generate energy to replace fossil fuel. As mass timber product consumption increases in the building sector in the future, the end-of-life treatment for structural timber may result in a greater recycle rate accompanied by much lower to near zero landfill rate. Several studies have evaluated end-of-life strategies for structural timber or EWPs to understand environmental benefits and economic impacts, using WBLCA and Life cycle cost (LCC) analysis (**Figure 5**).

In WBLCA, the end-of-life (EoL) phase refers to the impacts occurred in Stage C (**Figure 1**). At this stage, activities like deconstruction, waste management (recycling or landfilling), and waste transportation to the landfill are included to examine the total environmental and economic impacts of building materials. The assessments [21, 29–30] revealed a greater recycle/reuse rate for mass timber building products that would reduce not only the global warming potential but also the total life cycle costs of mass timber buildings. Studies [31–33] also have demonstrated the potential benefits at the EoL stage of recycling mass timber EWPs for the same or similar uses in subsequent building construction to extend the carbon storage time. Research [34] was conducted at the USDA Forest Service's Forest Products Laboratory (FPL) on timber recovered from the second Glulam structure built in the US at the FPL. After 75-years' service, minimal degradation of the Glulam beams structural performance was found and the reuse of the product in the same function was attested to be feasible.

The United States and many European countries are implementing stricter regulations for the disposal of wood construction materials in landfills. The reasoning is that while solid wood decomposes very slowly in landfills, it releases biogenic methane, a greenhouse gas with 25 times more global warming potential than carbon dioxide [35]. Thus, EoL strategies that lead to increasing the recycling of demolished wood materials from mass timber construction are expected to contribute positively to the global forest based circular carbon economy.

Carbon accounting at the EoL stage for mass timber EWPs should include CO_2 emissions from deconstruction and transportation of materials to landfills or remanufacturing facilities, and emissions from remanufacturing processes and landfill emissions. Carbon accounting also should include assessing the carbon benefits derived from storage in the products reused or repurposed, storage in landfills, and the substitution of fossil fuels for wood waste in energy recovery.

Previous mass timber building studies using WBLCA and LCC analyses [16, 23], including sensitivity analysis by different recycle rates, have demonstrated that a significant reduction of carbon emissions and total LCC can be achieved with improving EoL building management. With the greater reuse of mass timber construction products, demolition waste is minimized, and salvage values are amplified.

However, such reclamation practices for mass timber buildings also require front-end building materials designed for easy and safe disassembly. Therefore, EWPs and buildings designed with a goal of reuse or recycling are being recognized to have higher environmental and economic benefits. While the economic implications are still unclear, the WBLCA and LCC analyses strongly suggest

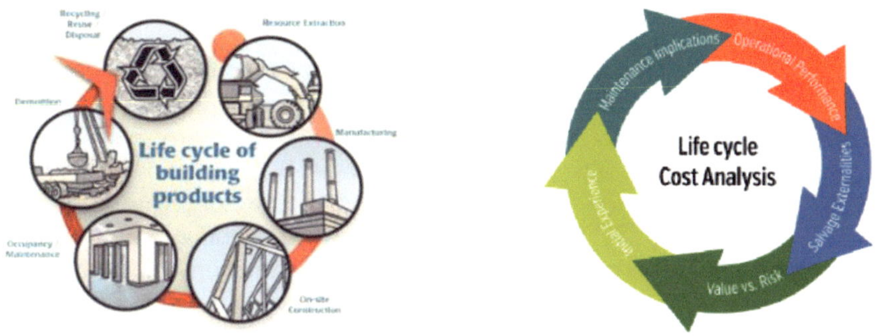

Figure 5.
Building life cycle stages for LCA and life cycle cost analysis.

landfills should be avoided, and recycling and reuse should be first principles in waste management for mass timber and EWPs.

6. Conclusions

With their inherent low carbon impact and carbon storage benefits, engineered wood products will play an important role in transforming the built environment from being a major contributor of greenhouse gas emissions to a central solution to the climate crisis. Whole building LCA studies revealed buildings constructed with mass timber, or with a large quantity of engineered wood products, yield significant embodied carbon reductions and are accompanied by large amounts of long term carbon storage, which aids in the mitigation of climate change. Potential avenues for mass timber products end-of-life treatment in construction were examined with LCA and LCC tools to assist policy makers in determining strategies for circularities in material and economy. The increased use of engineered wood and mass timber products in building construction is projected to offer considerable GHG mitigation potential, though the estimated GHG reduction benefits differed widely among studies due to different system boundaries, mass timber products used, and types of building analyzed. Such variability in the results suggests that the carbon benefits of mass timber buildings should not be generalized but should be evaluated on case-by-case basis.

Author details

Hongmei Gu*, Prakash Nepal, Matthew Arvanitis and Delton Alderman
USDA Forest Service Forest Products Laboratory, Madison, WI, USA

*Address all correspondence to: hongmei.gu@usda.gov

References

[1] Erol K, Brad D, editors. CLT Handbook. US edition. Library and Archives Canada Cataloguing in Publication; 2013. 572 p. ISBN 978-0-86488-553-1.

[2] United Nations Climate Change. The Paris Agreement. 2016. [Internet]. Available from: https://unfccc.int/process-and-meetings/the-paris-agreement/the-paris-agreement [Accessed: 2021-05-01]

[3] Architecture2030.Why the building sector? [Internet]. Available from: https://architecture2030.org/buildings_problem_why/ [Accessed: 2021-05-01]

[4] Rita S, Philip W, Editors. Environmental Life Cycle Assessment – measuring the environmental performance of products. Published by American Center for Life Cycle Assessment; 2014. 315 p. ISBN 978-0-9882145-5-2.

[5] ISO (2006a) 14040: Environmental Management – Life Cycle Assessment – Principles and Framework. International Organization for Standardization, Geneva, Switzerland.

[6] ISO (2006b) 14044: Environmental management – Life cycle assessment – Requirement and guidelines. International Organization for Standardization, Geneva, Switzerland.

[7] ISO 21930:2017. Sustainability in buildings and civil engineering works. Core rules for environmental product declarations of construction products and services. International Organization for Standardization. Geneva, Switzerland.

[8] EN 15978: 2011 Sustainability of construction works – Assessment of environmental performance of buildings – Calculation method. European Committee for Standardization. 60 pp.

[9] Lippke B., E Oneil, R Harroson, K Skog, L. Gustavson, R. Sathre: Life cycle impacts of forest management and wood utilization on carbon mitigation: Knowns and unknowns. Carbon Management. 2011;2:303-333. DOI: 10.4155/cmt.11.24

[10] Skog, KE, DC McKinley, RA Birdsey, SJ Hines, CW. Woodall, ED Reinhardt, JM Vose: Managing Carbon. In: Climate change and United States forests (advances in global change research, 57). 14th ed. Springer; 2014. p. 151-182. DOI: 0.1007/978-94-007-7515-2_7

[11] Brander, M, R tipper, C Hutchison, G Davis. Consequential and attributional approaches to LCA: A guide to policy makers with specific reference to greenhouse gas LCA of biofuels. Technical paper TP090403A. London, UK: Econometrica Press. 14 p.

[12] In: Smulski, Stephen, ed., Engineered wood products-A guide for specifiers, designers and users. ISBN-096556736-0-X. Madison, WI: PFS Research Foundation: pp. 1-1-1-39. Chapter 1. 1997.

[13] Tom W, Robert R, 2016. (Eds.) Proceedings: Mass Timber Research Workshop 2015. General Technical Report FPLGTR-241. Madison, WI: U.S. Department of Agriculture, Forest Service, Forest Products Laboratory. 364 p

[14] Samuel Z, Tom W, Karen M, Michael R, 2019. Research needs assessment for the mass timber industry: Proceedings of the 2nd North American mass timber research needs workshop. General Technical Report FPL-GTR-273. Madison, WI: U.S. Department of Agriculture, Forest

Service, Forest Products Laboratory. 41 p.

[15] WoodWorks. Wood Products Council. [Internet]. Available from: https://www.woodworks.org/why-wood/ [Accessed: 2021-05-01]

[16] Nepal P, KE Skog, DB McKeever, RD Bergman, KL Abt, RC Abt: Carbon mitigation impacts of increased softwood lumber and structural panel use for nonresidential construction in the United States. Forest Products Journal. 2016;66:77-87. DOI: doi:10.13073/FPJ-D-15-00019

[17] Hildebrandt J, N Hagemann, D. Thrän: The contribution of wood-based construction materials for leveraging a low carbon building sector in Europe. Sustainable Cities and Society. 2017;34:405-418. DOI: 10.1016/j.scs.2017.06.013

[18] Smyth C, G. Rampley, TC Lempriere, O Schwab, WA Kurz: Estimating product and energy substitution benefits in national-scale mitigation analyses for Canada. GCB Bioenergy. 2017;9:1071-1084. DOI: 10.1111/gcbb.12389

[19] De Wolf C, Yang F, Cox D, Charlson A, Hattan S, Ochsendorf J, 2016. Material quantities and embodied carbon dioxide in structures. Proceedings of the Institution of Civil Engineers - Engineering Sustainability, 169, 150-161.

[20] Hart J, D'Amico B, Pomponi F, 2021. Whole-life embodied carbon in multistory buildings: Steel, concrete and timber structures. Journal of Industrial Ecology 25(2):403-418. DOI: 10.1111/jiec.13139

[21] Liang S, Gu H, Bergman R, Kelley S, 2020. Comparative life-cycle assessment of a mass timber building and concrete alternative. Wood and Fiber Science. 52(2): 217-229.

[22] Chen Z, Gu H, Bergman R, Liang S, 2020. Comparative life-cycle assessment of a high-rise mass timber building with an equivalent reinforced concrete alternative using the Athena impact estimator for buildings. Sustainability 2020, 12, 4708. DOI:10.3390/su12114708.

[23] Simonen K, Rodriguez X, De Wolf C, 2017. Benchmarking the embodied carbon of buildings. Technology|Architecture + Design, 1, 208-218.

[24] United Nations Environment and International Energy Agency (2017): Towards a zero-emission efficient, and resilient buildings and construction sector. Global Status Report 2017. Pp. 48.

[25] US Environmental Protection Agency. Greenhouse Gases Equivalencies Calculator – Calculations and References. [Internet]. Available from: https://www.epa.gov/energy/greenhouse-gases-equivalencies-calculator-calculations-and-references [Accessed: 2021-05-20]

[26] Michael Green. How Timber Will Define the Skylines of Our Future. [Internet]. 2020. Available from: https://www.woodsolutions.com.au/webinars-podcasts/woodsolutions-weekly-webinars-2020/how-timber-will-define-skylines-our-future-michael-green#:~:text=How%20Timber%20Will%20Define%20the%20Skylines%20of%20Our%20Future%20-%20Michael%20Green,-Submitted%20by%20100content&text=Timber%20is%20an%20important%20ingredient,beautiful%2C%20healthy%20cities%20for%20tomorrow. [Accessed: 2021-05-11]

[27] Southwood Lumber Board. FPInnovations, Ben Romanchych Consulting. Oct. 2020. Mass Timber Outlook. https://softwoodlumberboard.org/wp-content/uploads/2021/03/SLB-Mass-Timber-Outlook-2021-Final-Condensed.pdf [Accessed: 2021-05-01]

[28] U.S. Environmental Protection Agency. Advancing Sustainable Materials Management: 2018 Fact Sheet. [Internet]. 2020. Available from: https://www.epa.gov/sites/production/files/2021-01/documents/2018_ff_fact_sheet_dec_2020_fnl_508.pdf [Accessed: 2021-05-01]

[29] Liang S, Gu H, Bilek T, Bergman R. 2019. Life Cycle Cost Analysis of a Mass Timber Building: Methodology and Hypothetical Case Study. Res. Paper FPL-RP-702. Madison, WI: U.S. Department of Agriculture, Forest Service, Forest Products Laboratory: 1-11.

[30] Gu H, Liang S, Bergman R, 2020. Comparison of building construction and life-cycle cost for a high-rise mass timber building with its concrete alternative. Forest Products Journal 70(4):482-492. DOI:10.13073/FPJ-D-20-00052

[31] Chen X, Pierobon F, Ganguly I, 2019. Life cycle assessment (LCA) of cross-laminated timber produced in western Washington: The role of logistics and wood species mix. Sustainability, 11: 1278.

[32] Chen X, 2019. Environmental Assessment of the Production and End-of-Life of Cross-Laminated Timber in Western Washington. Dissertation. University of Washington.

[33] Robertson A, Lam F, Cole R. 2012. A comparative cradle-to-gate life cycle assessment of mid-rise office building construction alternatives: Laminated timber or reinforced concrete. Buildings 2012, 2, 245-270; DOI: 10.3390/buildings2030245

[34] Rammer D, Moura J, Ross R, 2014. Structural Performance of the Second Oldest Glued-Laminated Structure in the United State. Proceedings paper at Structures Congress 2014 © ASCE 2014 https://www.fpl.fs.fed.us/documnts/pdf2014/fpl_2014_rammer001.pdf [Accessed: 2021-05-01]

[35] IPCC Climate Change 2007: The Physical Science Basis. Contribution of Working Group I to the Fourth Assessment Report of the Intergovernmental Panel on Climate Change. S. Solomon, D. Qin, M. Manning, Z. Chen, M. Marquis, K.B. Averyt, M. Tignor and H.L. Miller (Eds.)]. Cambridge University Press. Cambridge, United Kingdom 996 pp.

Section 2

Recent Research and Development of EWPs

Chapter 9

Review of Existing Methods for Evaluating Adhesive Bonds in Timber Products

Maryam Shirmohammadi and William Leggate

Abstract

Gluing is an integral part of the majority of production processes in the timber industry. The effectiveness of adhesive application, glue bond development and glue penetration into the wood structure is becoming more and more important as more structural glued timber products are used in construction and other applications. The continued increase in utilisation of mass timber products (MTPs) such as CLT, glulam and LVL in tall timber buildings requires an accurate and in-depth understanding of adhesive roles and their performance effectiveness during the life span of any of those products in relation to the type of loading applied, environmental effects (e.g. RH and temperature) and in-service condition of elements (e.g. exposure to major wet events and degradation from decay). This review aims to provide a comprehensive summary of existing imaging and other visualisation methods used to assess the glue line properties and examine the performance of glue lines in relation to factors such as species, product type and environmental conditions during manufacture and in-service life.

Keywords: glue line thickness, glue penetration, adhesion, timber species, image analysis, SEM, light microscopy, μCT, contrast agent, chemical imaging

1. Introduction to adhesive effectiveness

Gluing is an important and integral part of the majority of timber processing and new wood product designs and developments. Investigations into the effectiveness of the gluing process, formation of the glue line and structural strength of the product relies on an in-depth understanding of glue properties, the interaction between glue and timber species and type of product. This book chapter aims to review existing image analysis and other visualisation methods used to investigate the effectiveness of the gluing process and glue line formation in timber products at the cellular level.

Wood is a non-homogenous medium with a complex cellular structure and a range of chemical compositions that varies with type (hardwood vs. softwood), species, density and surface preparation. High extractives content, sensitivity of extractives to a high temperature environment and the surface characteristics of species affects the adhesion bond formation and strength. Adhesive bonding is currently considered to be mainly a combination of three mechanisms: mechanical interlocking (through wetting of the wood surface, glue penetration and distribution and), physical attraction and covalent chemical bonding [1, 2]. **Figure 1** shows a diagram of glue bond development and importance of the mechanisms leading to development of a strong glue line and good glue penetration into wood cellular structure.

The natural resinous and oily extractive content of some timbers in combination with different drying methods, or added chemicals during processing could affect the adhesive performance due to possible changes in wood surface chemistry. Adhesion is a surface phenomenon and its penetration varies with different wood surfaces, extractive contents and spatial scales of surfaces. The actual measurement and assessment of mechanical interlocking of glue in the wood industry requires more investigation and is still developing [1]. The physical attraction between the wood and adhesive in many adhesive types is considered as the primary bonding process through developing van der Waals forces and hydrogen bonds [1]. The initial surface wetting and absorption into the wood structure are fundamentally important in development of a strong adhesive profile though physical attraction. The covalent bonding between wood fibre and adhesives occur through electron sharing however, in waterproof adhesives where there are sufficient intermolecular physical attractions this electron sharing system may not be necessary [1].

The development of adhesive bonds requires wetting of the surface, flow of the adhesive into the wood structure and penetration into lumens and cell walls of wood (for some adhesive types only). A good wetting process requires formation of a low contact angle with the surface. The flow however refers to spread of adhesive liquid over the wood surface and better coverage of the surface will lead to better bond development. Penetration is the movement of the adhesive into the depth of the wood structure [3]. The overall penetration of adhesive in wood follows Darcy's law where liquid volume flow (Q in $m^3.s^{-1}$) is defined as function of the specific permeability of wood (K in m^2) and area perpendicular to the liquid flow (A in m^2), length in flow direction (L in m), dynamic viscosity of the adhesive liquid (η in Pa. s) and pressure gradient (ΔP in Pa) [4].

$$Q = K \times \frac{A}{L} \times \frac{1}{\eta} \times \Delta P \quad (1)$$

Permeability plays an important role in glue bond development and low permeability of wood types (such as Douglas-fir heartwood) can cause very low penetration of resin into radial and tangential surfaces while high permeability of the wood surface can lead to bond line starvation [5, 6].

Adhesive penetration into wood can be categorised into: gross penetration and cell wall penetration [6] (see **Figure 2**). In gross penetration the adhesive flows into the wood porous structure filling the lumens, this type of penetration occurs in most of the resin types at low viscosity. The cell wall penetration which is the flow of the adhesive into the cell walls of woody structure is only achieved with resins with small molecular weight (MW) components. A previous study identified that a MW of 3000

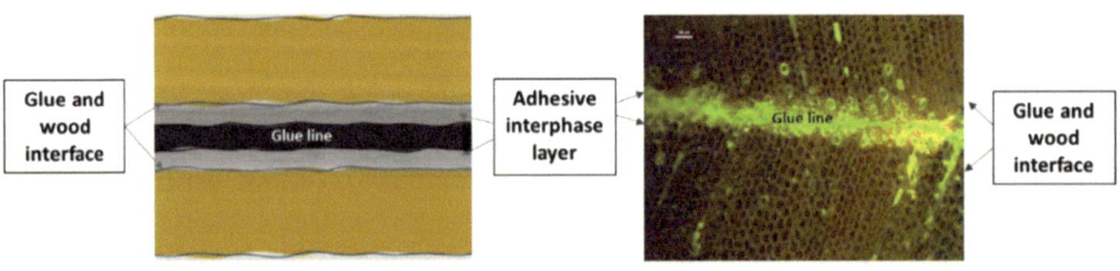

Figure 1.
Adhesive bonding mechanism including the glue and interaction between glue and timber.

DOI: http://dx.doi.org/10.5772/intechopen.99237

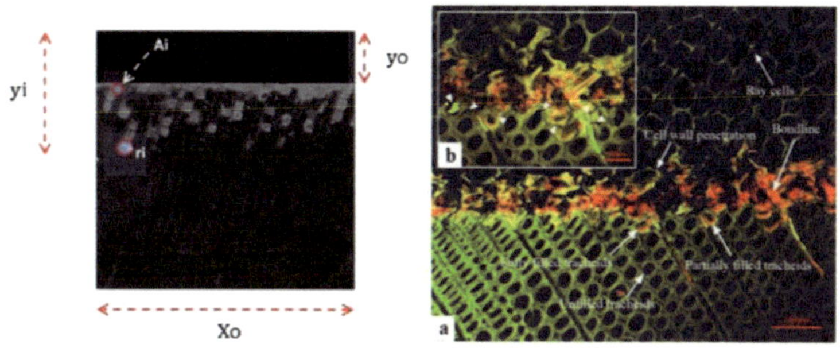

Figure 2.
(Left) Imaging of effective penetration (EP) and maximum penetration (MP) of gross penetration of into structure [8] and (right) penetration of urea formaldehyde (UF) into wood cell walls shown in green colour [9].

for polyethylene glycol (PEG) is a critical factor at room temperature to achieve cell wall penetration in Sitka spruce. Higher temperatures could cause larger MW due to the changes in cell wall polymer and intermolecular volumes [6, 7]. Both effective and maximum penetrations are critical in achieving a good adhesive bonding [4, 6].

Effective penetration (EP) is defined as the total area of adhesive in the interphase region of the glue line divided by the width of glue line. Maximum penetration (MP) is measured as the average value of five distance measurements taken of the five most remote adhesive objects observed in the image field [8].

$$EP = \frac{\sum_i^n A_i}{X_0} \; and \; MP = \frac{\sum_i^5 (y_i + r_i - y_0)}{5} \tag{2}$$

As shown in **Figure 2(a)**, A_i, X_0, y_i, r_i and y_0 are the area of adhesive studied in μm^2, width of image or maximum rectangle determining the area in μm, centroid of adhesive object i, mean radius of object i and reference y coordinate in μm [8].

Resin viscosity, molecular weight (MW), molecular weight distribution, resin solid content and surface tension are properties of an adhesive that affect the bond development and adhesive penetration [5, 6]. Literature suggests the lower MW leads to deeper penetration of adhesive into wood structure in comparison with higher MW resins [10]. Operational factors such as open processing time, pressing time, temperature and consolidation pressure also affect the glue bond development and penetration [4–6]. Adhesive interaction with the wood surface is a function of length scale and defined as interlocking/entanglement and charge. The interlocking or entanglement interactions occur over a longer length scale in comparison with charge interaction which happens in the molecular or nano-length scale [4, 11].

Adhesive penetration is defined by the maximum depth of adhesive in wood, the thickness of the glue bond and the surface condition that affects the penetration measurement and direction. In softwood, tracheids filled with glue are visible in an interconnected zone while the process in hardwoods is more complex. In hardwood, the vessels are filled with glue but they are usually scattered and not close to the actual glue line. In softwood samples the maximum glue penetration is usually measured by trigonometric relations between the uni-directional tracheids near or around the borderline (see **Figure 3**). Hass et al. [12] applied a new approach in defining adhesive penetration in hardwood samples (beech) by considering pore space, ray distribution, viscosity of adhesive and bond line morphology. The study showed that the position of measurement has direct and significant effects on bond line characteristics [12]. The anatomical structure of samples, its wood type, adhesive type and their properties (viscosity for example) could all affect the penetration measurement and its replicability for various studies and products.

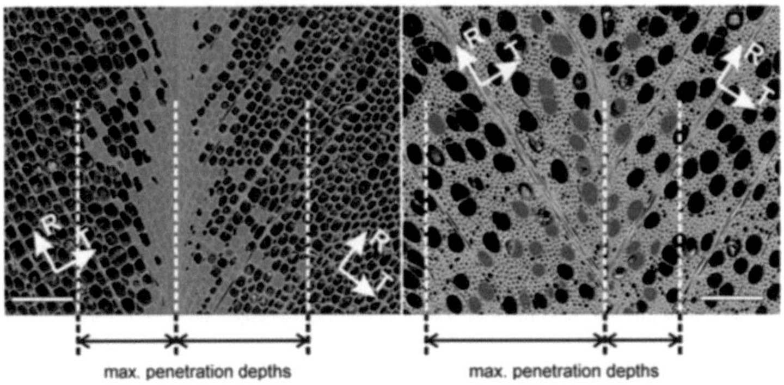

Figure 3.
SEM imaging of maximum adhesive penetration in (left) softwood (spruce) and (right) hardwood (beech). The length of the scale bar in images is 200 um [12].

2. Adhesive systems and types

The adhesive type used in each application need to be suitable for the end use of the product as well as being compatible with the wood type and characteristics and bonding conditions [3]. Wood adhesives can be categorised by their structural, semi-structural and non-structural applications as well as the strength and durability of the bond they develop.

Polymer formation can be used to categorise adhesive types. Linear polymers develop links that are like strings of beads; polyethylene and polypropylene are two linear polymers. The other type of polymer develop branches of linear chains and the properties of formed polymers changes intensely as the branches change. As the density and length of polymers developed changes the melting point, flexibility and strength of adhesive bond changes as well [3, 13].

2.1 Self-adhesion

Self-adhesion in wood products can be achieved in certain conditions however for better strength usually adhesives are needed. Hydrogen bonding, auto-cross-linking, wood welding and rotational welding are two of the self-adhesion types. Hydrogen bonding and auto-cross-linking of the lignin and hemicellulose components in high-density fibreboard requires little to no adhesive. In high moisture and temperature conditions, softening of lignin and the hemicellulose content of wood leads to hydrogen bonding of hemicellulose joining fibres and lignin to develop chemical bonds. Through vibrational welding at high temperature and cellular distortion, bonds are developed. The bond developed has shown good strength properties in dry environments, but the strength is very low in wet conditions. Rotational welding happens when a wood dowel is driven into another piece of wood. Self-adhesion can be improved by modifying the environment (moisture, heat etc.) as well as the addition of chemicals for better bond development [14–18].

2.2 Formaldehyde adhesives

Formaldehyde adhesives (resorcinol formaldehyde (RF), phenol resorcinol formaldehyde (PRF, urea formaldehyde (UF) and Mixed Urea Formaldehyde (MF-including melamine urea formaldehyde MUF) are usually water borne resins and the curing procedure involves polymerisation and loss of water. The loss of water in the bond line delays the reaction of adhesives with wood due to the reduction in the wettability and movement of resin. This will limit the collision required for polymerisation process and heat transfer.

Thermosetting phenol-formaldehyde (PF) or UF are the polymers used more commonly in structural veneer-based wood product applications. For exterior veneer based wood product applications usually PFs are used and UFs are mainly used for interior applications [1]. Formaldehyde adhesives develop a rigid bond and do not creep due to combined development of polymeric chains and cross-linking groups [3, 13].

2.3 Isocyanates in wood

Isocyanates in wood (Polymeric Diphenylmethane Diisocyanate, Emulsion Polymer Isocyanates, Polyurethane Adhesives).

Isocyanates are used in wood adhesion due to their reactive characteristics to compounds with reactive hydrogen. These adhesives however can react very fast with wood moisture which will compete against the required reaction with the hydroxyl group in wood's cellulose and hemicellulose and phenol and hydroxyl groups in lignin sections. The other drawback of these adhesive is their high reactivity level with the human body that could cause safety concerns during the gluing process. The most used type of isocyanates is polymeric diphenylmethane diisocynate (pMDI) in manufacturing oriented strand board (OSB) [3, 19]. PUR adhesives are also now widely in use for a wide range of application in timber products including glulam and cross laminated timber (CLT).

2.4 Epoxy resins

Epoxy resins are compounded with ketimines that assists with releasing the curing agent when the adhesive is exposed to moisture. Similar technology is already used in coating products [1]. Epoxies are produced with a range of curing times which can influence the degree of cure and mechanical strength of adhesive layer [3].

2.5 Polyvinyl and ethylene acetate (PVA) and dispersion adhesives

These are waterborne adhesives that are cost effective and do not require a heat curing operation and they are mainly used in furniture construction. These adhesives commonly exhibit good flow into the cell lumens that are exposed to glue however due to the high molecular weight they do not usually penetrate into wood cell walls [3]. Polyvinyl acetate (PVA) is commonly used for wood gluing in non-structural and furniture making however it lacks water resistance and has low load bearing properties.

2.6 Bio-based adhesives (protein Glues, tannin adhesives, lignin adhesives)

The protein driven from wheat grain (gluten) can react with aldehydes in a similar way to urea. Gluten has a high level of amine groups (lysine and arginine) which react similarly to the ones in melamine and phenols [20]. The availability of gluten from grain is an advantage for its application in wood adhesion. However, the powder form of the gluten limits its applicability to be used in current industrial manufacturing operations.

Lignin has a phenolic structure which makes it a potential replacement for phenol in phenolic resins used for wood adhesion [21]. Lignin based adhesives can be considered in two major categories including phenol formaldehyde and formaldehyde free adhesives. Initial investigations into using unmodified lignin in phenolic adhesives showed a reduction in glue strength and an increase in press time, so chemical modification of lignin has been suggested as a solution [22]. The use of Kraft lignin and polyethylenimine (PEI) for development of a formaldehyde free adhesive showed a very high shear strength and water resistance in the glue developed [23].

2.7 Miscellaneous composite adhesives

In this group of adhesives depending on the role of timber in the composite there are three different product types known including: wood-fibre cement boards, wood-plastic composites and wood filler for plastics [3]. Wood-fibre cement products use the plant fibre to reinforce the panels and reduce the possibility of fracture development, this field is still under further studies. Wood-plastic combination is used to reduce the product weight for industries such as automotive industry. These products require good polymer- fibre interaction otherwise exposure of fibre to moisture and under stress the interface can fail.

2.8 Construction adhesives

These adhesives are used in construction for attaching floors and wall coverings which provide better rigidity in comparison with using nails or screws only. These adhesives do not require fast curing as the nails or screws hold the connection together while curing is completed. These adhesives have high molecular weight with minimum amount of solvent and are applied at room temperature. They are used to cover the gaps in the connected joints. The adhesives used in construction are usually elastomers to provide some deformability for small range displacement so the bond lines do not fail/crack as timber expands or shrinks. These adhesives are not designed for larger scale movements due to their high molecular weight and require checks and maintenance of the joint over time [3].

2.9 Hot melts

These adhesives develop bonds quickly and are used for furniture, cabinet making, windows and edge banding of laminates. The high viscosity of these adhesive limits the wetting properties of them [3].

2.10 Pressure sensitive adhesives (PSAs)

These adhesives are high molecular weight polymers that have application in decorative laminates, tapes and labels. They are usually used in gluing plastics to wood and have different ranges for products used in indoor and outdoor conditions. Their low flow characteristics makes the pressure application important so that the force applied creates required deformation in the elastomeric adhesive [3]. In these adhesives debonding or failure of the bond developed can happen under large dimensional changes and is dependent on the stress–strain properties of the adhesive.

2.11 Contact adhesives

These adhesives are polymers in solvent and are applied on gluing surfaces and left for the solvent to evaporate before the two surfaces are brought together and pressed. These adhesives are used in bonding wood to plastic laminates.

2.12 Polymerizable acrylic

These adhesives are expensive and are not commonly used in bonding wood. Structural acrylic and cyanoacrylate instant adhesives are two types of these adhesives that can be used for gluing wood. They require a smooth surface and are used in products that require rapid curing and high strength bonds. They are used in products such as electronics assembly or decorative layers in panel manufacturing [3].

2.13 Film adhesives

These adhesives are used in applications where the use of liquid glue is difficult or limiting for the type of product. They are applied using an applicator such as fibreglass mat or tissue paper in applications such as gluing very thin layer wood veneers [3].

2.14 Formulation of adhesives

Adhesives are developed for specific applications where altering the properties of the adhesive can be required. Parameters such as surface roughness, moisture level in wood structure, cost, time between application and curing, type of production process and conditional changes (season) can influence the customisation required for formulation of adhesives [3].

3. Techniques used in imaging wood products

The conventional microscopic imaging has been the most common and primary tool for capturing wood structure using optical waves and high magnifications to clearly capture the structure of the sample. Modern optical microscopes can generate images with 1500 times magnification with a 0.2 um limit in spatial resolution. The light transmitting through transparent sections and reflecting on the other parts of object helps in differentiating the structural components and developing clear images of the surface. Use of bright, dark, polarised, phase difference and fluorescence enhances the imageability of objects depending on the differences between the structural components [24]. It is not easy to create perfect images of a natural object using optical wavelengths due to variations in optical waves passing through the lens of the microscopes [25]. However, use of digital multimedia and digital processing technology to enhance the light microscopic system allows the collection of imaging outputs, improves the image quality and provides more image post-processing options.

3.1 Microscopy (visible light, UV, IR)

Using light microscopy, the position and thickness of adhesive in lumens and other parts of the wood structure, can be captured and measured depending on wood species and adhesive types used [26]. Light microscopy can be used in generating 2 and 3D images of structures. 3D imaging of a sample's volume can be achieved by further processing of either microtome sectioning and stacking of images using visualisation software or using optical techniques (such as light-sheet microscopy, confocal laser scanning microscopy or optical projection tomography) to mount the images taken into volumetric models [27–29]. Florescence microscopy is used to separate wood cells from the glue line [30] using short wavelength light to brighten sections of the sample [24]. Fluorescence microscopy uses a high voltage mercury lamp [24] that can improve the image quality by permitting the light excitation to irradiate the specimen and separate the weaker re-radiating fluorescent light from the brighter excitation light (**Figure 4**) [31].

The fluorescence presence in the adhesive needs to be sufficient to be able to capture a clear image of the bond and glue line (see **Figure 1** for more information). In order to enhance the image quality fluorescence staining can be used however sufficient power is required to produce fluorescent light [24]. Specific optical filters (each has a certain wavelength) are needed to generate images with suitable wavelength of light emitted resulting in maximum fluorescent light outputs. An aqueous mix of 0.2% of acridine yellow which is absorbed by wood but not the PF glue is used to enhance the fluorescence presence in the imaging process [31]. However the staining methods used in the literature are limited and require extra experimental steps for sample preparation, and majority of reported staining trials applied stain to the resin prior to gluing. The application of stain in industrial scale production line can be costly and cause undesirable colour changes in the product [32]. Combining images taken in visible and florescence modes (FM) can enhance the visibility of the glue line. Mahrdt et al. [32] used a merging method to combine images taken in visible and fluorescence modes using gentian violet and brilliant sulphaflavine dyes respectively (**Figure 5**). The merging technique improved the post-processing required for quantifying glue line parameters by allowing

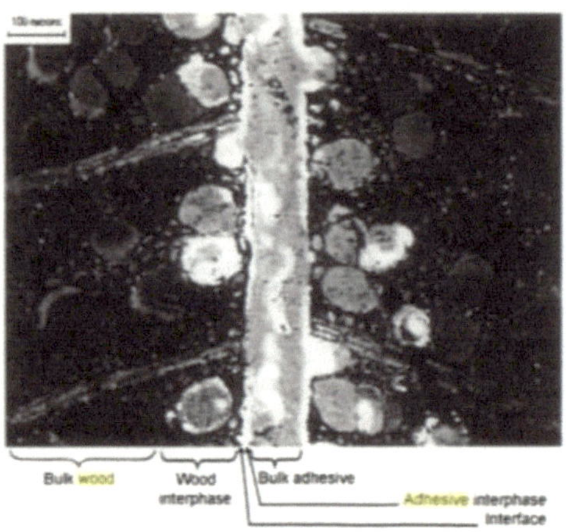

Figure 4.
Adhesive bond line fluorescence microscopy of an epoxy glue showing regions of glue penetration [30].

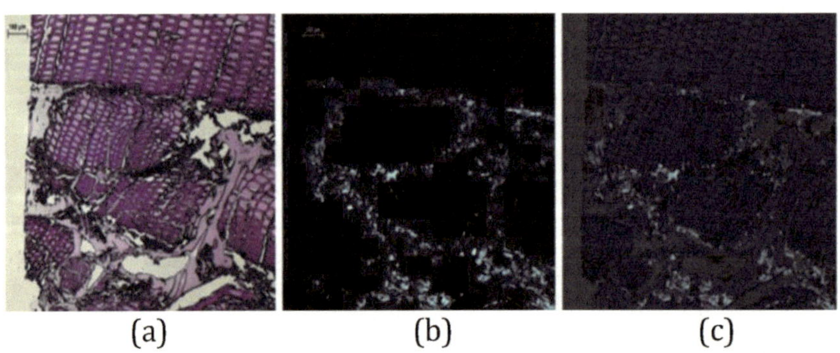

Figure 5.
Image of cross section and glueline using (a) visible, (b) fluorescence modes and (c) merged image [32].

semi-automatic analysis of images. The most common disadvantage associated with using FM is the limitation in capturing depth which is an important parameter in glue line investigation [33].

Based on the optical path used the fluorescent microscopy can be a transmission fluorescence microscopy (TFM) or an epifluorescence microscopy (EPI) [24]. TFM usually uses a dark field passing light through a condenser to excite the sample to emit fluorescent light in different directions. TFM provides generation of strong fluorescent light in low magnifications however once the magnification increases the fluorescent light reduces making the method more applicable for larger specimens [24]. EPI uses the objective lens as a well corrected condenser first then as a light gatherer to form an image (see optical path illustration in **Figure 6**).

The images produced by TFM are usually darker in comparison with images taken by EPI. Some of the other advantages of EPI systems are increased fluorescent light intensity, reduction in light loss, no loss of fluorescence intensity [24].

Imaging of adhesive penetration into wood based composites using EPI is shown in **Figure 7** [34]. The maximum penetration depth in fibres showed a rising trend as the moisture content increased. A study on the effects of moisture curing of polyurethane on the segment content (hard and soft content) of adhesive bonds showed that increasing hard phase content increased the intermolecular interactions, liquid viscosity of adhesive,and the soft phase glass transition temperature [35]. This study showed that glue bond line thickness increased as hard phase percentage increased while the hard mass percentage had a negative relation with the effective glue penetration (**Figure 8**).

Penetration of PF resin into poplar was quantified using EPI imaging at 2, 6 and 10% moisture content [34]. For imaging, a 50% phenol formaldehyde (PF) resin solution in water was sprayed onto the uniform oriented strand board. The mixture made of poplar and 2% potassium carbonate (K2CO3) as the catalyst was added to the strand boards [34, 36].

Using ultraviolet (UV) microscopy can generate better resolutions and higher magnification for images of glue lines. UV incident light microscopy imaging (UVLMI) was used to study the glue line in LVL samples of beech shown in **Figure 9** after cyclic temperature and RH conditioning [37]. The images taken showed a clear illustration of the modified melamine–formaldehyde resin (MUF) and PUR penetration in the LVL structure. High-resolution episcopic microscopy

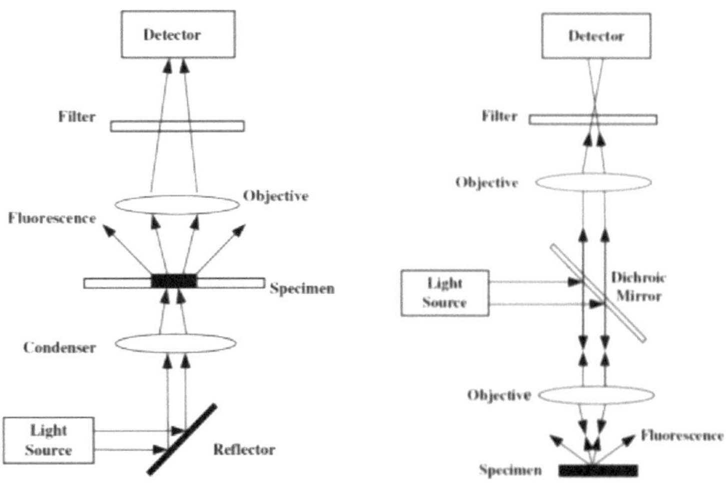

Figure 6.
Optical path differences between (left) TFM and (right) EPI [24].

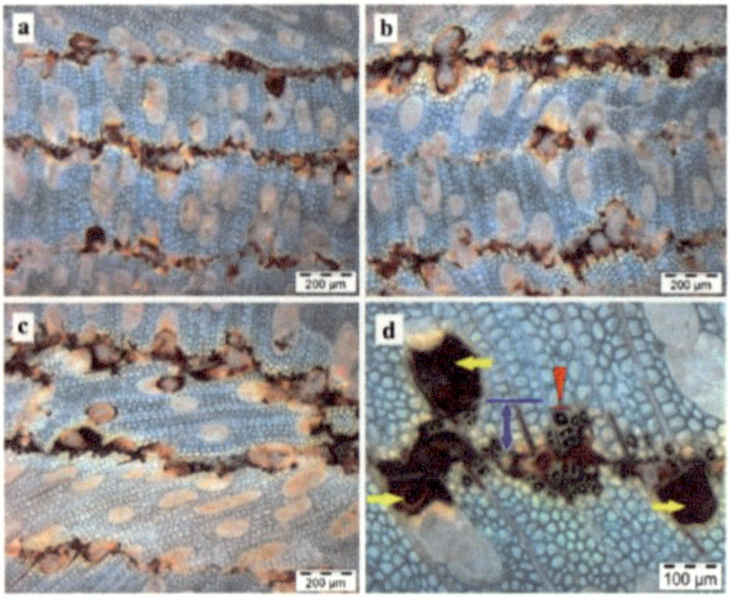

Figure 7.
Penetration of PF resin into poplar strands at different moisture contents of (a) 2%, (b) 6%, (c) 10% and (d) the discontinuity of glue in 6% moisture content samples [34].

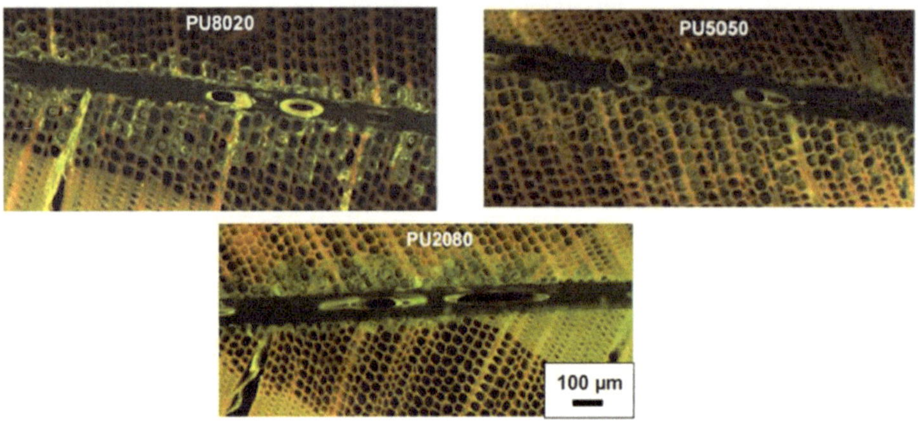

Figure 8.
Florescence microscopy imaging of PUR bond line and penetration development for top: PU8020 with 53.5%, middle: PU5050 with 65.4% and bottom: PU2080 with 72.5% hard phase mass [35].

((HREM) uses microtome sections taking images of thin layers of the sample block (1–5 µm thick). The images are then stacked using visualisation software to develop a 3D model of the sample block. A florescence stereomicroscope and a digital camera are used to capture the dye mixture embedded in the sample block in visualising the structural components [29]. The HREM method has been successfully used in various medical applications providing better resolution imaging in comparison with X-ray computed tomography, MRI, and optical projection tomography (OPT) [28, 38]. Samples require detailed processing of dehydrated tissues prior to HREM imaging using 4% paraformaldehyde (w/v) for 1–2 hours under vacuum and then incubated at 4°C overnight before being washed by ethanol and embedded in embedding solution (**Figure 10**) [29].

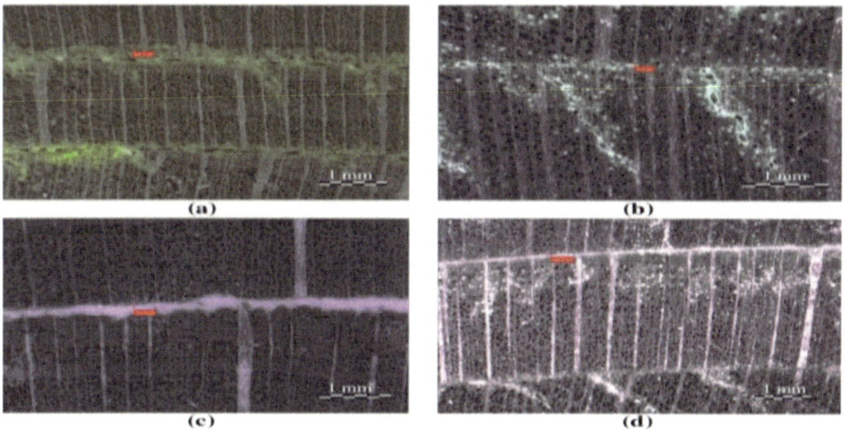

Figure 9.
Images of layered structure of LVL specimens conditioned in different temperature and relative humidity [37].

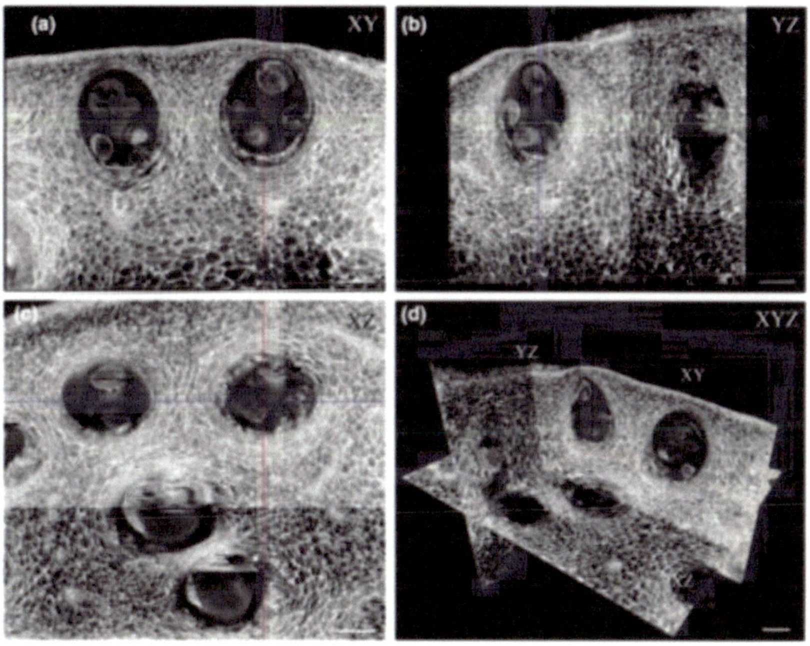

Figure 10.
Section images of mature Citrus limon *fruit using HREM [29].*

3.2 Electron microscopy (SEM, ESEM, transmission electron microscopy (TEM))

To enhance the spatial resolution of microscopes, the use of electron beams is becoming a more preferred technique replacing the visible light source. Electron microscopy can reach a maximum of more than 300 million times magnification. Moving electrons can act similar to optical wavelength variation. The concept of detecting the energy release and wavelength as electrons move are the foundation for development of transmission electron microscopy (TEM) in 1938 and scanning electron microscopy (SEM) in 1952 [24]. Imaging with a higher level of magnification and in greater depth is possible using electron microscopy. However the grey colour of the images produced limits the imaging of different glue lines and sample preparation and imaging in a vacuum environment are some of the limitations for electron microscopy [24, 33].

3.3 X-ray computed tomography (XCT)

Computed tomography (CT) was initially developed in 1970s for medical imaging, the CT scanner provides a range of image sizes including: large up to 250–500 μm or smaller μ-CT (min spatial resolution of 50 to 325 nm) and n-CT (min spatial resolution of 200 nm). For smaller size samples, reduction of the X-ray flux is required which would increase the required scanning time [27]. X-ray computed tomography (XCT) is used in microscale to image the cellular structure of samples in 3D mode providing spatial details of the complex xylem network [39]. XCT uses radiation that penetrates into the sample and then detects photons after being traversed through the specimen. XCT uses two different mechanisms to develop images of the structure: absorption contrast and phase-contrast tomography. In absorption contrast methods, the differences in material linear attenuation coefficients, μ, affects the absorption contrast resulting in sufficient contrast to map out the edges of cells in the sample structure. The linear attenuation coefficients are dependent on material density and atomic number (Z). Phase-contrast tomography uses photons incident beam phase shift at an interface between two different materials. Phase contrast tomography is suitable for materials with low absorption contrast such as soft material with a lower atomic number [40]. In wood adhesive penetration detection, the similarities between the attenuation PF and cellobiose (chosen polymer in this study to represent cell walls in wood structure) showed that XCT is not a suitable detecting method if the PF glue was tested without any contrast agent added [41]. However, this may not apply to all glue types and timber products depending on wood species, their characterises and densities (**Figure 11**).

The lack of x-ray absorption contrast by the adhesive and the wood cells makes the use of micro CT (μCT) technology in bond line imaging and adhesive penetration challenging. The insufficient differences between the natural density of wood and the glue types used require addition of an enhancing agent to the glue for the purpose of imaging. Heavy metals have been used as contrast agent for imaging of glue line in wood products however their limited tag mobility and phase separation of those components made use of them less effective in studying the glue line properties [42]. The scanning is dependent on the absorbed contrast based on a materials linear attenuation coefficient (μ) differences. The linear attenuation coefficient is a function of material density and elemental composition. Based on Beer–Lambert's law, μ is calculated using the transmitted (I) and initial (I_0) radiation intensity and the material thickness (d) [43]:

$$I/_{I_0} = exp^{(-\mu d)} \tag{3}$$

Considering the differences between adhesives used in the timber industry and the density of commercial timber species, achieving the required contrast between the two materials will require addition of contrast agents to the wood or adhesive used. The use of contrast agent additives has been studied for wood adhesive penetration studies before [33, 40, 41, 44, 45] . From a range of tested contrast agents used with PF (see **Figure 12**) iodine was selected as the preferred agent, improving the clarity of differences between adhesive and cell walls. Rubidium was also tested as a contrast agent in combination with PF. RbPF (added in form of RbOH as a contrast agent to PF with molar ratio of 14.3) showed significantly clearer differences between the glue and cell walls. Use of Rb as contrast agent produced clear images and

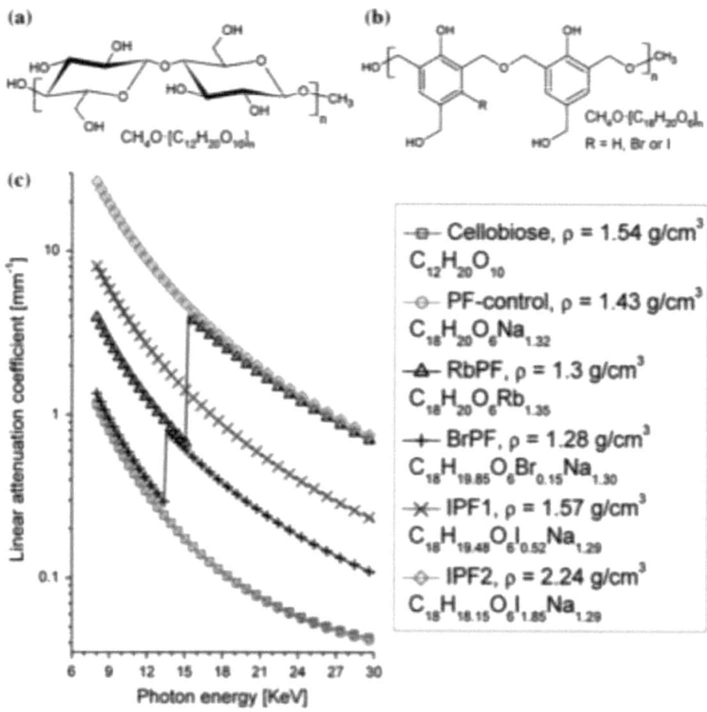

Figure 11.
Linear attenuation recorded for PF- control, brominated (BRPF), iodinated (IPF1 and 2) and rubidium (RBPF) adhesives and cellobiose as representor of wood cell walls polymer [41].

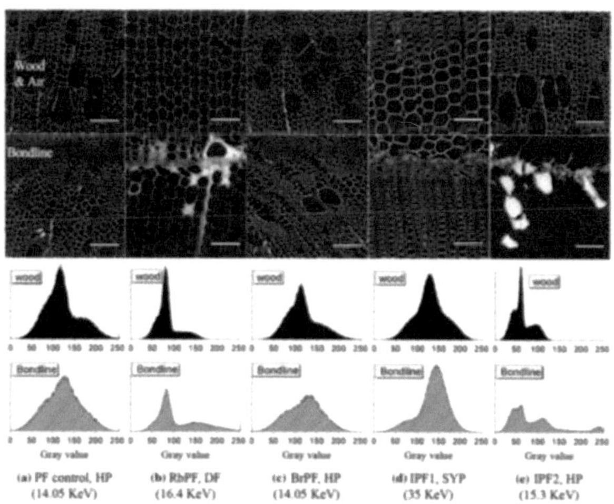

Figure 12.
Results of imaging adhesives with different contrast agents added [41].

adhesive with Rb travelled easier in species with higher permeability such as oak and poplar while Douglas-fir showed lower movement of Rb in sample structure. The images taken from samples also showed that Rb travelled into the wood structure independently of the adhesive and penetrated further than the adhesive which indicated the complexity of using Rb as a contrast agent [33].

Use of x-ray micro tomography (XMT) in capturing the PF adhesive penetration in Douglas-fir samples improved the clarity of glue penetration in the sample structure however, the grey-scale results did not illustrate clearly the interface between cell walls and resin filled lumens (**Figure 13**).

For clarity and image quality using XMT, samples need to be dried. The excess moisture - at saturation point can reduce the cell wall contrast and image quality. However, the oven drying of wood samples below FSP could increase the risk of cell wall collapse. Combining electron energy loss spectroscopy (EELS) or energy dispersive X-ray analysis (EDXA) with electron microscopy has shown advantages in quantifying the glue line parameters and depth. EELS uses an electron probe targeted with high capabilities in imaging dark modes (these couple weakly with the optical excitation making imaging more difficult in comparison with light mode) to map different locations of the sample [46]. The SEM-EDXA is suitable in investigating if the adhesive penetrated into the cell structure (**Figure 14**) [47].

Iodinated (IPF) and brominated (BrPF) PF resins were used in μCT imaging of the glue lines in wood composite panels. The study of glue viscosity and MW (after curing) showed no changes in BrPF. However, the MW of glue before curing increased as percentage of added contrast agent increased. The IPF had slightly lower viscosity compared to control PF samples. The powder density determined for BrPF was lower than controlled PF samples. Further testing of tagged adhesive using fluorescent micrographs and energy-dispersive spectroscopy (EDS) elemental

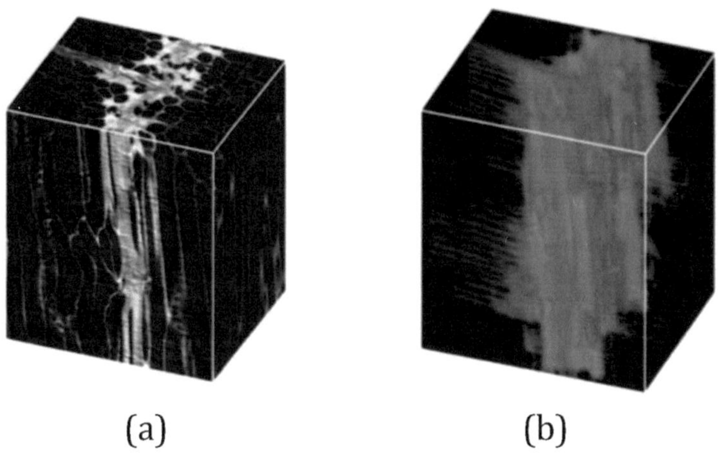

Figure 13.
3D images of Douglas-fir with phenol–formaldehyde adhesive in transverse surface on top (a) with the cell wall details and (b) removed cell wall details [33].

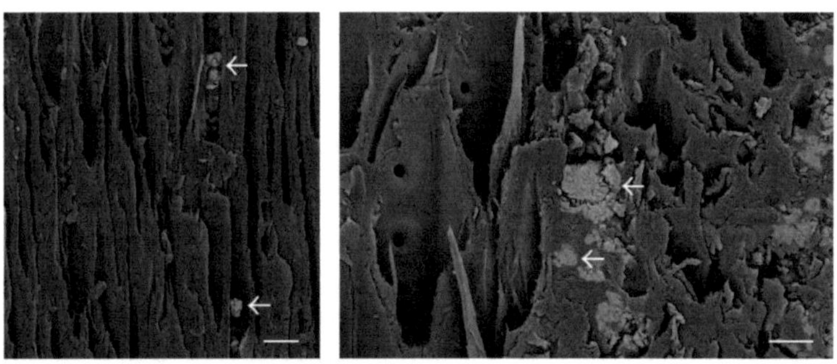

Figure 14.
EDX spectrum and backscattered SEM imaging of OSB using zinc borate as a contrast agent, arrows show the zinc borate particles in lumens of wood cell [47].

maps by Modzel [48] showed the presence of Rb in cell-walls in sections with no RbPF glue presence (**Figure 15**) [43].

In a study by Kamke et al. Iodinated PF (IPF) was made by adding 10% 3-iodophenol to plywood adhesives following commercial adhesive synthesis manufacturing process. The final adhesive had 39.5% weight iodine as the contrast agent to enhance the x-ray imaging results. Even though the viscosity of adhesive at 25C was 930 cP which could have changed the curing properties of the glue it did not compromise the behaviour of the bond. Images from this study are shown in [41]. The images generated with ionising radiation (called tomograms or slices) are obtained from the translation and rotation of the source and detectors. In this imaging technique, the attenuation coefficient of x-ray or gamma rays are recorded. In a non-homogenous medium such as wood structure it detects the attenuation coefficient depending on the quantum energy of the ionising radiation and the chemical composition of the sample [49]. The computer tomography (CT) method provides the option of imaging the full 3D structure of the sample including bark, knots, heart and sapwood borders etc. [44]. The scanning parameters measured by ionising radiation vary with wood species, size of the specimen, level of contrast in density for different defects, end use of scan information, speed of scanning required for imaging [49].

3.4 Magnetic resonance imaging (MRI) or nuclear magnetic resonance (NMR)

MRI or NMR was developed in 1970s for medical imaging and it was used later in the 1980s in imaging plant material [27, 50]. MRI is a non-destructive method to develop images of plant structure and has been used as an in vivo method to determine the water content and moisture movement in plant structure [51]. Images of a *Quercus serrata* branch were taken using MRI at different signal intensity specification to detect different components (shown in **Figure 16**) including water, annual ring, inner rings [52]. Although no studies have been reported in the literature on the use of MRI or NMR for studying adhesive bonds, both these technologies have considerable potential for this purpose.

Thermal technique uses the temperature to map out the surface of the object applying an active or passive heating procedure. The cyclic loading in structural timber for example can be used as an active source of heating to detect the effects of defects in structure on the mechanical strength and performance of products [49].

The passive heating process usually applies for detecting knots direction, slope of grain, moisture distribution and rupturing in wood structure. The passive heating process does not create the destructive effects on the samples due to low thermal stress however it requires fast recording methods to generate accurate images.

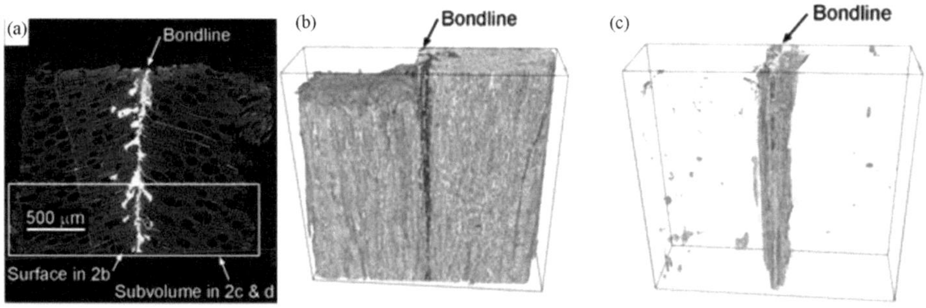

Figure 15.
X-ray images of poplar glued using iodinated PF, (a) cross section, (b) wood and bond line in 3D, (c) 3D bond line without wood sections [40].

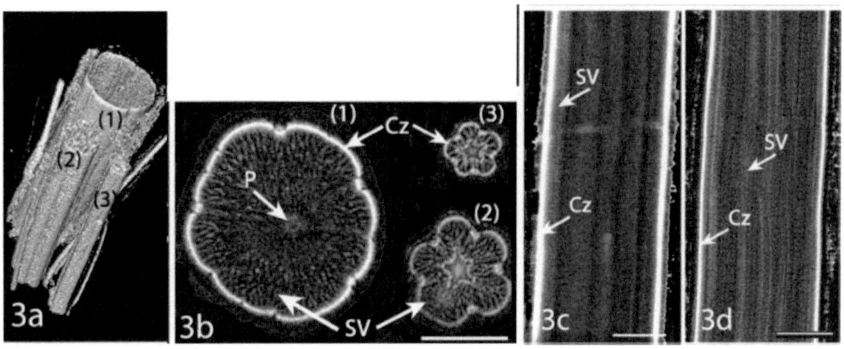

Figure 16.
Images of Quercus serrata *branch using MRI, (a) Surface-rendered image reconstructed from a 3D-FSPGR dataset. Bolts 1 to 3 correspond to the numbers in* ***Figure 3b****. (b) Proton density weighted crosscut images. TR, 2000 ms; TE, 25 ms. (c & d) A proton density weighted (TR, 2000 ms; TE, 20 ms) and a T1 weighted (TR, 540 ms; TE, 20 ms) radial image, respectively. Pixel size 156 μm. — Scale bar = 1 cm [52].*

Infrared cameras are the most common types of detecting systems using electronic detection of infrared emitted from the sample [49]. These methods are used in determining the integrity of the surface and sub-surface, for example in wood based composites [53].

Scanning thermal microscopy (SThM) uses two different modes to detect the changes in chemical composition of samples and map out the adhesive penetration in the structure using a thermal probe. The thermal contrast microscopy (TCM) measures the changes in temperature of the sample surface. Conductivity contrast microscopy (CCM) also measures the sample's surface conductivity as the temperature is kept constant [54]. Images of wood phenol-resorcinol-formaldehyde-adhesive bonds (PRF) using SThM are presented in **Figure 17**.

In microwave techniques, Dielectric properties of the sample structure are used to image the structure in microwave imaging methods. This method is commonly used for assessment of wood products after drying and gluing, for detecting any internal detects such as knots, spiral grains, discontinuation in structure of logs, lumber and wood based composites [49].

Ultrasonic methods are low cost methods for detecting the adhesive content of wood products with a high sensitivity to delamination, however achieving imaging with high resolution is difficult and coupling pressure dependency reproducibility

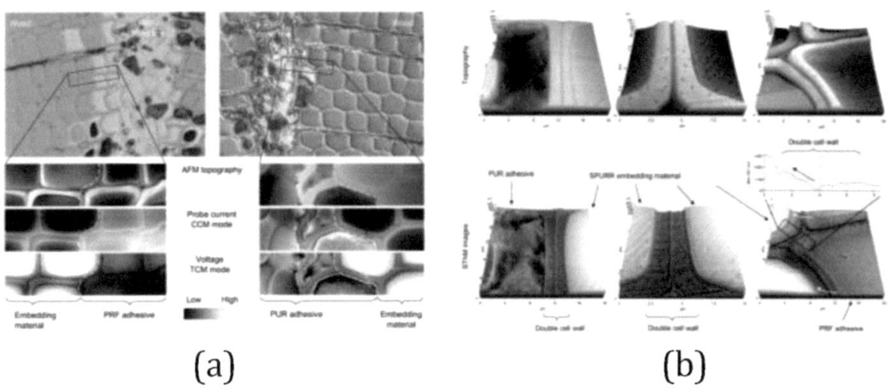

Figure 17.
(a) Images of adhesive bond in spruce samples using (top): Light microscopy and (bottom): SThM mapping the topography, probe current and voltage and (b) differences in topography for SPURR epoxy resin (Centre) and in contact with PUR adhesive (left) and PRF adhesive (right).

of this technique is poor [44]. There have been reported studies on the uncertainty around voltage values that the tranducer recorded (20–25% variation) where extra applied hand pressure is required for ultrasonic readings [55]. Air coupled ultrasonics systems however have shown promising results eliminating this variation which requires future investigation on its effectiveness in detecting complex glued sections (**Figure 18**) [44, 55].

FTIR imaging of chemical components (Fourier Transform Infrared) provides details of functional groups based on absorption, of chemical band intensity, band areas, and position. **Figure 19** shows the FTIR readings for wood and Polymeric diphenylmethane diisocyanate (pMDI) in uncured and cured conditions [56]. This study reported on the challenges in detecting glue formation in the cell walls using this method.

Chemical imaging is used to map out the intensity and distribution of different components in wood structure [57, 57]. The study used non-covalent interaction of

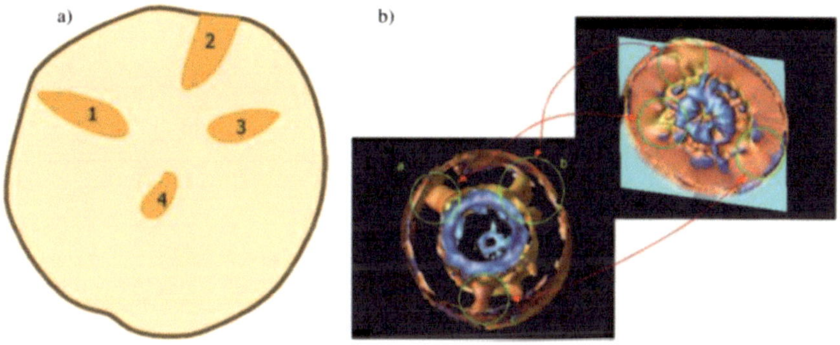

Figure 18.
Microwave technique used to detect knots in radial (a) and reconstructed image (b) [49].

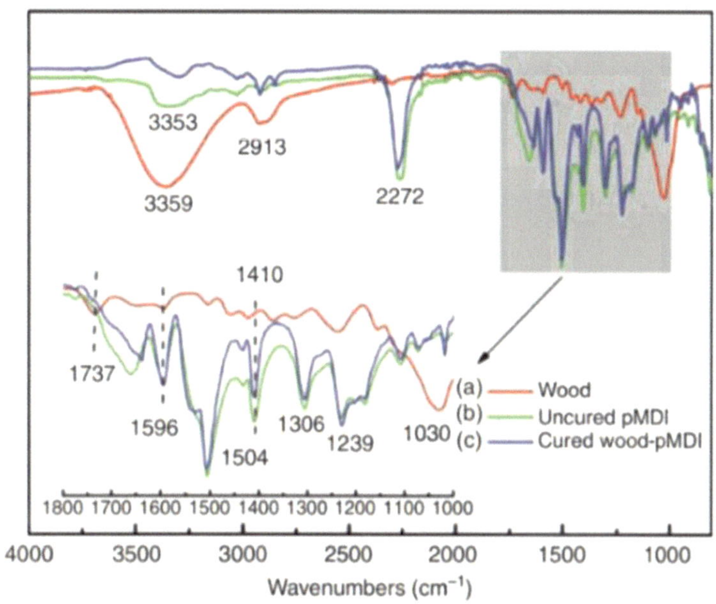

Figure 19.
*Results of imaging using FTIR spectra of loblolly pine (*Pinus taeda *L.) wood and pMDI, and wood-pMDI [56].*

acetylated nanocrystalline cellulose (AC-NCC) with polylactic acid (PLA) in material tested as an indicator for chemical imaging. **Figure 20** shows the readings for different chemical components in the samples including visible light image in left column, mCH2 absorption maps in middle column and integrated mC=O second derivative absorption peak areas in right column. The results showed that FTIR imaging can be used effectively to image the chemical links between a substrate and functionalised filler [57].

3.5 X-ray photoelectron spectroscopy (XFS)

As discussed earlier, X-ray computed tomography (XCT) was used after adding the bromine substituted phenol formaldehyde (BrPF) to increase the X-ray

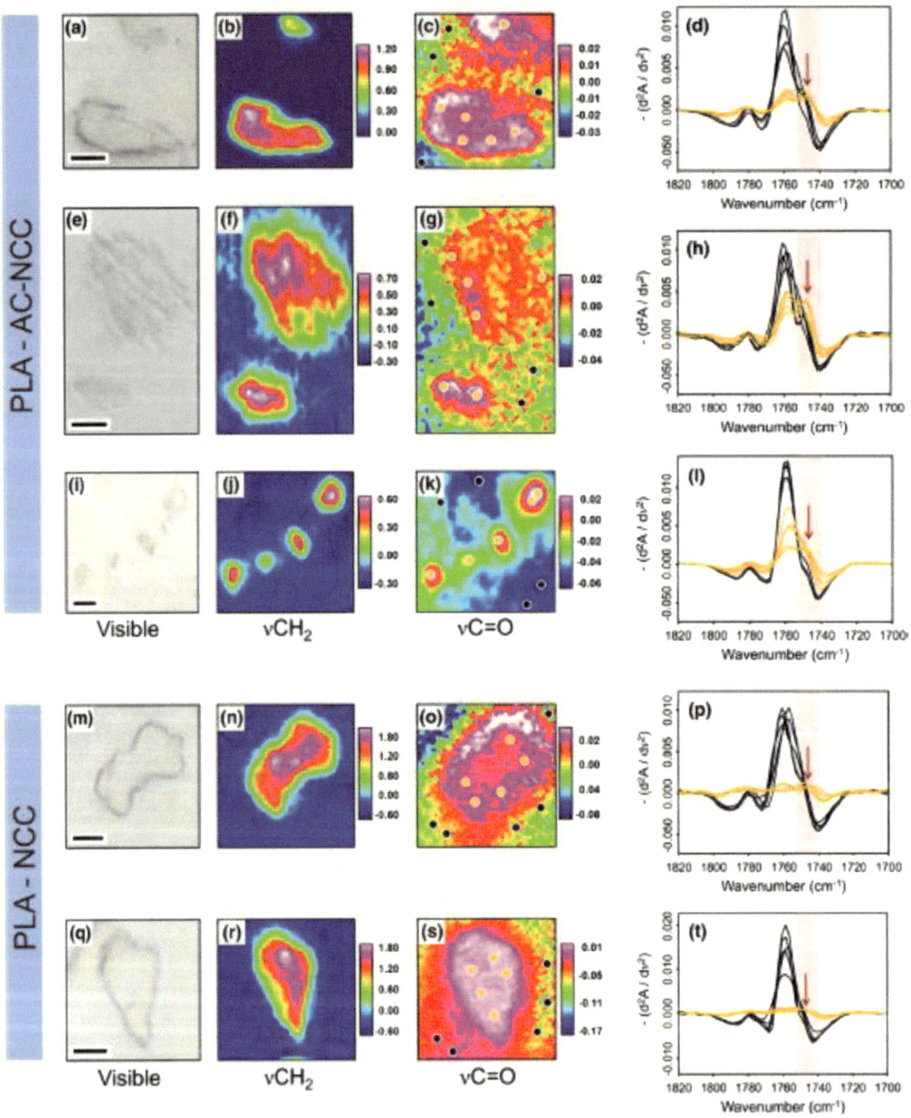

Figure 20.
Results of absorption map for different particles using chemical imaging and FTIR micro spectroscopy for composite materials [57].

attenuation and capture a brighter glue line which is distinguishable from wood cells [58].

This study also used X-ray fluorescence microscopy (XFM) to acquire images of the bond line [59]. This study investigated the required adhesive (BrPF) flow and infiltration to develop an effective glue bond in loblolly pine (*Pinus taeda*). Four materials including BrPF, BrPF-mixed with wood, wood and air in the porous structure of wood was observed in the images as shown in **Figure 21** (left image (b)). The study indicated some uncertainty around the wood composition with BrPF that was not included in the 3D reconstruction of the image. **Figure 21** shows the X-ray and X-ray fluorescence image and the sections in red circles that were not used in 3D reconstruction. The potential inclusion of wood cell walls infiltrated with BrPF or another material interface in the images could be a limitation in 3D imaging with XCT that could be overcome using XFM. Image on the right in **Figure 21** shows the consistency of XCT images.

The XCT method provides the option of visualising the flow of BrPF and penetration into the 3D wood structure over time. In sections with less clarity of BrPF mixed with wood, XFM provided a detailed observation for the glue presence inside the cell structure of the wood (**Figure 22**).

A study of PF in wood cell walls using XFM in combination with nanoindentation showed that using BrPF enhanced the cell wall matrix (see sections showed in **Figure 21**). The different molecular weights (MW) of BrPF tested showed that the lower MW had greater effects on preventing the softening of cell walls due to gained moisture. In **Figure 23**, the glue line is shown by arrows and the dashed white line shows the cells that were tested for mechanical strength using nanoindentation [59].

Nano indentation and scanning probe microscopy (SPM) imaging of year old Masson pine (*Pinus massoniana* Lamb.) showed the advantages that SPM could provide in investigating the cellular strength [60]. **Figure 24** shows the detailed images of cells and modified (PF resin (as thermosetting agent) impregnated samples at 15, 20, 25 and 30% treatments) cells. This method can be used in detecting the presence of glue in various sections of the cell wall and its effects on cell wall strength as the thickness of glue and the properties of bond line (thickness and depth of penetration) varies.

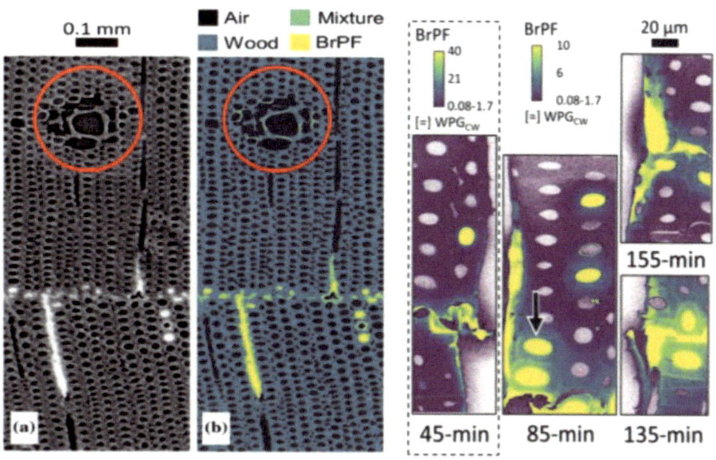

Figure 21.
Left- X-ray computed tomography (XCT) and right- X-ray fluorescence microscopy (XFM) images of glue line [58].

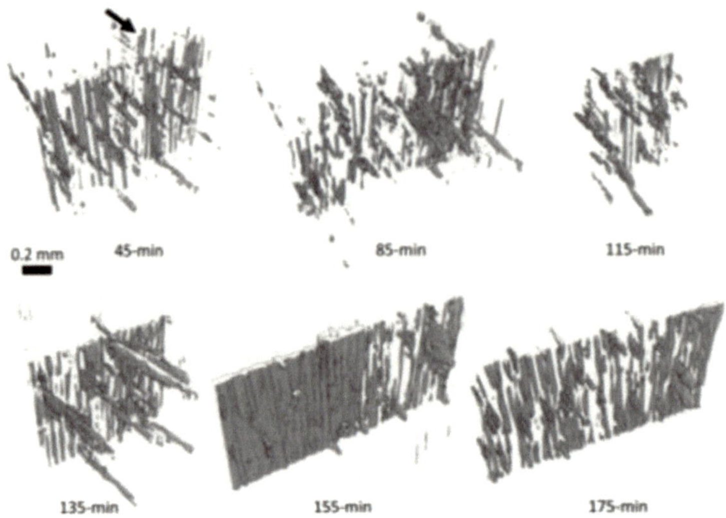

Figure 22.
Flow of BrPF into wood structure over time [58].

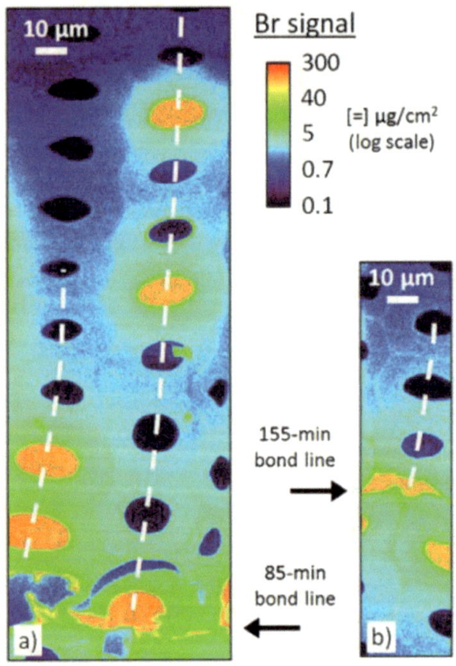

Figure 23.
XFM used to map Br penetration into wood cell structure, (a) after 85 min BrPF and (b) 155 min samples [59].

3.6 Epifluorescence microscopy (EPI)

EPI microscopy can provide imaging of the cell structure from the surface of samples used for imaging.

A study of adhesive penetration in particle board composites using EPI showed this microscopic method was effective for quantitatively assessing glue interaction

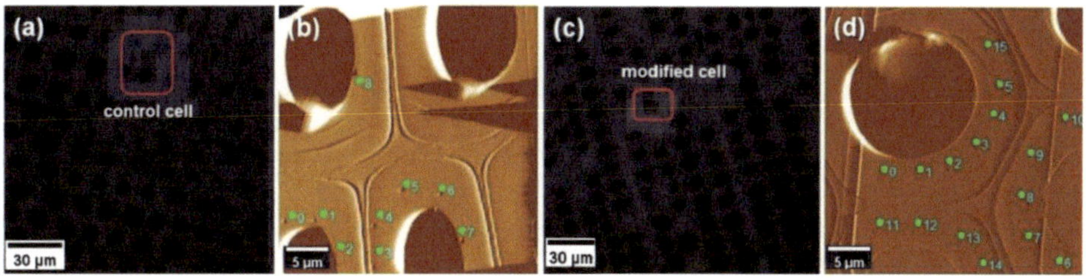

Figure 24.
Images of Masson pine (a and c) under microscope and (b and d) under SPM equipped with nano indenter probe [60].

with wood. The study determined adhesive filled fibre numbers (FFN), adhesive filled vessels number (FVN), maximum adhesive penetration depth in fibre (MPD_f) and in rays (MPD_r). **Figure 25(a)** shows the results of EPI microscopy in poplar samples at different moisture content (2, 6 and 10% MC) [34]. Penetration of resin was higher in higher moisture content wood. **Figure 25(b)** shows results of EPI imaging of three UF adhesives in poplar in radial and tangential directions. The images taken showed clear sections where glue penetrated into vessels, rays as well as showing the bond lines. Three UF adhesives were tested at different viscosities including UF I, 545 mPa.s < UF II, 745 mPa.s < UF III, 1644 mPa.s [36].

This study showed penetration into the radial direction was lower than penetration in tangential direction which could be related to the presence of pits (on radial walls) that facilitate the adhesive movement in tangential direction. Shear strength of specimens increased as penetration increased and was higher in adhesive with lower viscosity (due to better flow of adhesive) [36].

3.6.1 Confocal laser scanning microscopy (CLSM)

Confocal laser scanning microscopy (CLSM) provides an optical depth selection option to image the sections of a structure that are usually very hard to physically prepare. Imaging of glued sections using CLSM under FM provides a clear representation of the glue line and its borders in wood structure. **Figure 26(a)** represents the images taken from a coated wood surface in a study investigating the durability of the coating material used and penetration into surface cracks of radiata pine (*Pinus radiata*). The comparison with light microscopic imaging showed the advantages of using CLSM in showing the details of penetration into cell lumina and fine surface cracks [61].

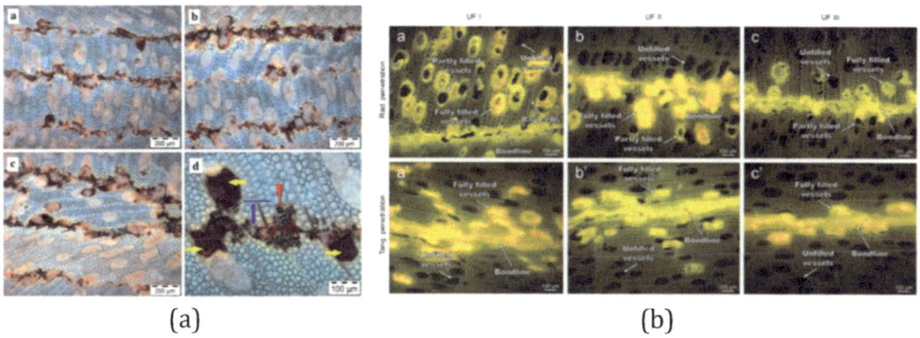

Figure 25.
(a) Images of UF penetration into wood structure using EPI [34] and (b) EPI imaging of adhesive penetration into different planes of poplar (a, b and c are in radial and a' b' and c' are in tangential directions) [36].

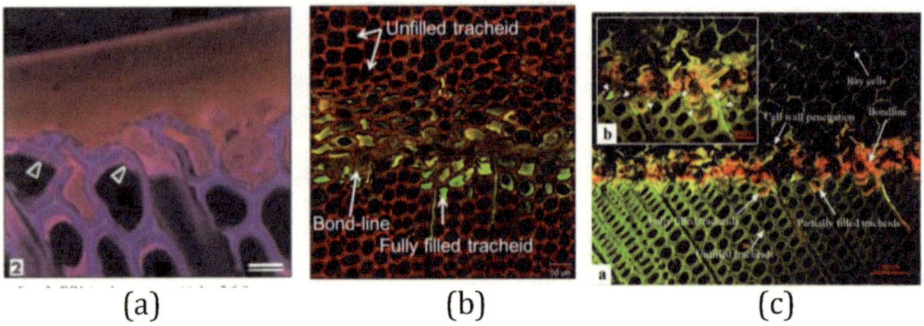

Figure 26.
CLSM used in (a) imaging coating surface and penetration of coating material into the wood structure (bar = 20 μm) [61], (b) urea formaldehyde (UF) resin penetration into wood structure [62] and (c) UF adhesive imaging showing wood structure and resin penetration [9].

The separation between urea formaldehyde (UF) resin and wood cells is not usually easy with common microscopic methods; **Figure 26**-right show images of UF resin using CLSM [9].

Figure 26(b) shows results of CLSM of UF resin penetration in radiata pine (*Pinus radiata D. Don*) veneer layers (2 mm thick) of plywood including fully adhesive filled and unfilled tracheids and the glue bond line developed [62].

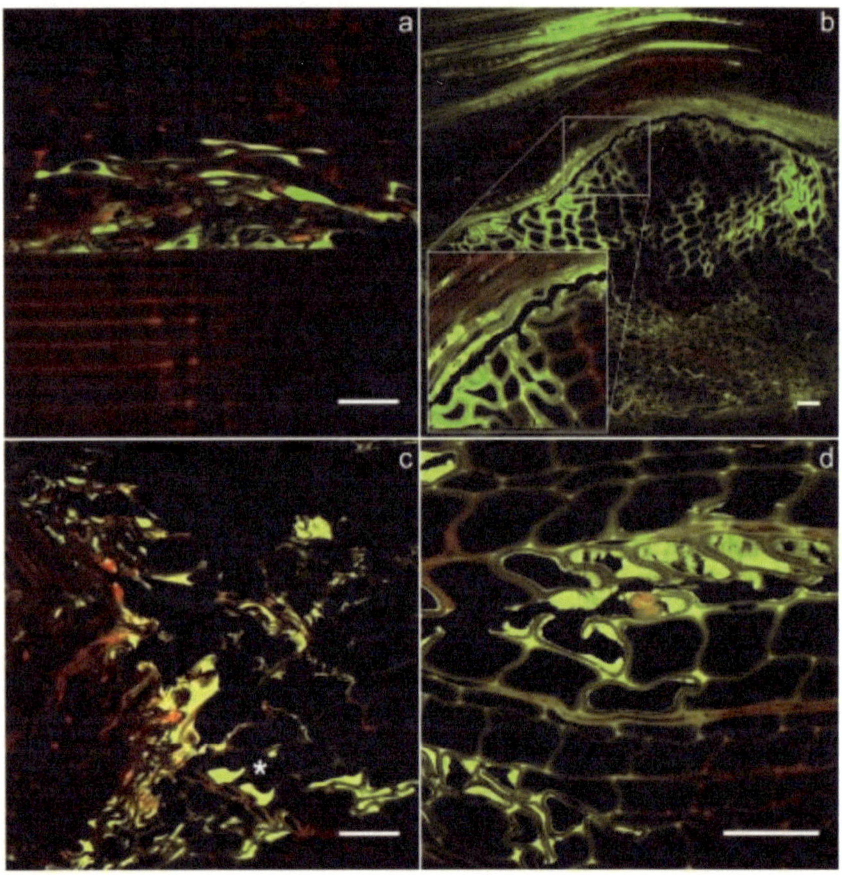

Figure 27.
Images of particleboard section glued using fluorescence-labelled UF resin, bar = 50 μm [63].

Figure 26(c) shows results of CLSM imaging of UF resin in samples of *Cryptomeria fortune* Hooibrenk. This study showed that UF resin penetrated in axial tracheids and ray lumens and cracks around the glue line. Imaging showed that glue penetrated easier in the direction of lowest resistance (shown in image by arrows) to the compression applied by the press machine [9].

CLSM has also been used in imaging different particle board samples targeting the UF resin component in board structure enabling the quantification of resin percentage in the total board area. This study which combined CLSM and fluorescence labelled UF (**Figure 27**) showed clear sections of cells and glue penetration into the structure of boards (made from Norway spruce (*Picea abies* (L.) Karst.) [63]. Singh et al. used [61, 64] confocal florescence micrographs to capture the adhesive penetration into the wood cell structure as well as detailed information on interlocking of cells by glue (see **Figure 28**). CLSM is a new imaging technique that

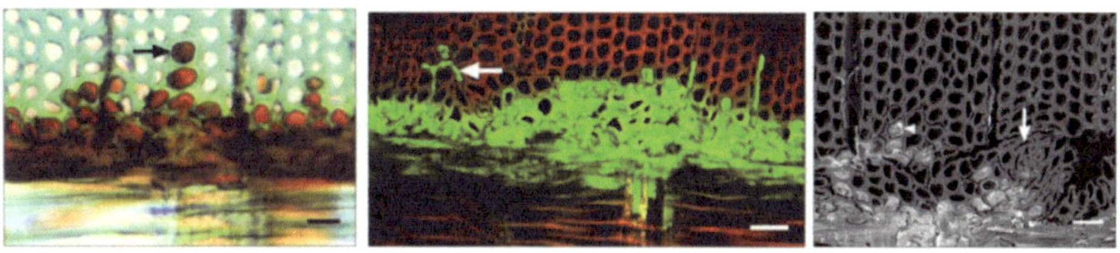

Figure 28.
Left- light microscopy image with high magnification, middle- image taken using confocal florescence micrograph and right- SEM image [64].

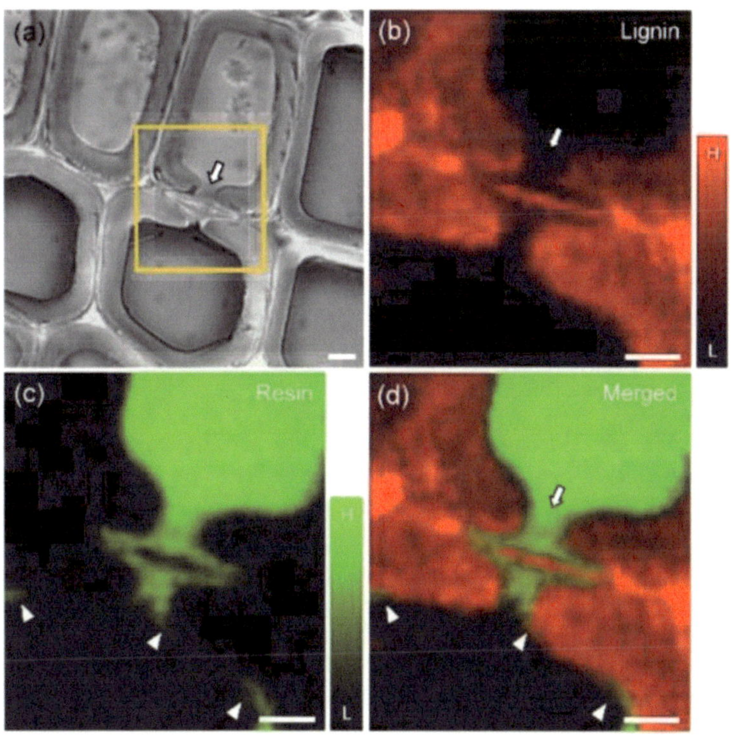

Figure 29.
Raman mapping on cross section of coated samples of Japanese cedar (a) Raman image (b) lignin, (c) alkyd resin and (d) merged information adrenal glands [65].

has capabilities in enhancing the contrast between the wood and coating or adhesive materials to improve the image quality and clarity.

3.7 Chemical mapping - FTIR microscopy

Chemical mapping and imaging of biological and biometric samples- FT-Raman spectroscopy- has recently been used in studying wood including coatings, degradation and modification of cell walls. Similar to NIR detecting systems, Raman spectroscopy uses the vibrational transitions [65]. However, Raman spectroscopy uses the changes in polarizability of functional groups while NIR uses differences in infrared absorption to detect chemical components. A study of Japanese cedar (*Cryptomeria japonica* D. Don) using confocal Raman microscopy showed clear details of the structure and chemistry of wood including distribution of alkyd resin and lignin (**Figure 29**). The study showed the great capability of Raman microscopy in detecting different chemical components of wood and wood coating materials at the cellular level of structure.

Similar work done by Wang et al. [60] showed the advantages that Raman spectroscopy can provide in studying the penetration of PF resin into the cellular structure of Masson pine. The images indicated the penetration of PF resin into cell lumens as well as into cell walls. The images also show the interaction of the PF with polymers in the cell walls. The nano indentation done on cell walls in combination with the Raman spectroscopy conducted showed that the added PF resin in the wood structure reduced the dimensional changes of the structure resulting from swelling and shrinkage [60].

4. Summary and future directions

The review focuses on existing studies and techniques used for visualising adhesive bonds in various timber products. This review highlighted the capacity, advantages, limitations and the potential of the different visualisation methods including Microscopy (visible light, UV, IR), Electron Microscopy (SEM, ESEM, transmission electron microscopy (TEM)), X-ray computed tomography (XCT), Thermal techniques, Microwave techniques, Ionising radiation, Ultrasonics, FTIR imaging of chemical components, X-ray imaging, Epifluorescence Microscopy (EPI), Confocal laser scanning microscopy (CLSM) and Chemical mapping. The majority of work published has focused on applications and types of wood used for specific products rather than developing/following a standard protocol in determining definitions for adhesive bond development and measurements. These studies are mainly qualitative and comparative only to samples tested in each study. The review suggests that there is a need for developing standard definitions for glue bond shape, thickness, penetration depth and scale in both 2 and 3D in order to be able to quantitatively assess the effectiveness of gluing process for various timber products. The effects of wood types, anatomical structure, gluing properties, curing mechanisms and glue types used for different applications/products need to be studied for different structural and non-structural timber elements. The review also highlighted the need for better definitions of glue lines, penetration in relation to wood grain direction and cellular details which could affect the penetration effectiveness depending on surface wetting properties, extractive content in cells and size of cells. The interaction between glue, cell walls and chemical extractives of different wood types before and after glue application and product manufacturing will need to be carefully studied and a summary of potential differences and similarities for each type of timber product can be developed for industry to access.

The existing literature shows the strength and capacity of different visualisation methods for analysing the adhesive effectiveness in different wood types, however more research is required to examine the potential of each method for a range of timber products and wood types (e.g. soft vs. hardwoods). The 3D imaging can provide detailed information on adhesive bonds in various conditions including where there is variation in MC, glue type (different curing requirements and curing time) and wood type (including low to high density and for soft and hardwood).

In the use of X-ray scanning where density differences between the wood and glue are the main visualisation factor, further studies are needed for investigating the effects of adding contrast agents (such as iodine) to adhesives and the potential effects on glue properties that could lead to bond line variations and changes in penetration patterns.

Further work is required to monitor potential effects of industrial processes and environmental conditional factors on bond formation and adhesive penetration during and after production and in the service life of timber elements specially for structural products such as glulam, CLT and LVL.

Acknowledgements

The authors acknowledge the Queensland Department of Agriculture and Fisheries and the Australian Centre for International Agricultural Research (ACIAR) for providing the funding which supported this work, as a component of the ACIAR Project FST/2016/151 “Advancing enhanced wood manufacturing industries in Laos and Australia”.

Author details

Maryam Shirmohammadi* and William Leggate
Department of Agriculture and Fisheries, Brisbane, Australia

*Address all correspondence to: maryam.shirmohammadi@daf.qld.gov.au

References

[1] Skeist I., Handbook of adhesives. 2012: Springer Science & Business Media.

[2] Frihart, C.R. and C.G. Hunt, Adhesives with wood materials: bond formation and performance. Wood handbook: wood as an engineering material: chapter 10. Centennial ed. General technical report FPL; GTR-190. Madison, WI: US Dept. of Agriculture, Forest Service, Forest Products Laboratory, 2010: p. 10.1-10.24.

[3] Frihart, C.R., Wood adhesion and adhesives. 2005: CRC Press, Boca Raton, FL.

[4] Kumar, R. and A. Pizzi, Adhesives for Wood and Lignocellulosic Materials. 2019: John Wiley & Sons.

[5] Siau, J., Wood: Influence of moisture on physical properties. Dept. Wood Sci. and For. Prod., Virginia Polytech. Inst. State Univ., USA, 1995.

[6] Kamke, F.A. and J.N. Lee, Adhesive penetration in wood—a review. Wood and Fiber Science, 2007. 39(2): p. 205-220.

[7] Tarkow, H., W. Feist, and C. Southerland, Penetration versus molecular size—interaction of wood with polymeric materials. For Prod J, 1966. 16(10): p. 61-65.

[8] Bastani, A., S. Adamopoulos, and H. Militz, Gross adhesive penetration in furfurylated, N-methylol melamine-modified and heat-treated wood examined by fluorescence microscopy. European Journal of Wood and Wood Products, 2015. 73(5): p. 635-642.

[9] Qin, L., L. Lin, and F. Fu, Microstructural and micromechanical characterization of modified urea-formaldehyde resin penetration into wood. BioResources, 2016. 11(1): p. 182-194.

[10] Stephens, R.S. and N.P. Kutscha, Effect of resin molecular weight on bonding flakeboard. Wood and fiber science, 2007. 19(4): p. 353-361.

[11] Gardner, D.J., C.E. Frazier, and A. W. Christiansen, Characteristics of the wood adhesion bonding mechanism using hydroxymethyl resorcinol. Wood Adhesives, November 2-4, 2005; San Diego, California, USA. Madison, WI: Forest Products Society, 2006: ISBN: 1892529459: pages 93-97.

[12] Hass, P., et al., Adhesive penetration in beech wood: experiments. Wood science and technology, 2012. 46(1–3): p. 243-256.

[13] Rowell, R.M., Handbook of wood chemistry and wood composites. 2012: CRC press.

[14] Young, R., M. Fujita, and B. River, New approaches to wood bonding A base-activated lignin adhesive system. Wood Science and Technology, 1985. 19 (4): p. 363-381.

[15] Karlsson, O. and U. Westermark. Auto-adhesive bonding by oxidative treatment of wood. in International Symposium on Wood and Pulping Chemistry: 09/06/2003–12/06/2003. 2003.

[16] Widsten, P., et al., Effect of high defibration temperature on the properties of medium-density fiberboard (MDF) made from laccase-treated hardwood fibers. Journal of adhesion science and technology, 2003. 17(1): p. 67-78.

[17] Felby, C., J. Hassingboe, and M. Lund, Pilot-scale production of fiberboards made by laccase oxidized wood fibers: board properties and evidence for cross-linking of lignin. Enzyme and microbial technology, 2002. 31(6): p. 736-741.

[18] Kharazipour, A., A. Hüttermann, and H. Luedemann, Enzymatic activation of wood fibres as a means for the production of wood composites. Journal of Adhesion Science and Technology, 1997. 11(3): p. 419-427.

[19] Frazier, C.E., Isocyanate wood binders. Handbook of adhesive technology, 2003. 2.

[20] Lei, H., et al., Gluten protein adhesives for wood panels. Journal of Adhesion Science and Technology, 2010. 24(8–10): p. 1583-1596.

[21] Ferdosian, F., et al., Bio-based adhesives and evaluation for wood composites application. Polymers, 2017. 9(2): p. 70.

[22] Sarika, P., et al., Bio-Based Alternatives to Phenol and Formaldehyde for the Production of Resins. Polymers, 2020. 12(10): p. 2237.

[23] Geng, X. and K. Li, Investigation of wood adhesives from kraft lignin and polyethylenimine. Journal of adhesion science and technology, 2006. 20(8): p. 847-858.

[24] Chen, H.-H., et al., Quantitative analysis of nanoparticle internalization in mammalian cells by high resolution X-ray microscopy. Journal of Nanobiotechnology, 2011. 9(1): p. 1-15.

[25] Chen, X., B. Zheng, and H. Liu, Optical and digital microscopic imaging techniques and applications in pathology. Analytical Cellular Pathology, 2011. 34(1, 2): p. 5-18.

[26] Kamke, F.A. and J.N. Lee, Adhesive Penetration In Wood—A Review. Wood and Fiber Science, 2005. 39(2): p. 205-220.

[27] Hesse, L., et al., Structural and functional imaging of large and opaque plant specimens. Journal of experimental botany, 2019. 70(14): p. 3659-3678.

[28] Geyer, S.H., T.J. Mohun, and W.J. Weninger, Visualizing vertebrate embryos with episcopic 3D imaging techniques. The Scientific World Journal, 2009. 9: p. 1423-1437.

[29] Cinnamon, Y., et al., High-resolution episcopic microscopy enables three-dimensional visualization of plant morphology and development. Plant direct, 2019. 3(11).

[30] Damico, D.J., Advances in adhesives, adhesion science, and testing. 2005: ASTM International.

[31] Jermal G. Chandler, R.L.B., and Charles R. Frihart Examination of adhesive penetration in modified wood using fluorescence microscopy in the adhesive and sealant council, Inc. 2005, Forest Products Laboratory, USDA Forest Service, Madison.

[32] Mahrdt, E., et al., Light microscopic detection of UF adhesive in industrial particle board. Wood Sci Technol, 2015 (49): p. 517-526.

[33] Modzel, G., F.A. Kamke, and F.D. Carlo, Comparative analysis of a wood: adhesive bondline. Wood Sci Technol, 2011. 2011(45): p. 147-158.

[34] Edalat, H., et al., A new quantitative method for evaluation of adhesive penetration pattern in particulate wood-based composites: Elemental counting method. Wood science and technology, 2014. 48(4): p. 703-712.

[35] Ren, D. and C.E. Frazier, Structure–property behavior of moisture-cure polyurethane wood adhesives: Influence of hard segment content. International Journal of Adhesion and Adhesives, 2013. 45: p. 118-124.

[36] Gavrilovic-Grmusa, I., et al., Influence of the viscosity of UF resins

on the radial and tangential penetration into poplar wood and on the shear strength of adhesive joints. 2012.

[37] Hirschmüller, S., et al., Adhesive properties of laminated veneer lumber poles for use in temporary soil nailing: bonding performance of curved structures in a high-alkaline environment. Wood science and technology, 2018. 52(5): p. 1289-1311.

[38] Geyer, S.H. and W.J. Weninger, High-Resolution Episcopic Microscopy (HREM): Looking Back on 13 Years of Successful Generation of Digital Volume Data of Organic Material for 3D Visualisation and 3D Display. Applied Sciences, 2019. 9(18): p. 3826.

[39] Brodersen, C.R., Visualizing wood anatomy in three dimensions with high-resolution X-ray micro-tomography (μCT)–a review. IAWA Journal, 2013. 34(4): p. 408-424.

[40] Paris, J.L. and F.A. Kamke, Quantitative wood–adhesive penetration with X-ray computed tomography. International Journal of Adhesion and Adhesives, 2015. 61: p. 71-80.

[41] Kamke, F., et al., Methodology for micromechanical analysis of wood adhesive bonds using X-ray computed tomography and numerical modeling. Wood and Fiber Science, 2014. 46(1): p. 15-28.

[42] Paris, J.L., F.A. Kamke, and X. Xiao, X-ray computed tomography of wood-adhesive bondlines: attenuation and phase-contrast effects. Wood Science and Technology, 2015. 49(6): p. 1185-1208.

[43] Paris, J.L., et al., Phenol formaldehyde adhesives formulated for advanced X-ray imaging in wood-composite bondlines. Journal of materials science, 2014. 49(2): p. 580-591.

[44] Sanabria, S.J., et al., Assessment of glued timber integrity by limited-angle microfocus X-ray computed tomography. European Journal of Wood and Wood Products, 2011. 69(4): p. 605-617.

[45] Lasaygues, P., et al. Non-destructive diagnosis of the integrity of green wood using ultrasonic computed tomography. 2007.

[46] Madsen, S.J. and M.L.B.A.R.S. Majid Esfandyarpour, Observing Plasmon Damping Due to Adhesion Layers in Gold Nanostructures Using Electron Energy Loss Spectroscopy, in ACS Photonics. 2018.

[47] Hill, C.A., Wood modification: chemical, thermal and other processes. Vol. 5. 2007: John Wiley & Sons.

[48] Modzel, G.G.R., Computed X-ray analysis of wood-adhesive bonds. 2009: Oregon State University.

[49] Bucur, V., Techniques for high resolution imaging of wood structure: a review. Measurement Science and Technology, 2003. 14(12): p. R91.

[50] Lauterbur, P.C., Image formation by induced local interactions: examples employing nuclear magnetic resonance. 1973.

[51] Robert, E.M., et al., Visualization of the stem water content of two genera with secondary phloem produced by successive cambia through Magnetic Resonance Imaging (MRI). Journal of Plant Hydraulics, 2014. 1: p. e006-e006.

[52] Kuroda, K., et al., Magnetic resonance micro-imaging of xylem sap distribution and necrotic lesions in tree stems. IAWA Journal, 2006. 27(1): p. 3-17.

[53] Wu, D. and G. Busse, Lock-in thermography for nondestructive evaluation of materials. Revue générale de thermique, 1998. 37(8): p. 693-703.

[54] Konnerth, J., et al., Adhesive penetration of wood cell walls investigated byscanning thermal microscopy (SThM). De Gruyter, 2008. 62(1).

[55] Sanabria, S.J., et al., Air-coupled ultrasound as an accurate and reproducible method for bonding assessment of glued timber. Wood science and technology, 2011. 45(4): p. 645-659.

[56] Wang, X., et al., Effect of the penetration of isocyanates (pMDI) on the nanomechanics of wood cell wall evaluated by AFM-IR and nanoindentation (NI). Holzforschung, 2018. 72(4): p. 301-309.

[57] Mukherjee, T., et al., Chemically imaging the interaction of acetylated nanocrystalline cellulose (NCC) with a polylactic acid (PLA) polymer matrix. Cellulose, 2017. 2017(24): p. 1717-1729.

[58] Jakes, J.E., et al., X-ray methods to observe and quantify adhesive penetration into wood. Journal of materials science, 2019. 54(1): p. 705-718.

[59] Jakes, J.E., et al., Synchrotron-based X-ray fluorescence microscopy in conjunction with nanoindentation to study molecular-scale interactions of phenol–formaldehyde in wood cell walls. ACS applied materials & interfaces, 2015. 7(12): p. 6584-6589.

[60] Wang, X., et al., Multi-Scale Evaluation of the Effect of Phenol Formaldehyde Resin Impregnation on the Dimensional Stability and Mechanical Properties of Pinus Massoniana Lamb. Forests, 2019. 10(8): p. 646.

[61] Singh, A.P. and B.S. Dawson, Confocal microscope—A valuable tool for examining wood-coating interface. JCT research, 2004. 1(3): p. 235.

[62] Jeong, B. and B.-D. Park, Effect of molecular weight of urea–formaldehyde resins on their cure kinetics, interphase, penetration into wood, and adhesion in bonding wood. Wood Science and Technology, 2019. 53(3): p. 665-685.

[63] Altgen, D., et al., Analyzing the UF resin distribution in particleboards by confocal laser scanning microscopy. Composites Part A: Applied Science and Manufacturing, 2019. 125: p. 105529.

[64] Singh, A., et al., Light, confocal and scanning electron microscopy of wood-adhesive interface. Microsc Anal, 2008. 22(3): p. 5-8.

[65] Gierlinger, N. and M. Schwanninger, Chemical imaging of poplar wood cell walls by confocal Raman microscopy. Plant physiology, 2006. 140(4): p. 1246-1254.

Chapter 10

Size Effect of Core Strands on the Major Physical and Mechanical Properties of Oriented Strand Boards from Fast Growing Tropical Species

Wan Mohd Nazri Wan Abdul Rahman, Nur Sakinah Mohamed Tamat, Nor Yuziah Mohd Yunus and Jamaludin Kasim

Abstract

Oriented strand board (OSB) is generally used for sheathing in residential walls, floors, and roofs. Because of its low pricing and utilisation of tiny diameter logs from fast-growing trees and thinning logs as raw materials, OSB is anticipated to gain popularity. In chapter, board properties of OSB using smaller strand size of *Leucaena leucocephala* as core layer had been studied. Small strand size of S3 (length = 75 mm, width = 3.2 to 6.3 mm) was located in the middle layer of the board while bigger strand sizes of S1 (length = 75 mm, width = 12.7 to 19.0 mm) and S2 (length = 75 mm, width = 6.3 to 12.7 mm) were located at the face and back layers. Utilization of smaller strands (S3) in the middle layers may yield boards that have better physical and mechanical properties. Except for MOR in the minor axis, board density and resin content were shown to have a substantial impact on physical and mechanical properties. Except for MOR in the major axis, strand size had little affected on physical and mechanical properties. The effects of board density on mechanical properties were discovered to affect significantly different. With a positive correlation, board density had a significant effect on thickness swelling. Between S1+S3 and S2+S3 strand size, there is no significant effect on bending properties, internal bond strength and thickness swelling. The effect of resin content on bending properties revealed a significant difference of MOR in major axis, as well as MOE values in both major and minor axes. Even when the resin content was as low as 5%, all treatments of OSB passed the general requirement of general purpose OSB.

Keywords: oriented strand board, strand size, core layer, physical properties, mechanical properties

1. Introduction

Leucaena leucocephala is a fast-growing tree species and can be cultivated in Malaysia. This species has potential as an alternative resource for the Malaysian

IntechOpen

furniture and panel board industries in the coming years because of a shortage of the current frequently used raw material of rubberwood [1]. The use of a fast-growing species for wood composite production may offer some advantages, such as a shorter time required to activate production compared with other woody plants [2]. In Malaysia, the study on OSB from rubberwood was successfully carried out at the laboratory for the first time in 1996 by researchers from the Forest Research Institute Malaysia (FRIM), showed a good sign of using plantation species [3].

OSB is an engineered wood–based panel material in which long strands of wood are crushed together in layers and bonded together with a synthetic resin glue. Despite its strength, the effective performance of OSB begins immediately after it is created; OSB panels are light in weight and simple to handle and instal. Due to the coordination between the board and the adhesive to generate a strong, dimensionally stable panel, the panels also demonstrate outstanding fastener–holding, resist deflection, delamination, and warping. OSB is commonly used for flooring, roof and decking, and wall sheathing because it is uniquely acceptable for load–bearing applications in construction [4].

In order to efficiently used OSB products, it is important to understand the material and manufacturing variables that affect properties of the boards. As a result, this study will investigate the properties of OSB from *L. leucocephala* wood. This chapter describes the findings of the study conducted to evaluate the properties of the OSB produced from smaller strand size. The strand size of S3 as core material in the manufacture of OSB using eight-year-old *L. leucocephala* wood (**Figure 1**) strands was conducted in the study. Smaller strand size of S3 or fines is an additional component to this study. Any wood particles that pass through a 6.3 mm screen but remain at 3.2 mm screen are considered fines. They are a natural by-product of the stranding process and are used as filler material in the board. They add to the board's mass but are assumed to have no effect on the board's final volume. An optimal manufacturing process would use a balance of wood strands and fines to achieve the strength requirements while reducing the cost of raw materials. The S3 strand size (length = 75 mm, width = 3.2 to 6.3 mm) was found in the middle layer of the board, while the S1 (length = 75 mm, width = 12.7 to 19.0 mm) and S2 strand sizes (length = 75 mm, width = 6.3 to 12.7 mm) were found on the face and back layers (**Figure 2**). The focus of this study is to maximise recovery and to find the best combination of treatments. Perhaps, by utilisation of smaller strands (S3) in the middle layers may yield boards that have better physical and mechanical properties. By utilising S3; residue utilisation, only a small portion will be burned as waste or sent to a landfill. **Table 1** exhibit the properties of OSB according to board density, strand size and resin content.

Boards made from S1S3 with 9% resin content gave the best value of physical and mechanical properties in the fabrication of boards with a target board density of 700 kgm^{-3} (MOR major axis; 43.57 MPa, MOE major axis; 7377 MPa, MOR

Figure 1.
Leucaena leucocephala *logs.*

DOI: http://dx.doi.org/10.5772/intechopen.99953

Figure 2.
Strand size.

				Major axis		Minor axis			
Target density (kgm^{-3})	**strand size**	**Resin (%)**	**Actual density (kgm^{-3})**	**MOR (MP)**	**MOE (MPa)**	**MOR (MP)**	**MOE (MPa)**	**IB (MPa)**	**TS (%)**
700	S1S3	5	699	38.11	6455	15.94	1444	0.64	15.75
700	S1S3	7	695	40.70	6703	16.84	1485	0.79	14.11
700	S1S3	9	707	3.57	7377	16.26	1597	1.20	12.05
700	S2S3	5	697	36.00	6684	14.41	1444	0.57	20.43
700	S2S3	7	695	39.73	6929	14.29	1483	0.90	15.19
700	S2S3	9	703	41.05	7013	14.71	1569	0.98	12.30
800	S1S3	7	793	52.12	7324	18.22	1624	0.92	19.42
800	S1S3	9	797	56.24	8555	20.66	2024	2.06	12.93
800	S2S3	7	792	46.08	7663	18.67	1804	1.34	15.34
800	S2S3	9	799	51.03	7783	19.76	2003	1.92	15.07
	Min. req.			18 MPa	2500 MPa	9 MPa	1200 MPa	0.28 MPa	<25%
Type OSB/1: General purpose EN Standard				EN 310: 1993	EN 310: 1993	EN 310: 1993	EN 310: 1993	EN 319: 1993	EN 317: 1993

Note: MOR = Modulus of Rupture, MOE = Modulus of Elasticity, IB = Internal Bond, TS = Thickness Swelling.

Table 1.
Properties of OSB.

minor axis; 16.26 MPa and MOE minor axis; 1597 MPa, Internal Bond; 1.20 MPa and thickness swelling; 12.05%). However, boards produced from S2S3, with a density of 700 kgm^{-3} and 5% resin content, had the lowest mechanical properties in both major and minor axes, with the exception of MOE in major axis. The board also had the highest percentage of thickness swelling (20.43%), yet it still met the

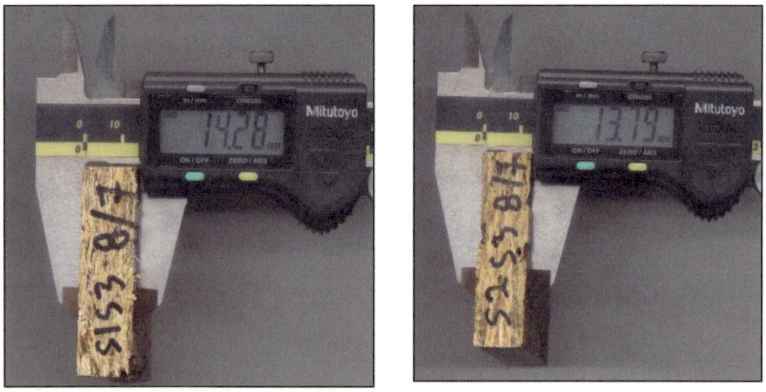

a) S1S3 b) S2S3

Figure 3.
Spring back phenomena.

EN standard's minimal requirements. The highest MOR values in the major axis were 2.4 times higher than the EN standard specification. Furthermore, MOR in the major axis performed 37.32% better than MOR in the minor axis. MOE values in the major axis perform tremendously with 3 times greater than minimum requirement. By raising the resin content from 5–9%, the internal bond values of OSB board improved significantly. Internal bond improvements were 1.7 to 1.8 times better than the internal bond with 5% resin content. The boards with 9% resin content had the best thickness swelling overall. Physical and mechanical properties were determined to comply with EN 300 standard [5] for general purpose (type OSB/1) even at the lowest resin content of 5%, according to the study's findings.

In the major and minor axes, the boards with a target board density of 800 kgm^{-3} and a combination of strand sizes S1S3 at 9% resin content had the highest MOR and MOE values. In the major axis, the MOR and MOE values are 56.24 MPa and 8555 MPa, respectively. MOR had the highest minor axis value of 20.66 MPa and MOE had the highest minor axis value of 2024 MPa. The board also had the highest internal bond values, at 2.06 MPa, which was 7 times higher than the EN standard's specification. The board's thickness swelling was also lower, at 12.93%. Boards constructed from target density of 800 kgm^{-3} with strand size of S1S3 and 7% resin content, on the other hand, had the lowest values for mechanical properties in major and minor axes, with the exception of MOR in major axis. This board likewise has the highest percentage of thickness swelling (19.42%), yet it still meets the maximum requirement of 25% and below. The findings of this study revealed that boards with a resin content of 7% exhibit spring back phenomena, resulting in a thickness increase of more than 12 mm (**Figure 3**). The 'spring back' effect was reduced by reducing the board density and increasing the resin amount. As a result of the performance at 7% resin content, the fabrication of boards with 5% resin content was not conducted. Because the board specimens blow in the core layer, no data from the treatment of S1S3 and S2S3 at board density of 800kgm^{-3} with 5% resin content was acquired. In general, boards with strand sizes of S1S3 perform better than boards with strand sizes of S2S3, and all boards fulfil the EN 300 minimum criteria.

2. Statistical significance

The analysis of variance (ANOVA) and correlation analyses are presented for discussion. The ANOVA of the effect of board density, strand size and resin content

		Major axis		Minor axis			
SOV	Df	MOR (MPa)	MOE (MPa)	MOR (MPa)	MOE (MPa)	IB (MPa)	TS (%)
Density (D)	1	51.90**	40.57**	22.97**	21.03**	115.48**	6.76*
Strand Size (S)	1	5.66*	0.0 ns	1.91 ns	0.08 ns	1.58 ns	0.69 ns
Resin Content (RC)	1	5.80*	17.51**	0.84 ns	6.19*	74.26**	13.25**
D x S	1	3.36 ns	0.35 ns	1.29 ns	0.24 ns	6.90*	2.38 ns
D x RC	1	0.06 ns	0.10 ns	1.03 ns	1.37 ns	21.72**	0.24 ns
S x RC	1	0.35 ns	10.01**	0.02 ns	0.38 ns	7.67**	1.33 ns
D x S x RC	1	0.41 ns	0.01 ns	0.55 ns	0.19 ns	0.14 ns	0.99 ns

Note: SOV = Source of variance, Df = Degree of freedom, ns = not significant at p > 0.05.
** significant at p < 0.05.*
***highly significant at p < 0.01.*

Table 2.
Summary of the ANOVA on the properties of S1S3 and S2S3 board.

and their interactions on the OSB properties are shown in **Table 2**. Mechanical properties and thickness swelling were both affected by density, according to statistical analysis. Except for MOR in the major axis, strand size did not affect mechanical properties and thickness swelling. Except for MOR in the minor axis, all mechanical properties and thickness swelling were found to be affected significantly by resin content. Except for internal bond, there is no significant difference in the interaction between board density and strand size. The interaction between board density and resin content showed a similar pattern. Despite this, there was no significant difference in the interaction between strand size and resin, with the exception of MOE in the major axis and internal bond strength. In the interaction of all main factors, no significant effect was found.

3. Effects of board density

The effect of board density on mechanical properties are given in **Figure 4**. The mechanical properties showed higher value with increase in board density. The mechanical properties are significantly different at $p < 0.05$, according to the t-test comparison. With increasing board density, the values of MOR and MOE in major and minor axes of each board increased almost linearly. The MOR in the major axis was approximately 2.6 times that of the minor axis. In MOE, a similar trend in density effects was seen. According to [6] the values of MOR and MOE determined in the parallel direction are approximately 40–50% greater than the values determined in the perpendicular direction. MOR major axis, MOE major axis, MOR minor axis, MOE minor axis, and internal bond increased with increased board density ($r = 0.68^{**}$, 0.59^{**}, 0.56^{**}, 0.50^{**}, and 0.61^{**}, respectively) according to correlation analysis (**Table 3**). According to [7], board density was one of the most critical elements influencing particleboard mechanical properties. In wood composites, board density is the most important component for board structure to sustain load, so increasing density means increasing material resistance to outside forces and achieving strong composite unit contacts for improved bonding strength. Study by [8] investigated three different board densities (0.53, 0.66 and 0.78 gcm^{-3}) and found that mechanical properties increased as panel specific gravity increased for douglas-fir flakeboards.

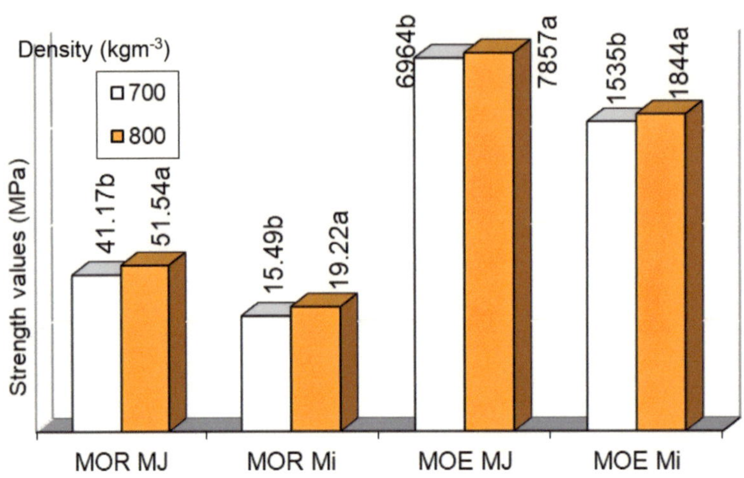

Figure 4.
Effect of board density on bending properties.

	Major axis		Minor axis			
Variable	**MOR (MPa)**	**MOE (MPa)**	**MOR (MPa)**	**MOE (MPa)**	**IB (MPa)**	**TS (%)**
Density	0.68**	0.59**	0.56**	0.50**	0.61**	0.30*
Strand Size	−0.18 ns	−0.04 ns	−0.18 ns	0.06 ns	−0.01 ns	−0.02 ns
Resin	0.27*	0.40**	0.09 ns	0.29*	0.52**	−0.40**

Note: ns = no significant correlation.
**Correlation is significant at the 0.05 level.*
***Correlation is significant at the 0.01 level.*

Table 3.
Correlation coefficients of the effect of Strand size, resin content and density on board properties for S1S3 and S2S3.

The internal bond follows a similar pattern to bending properties and has a significant effect (**Figure 5**). A positive correlation between board density and internal bond strength (r = 0.61**) was revealed in the correlation analysis (**Table 3**). Because the core layer's strand size (S3) is smaller, it's easier to increase the core density, which leads to more intimate contact between strands and hence stronger internal bonding. Higher amount of materials in a board density of 800 kgm^{-3} resulted in a stiffer board. Board density has been studied as one of the elements that affect the internal bonding of strand composite panels, according to [9].

Figure 5 also revealed a significant difference in board thickness swelling. Thickness swelling has a positive correlation with board density (r = 0.30*) according to correlation analysis (**Table 3**). A probable reason for this behaviour is that when immersed in water, the greater density zone and finer particles in the core rebound to swell more. When both were constructed with the same wood supply, greater density boards had more compression set than lower density boards, but expanded more following immersion in water, according to [10]. According to [11] in their study found a good linear correlation between density and thickness swelling values. According to [12], OSB thickness swelling not only causes aesthetic problems in some applications, but it is also linked to a reduction in the material's strength and stiffness. However, decreasing board density could increase board dimensional stability [8].

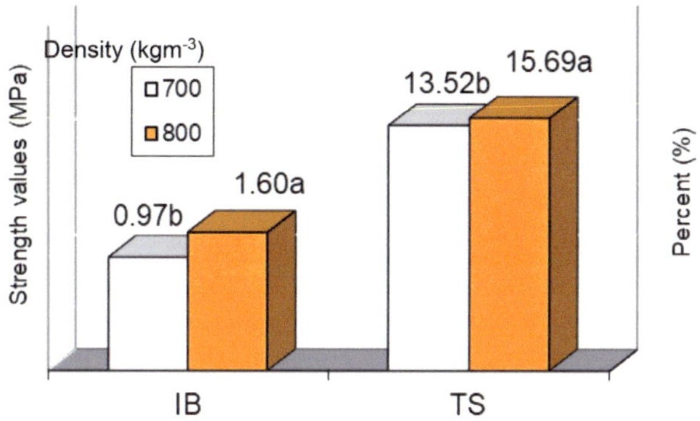

Figure 5.
Effects of board density on internal bond and thickness swelling.

4. Effects of strand size

Figure 6 shows the effect of strand size on bending properties. There is no significant difference on bending properties between combination of strand size of S1S3 and S2S3. Study by [13] discovered that the length of flakes between 5.0 and 7.5 cm had no significant effect on MOR and MOE. This is because smaller strand sizes provide denser structures between strands in the core part, whereas larger strand sizes of S1 and S2 act as stress supporters on the board's face and back. The lower value of bending properties could be due to the distribution of large wood strands. Large strands that should improve bending strength may be heavily disoriented, according to [14], whereas smaller strands that have less influence on bending strength may be well orientated. Due to additional space or void in the core layer and adjacent layers of face and back (**Figure 7**), boards with S1S3 strands have a higher spring back. Boards with S2S3 strands, on the other hand, had less spring back because the S2 strand size was smaller, resulting in less void between the S2 and S3 strands. According to [15], the presence and distribution of macro-voids are governed by the bigger and longer strand, which could be filled by smaller strand size to close the gap. Increased proportions of smaller strands had a negative impact on the mechanical properties of the board in general. The mechanical properties had an insignificant ($r = -0.18$ ns, -0.01 ns, -0.18 ns, and 0.06 ns) correlation

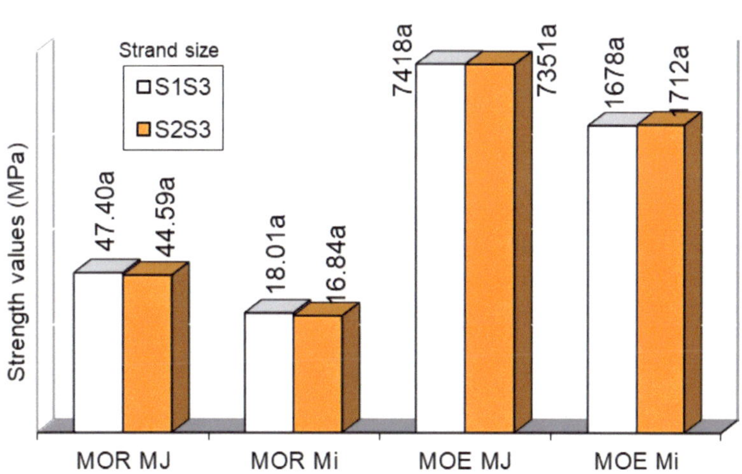

Figure 6.
Effects of Strand size on bending properties.

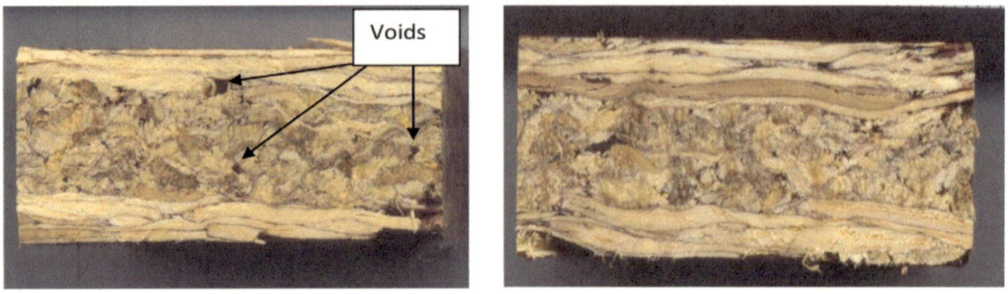

Figure 7.
Voids within the board.

with strand size, according to the correlation analysis (**Table 3**). Study by [16] reported that the MOR and MOE of the boards were not affected greatly with increase in strand size.

One of the important mechanical properties is internal bond strength. The majority of failure in the internal bond test on boards manufactured from S1S3 and S2S3 strand size happened in the wood part, not at the adhesive or glue line (**Figure 8**). The strand size used in the study had no effect on the internal bond strength of the boards, according to the statistical analysis (**Figure 9**). The internal bond strength showed an insignificant ($r = -0.01$ ns) correlation with decreasing strand size, according to the correlation analysis (**Table 3**). Internal bond strength did not increase for boards with 30% fines content, and even reduced as fines content was increased to 45%, according to [17]. This is likely owing to poor bonding, resulting in less resin covering on the surfaces of wood fines. According to [18], wood elements are deposited randomly over the horizontal plane in a more or less layer-by-layer way throughout the hand-forming process, and voids can result between any adjacent elements in any layer. The number of these voids decreases in a unit area within one layer as flake size rises, whereas void size increases.

The strand size showed no significant effect on thickness swelling readings after a 24-hour soak at a 95% significant level (**Figure 9**). The results demonstrate that S2S3 boards with a combination of strand sizes outperformed S1S3 boards with a

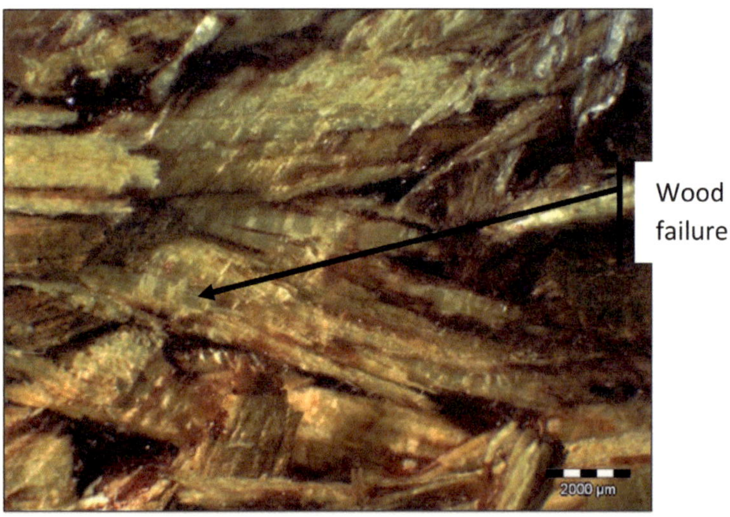

Figure 8.
Failure occurred in the wood portion.

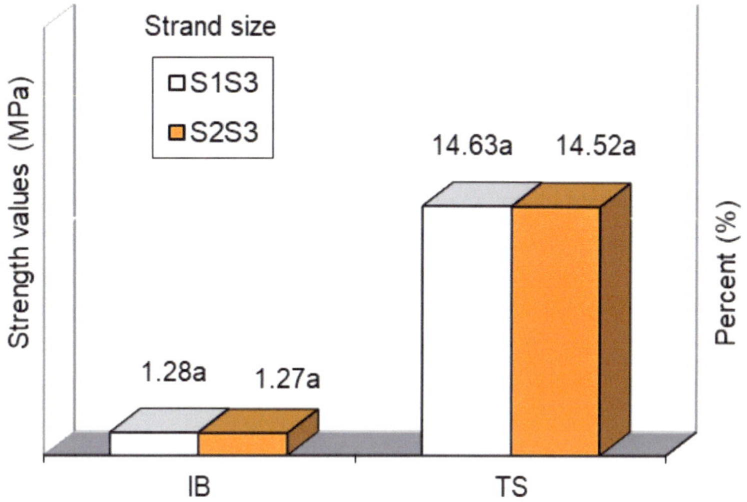

Figure 9.
Effects of Strand size on internal bond and thickness swelling.

variety of strand sizes. The statistical analysis revealed that bigger strand size of S1 and S2 strand sizes in the face and back layers had an excellent role in preventing the board from swelling further, with no significant different between boards manufactured with S1S3 and S2S3 strands. The use of strand sizes of S1, S2, and S3 to maximise recovery without jeopardising board attributes was demonstrated in this treatment technique. The thickness swelling values showed insignificant ($r = -0.02$ ns) correlation with strand size, according to the correlation analysis (**Table 3**). The thickness swelling of smaller strand size at the core center of the OSB was higher than that in the surface region, according to [19].

5. Effects of resin content

The influence of resin content on bending properties is shown in **Figure 10**. MOR values in major axis and MOE values in both major and minor axes shows significant difference according to the statistical analysis. With increased resin content, however, MOR exhibits no significant difference in minor axis. The bending properties of MOR in major axis, MOE in major axis, MOR in minor axis, and MOE in minor axis all show a positive association with increased of resin content ($r = 0.27^*$, $r = 0.40^{**}$, $r = 0.09$ ns, and $r = 0.29^*$, respectively) according to the correlation analysis (**Table 3**). Insufficient resin is available to bond smaller strand sizes, resulting in lower performance by 7% resin content, especially for S2 and S3. Because smaller strand sizes absorb more resin and create uneven resin distribution among smaller strands during the blending process, higher board density and the presence of smaller strand sizes resulted in less bonding contact. Smaller strand sizes or particles resulted in insufficient resin coverage on the surface, resulting in poor strength [20–23]. Nonetheless, in terms of rupture property, the boards with 7% resin content are comparable to boards with 9% resin content. As a result, it might be possible to make the boards with a lower resin level. According to [24], high-cost resin adhesives must be applied at appropriate rates to ensure both the product's exceptional qualities and its economic viability. When considering reduced application rates, the composite's improved performance and quality must be maintained.

Figure 11 indicates that increasing the resin content shows a significant difference on internal bond strength. A positive correlation between resin content and

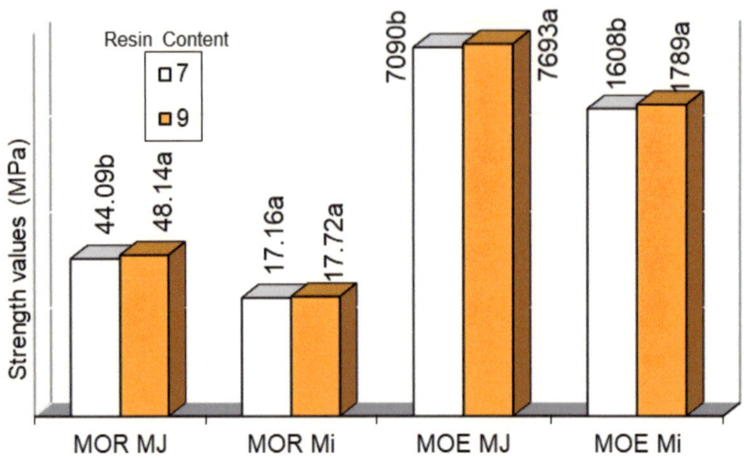

Figure 10.
Effects of resin content on bending properties.

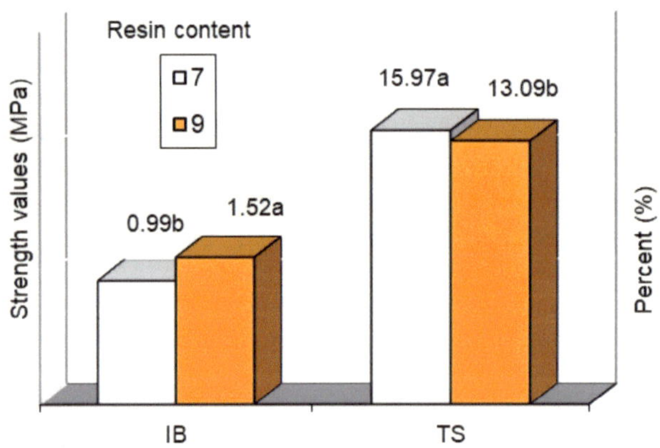

Figure 11.
Effects of resin content on internal bond and thickness swelling.

internal bond strength (r = 0.52**) was also discovered in the correlation analysis (**Table 3**), indicating that increasing resin content improved bonding performance. According to the findings, a high resin content is required for successful bonding, particularly for small strand geometry (S3; core layer). When insufficient resin content is employed, high levels of wood fines in the core layer might contribute to lower internal bond strength, according to [17]. According to a study by [20], finer particles in high-density surface zones required far less resin than coarser particles in lower-density core sections. This finding highlights the need of good bond formation, which may be achieved with the right board forming parameters.

After 24 hours of immersion in water, boards with 9% resin content show significantly lower thickness swelling (**Figure 11**). The thickness swelling was reduced by increasing the resin content. When the resin content was increased to 9%, the dimensional stability of the boards was significantly improved (19%). The correlation analysis (**Table 3**) also demonstrated that resin content and thickness swelling have a negative relationship (r = −0.40**). When the board is immersed in water, it becomes dimensionally unstable, and increasing the resin content improves the board's resistance to water exposure. According to [25] normally, board decreases in thickness-swell with the increase of resin content.

6. Conclusions

Based on the findings of this study, OSB made from smaller strands size of *Leucaena leucocephala* as core layers demonstrated good physical and mechanical properties. All boards produced were complied with the standard requirement values MOE, MOR, internal bond strength and thickness swelling in accordance to EN 300 Standard for general purpose (type OSB/1) which are used as the sheathing materials for walls, floors, roofs, general construction and renovation work.

Acknowledgements

Acknowledgement given to Center of Wood Industries of Universiti Teknologi MARA Cawangan Pahang for providing raw materials and use of facilities for the experimental work.

Author details

Wan Mohd Nazri Wan Abdul Rahman, Nur Sakinah Mohamed Tamat*,
Nor Yuziah Mohd Yunus and Jamaludin Kasim
Faculty of Applied Sciences, Department of Wood Industry, Universiti Teknologi MARA Cawangan Pahang, Bandar Tun Abdul Razak, Pahang, Malaysia

*Address all correspondence to: sakinah292@yahoo.com

References

[1] Wan Mohd Nazri WAR, Hilmi AR, Nor Yuziah MY, Yani J, Mohd Nizar K.: *Leucaena leucocephala*: A fast-growing tree for the Malaysian particleboard industry. Bioresources. 2020;15: 7433-7442.

[2] Oggiano, Angelini LG, Cappelletto, P.: Pulping and paper properties of some fibre crops. Ind. Crops Prod. 1997; 7:59-67.

[3] Mohd Nor MY, Horie, H.: The manufacture of oriented strand boards from Rubberwood. Journal of Tropical Forest Products. 3;1:43-50.

[4] Abeeb OF. Properties of oriented strand board (OSB) made from oil palm trunk strand bonded with phenol formaldehyde resin for structural and aerospace application [thesis]. Selangor: Universiti Putra Malaysia; 2019.

[5] European Standard (EN) 300 (2006) Oriented Strand Boards (OSB) – Definitions, classification and specifications. European Committee for Standardization; Brussels, Belgium.

[6] Zhou D: A study of oriented structural board made from hybrid poplar: Physical and mechanical properties of OSB. Holz als Roh- und Werkstoff. 1990;48:293-296.

[7] Muszynski Z, McNatt JD.: Investigations on the use of spruce bark in the manufacture of particleboard in Poland. Forest Products Journal. 1984;34:28-35.

[8] Vital BR, Wilson JB, Kanarek PH.: Parameters affecting dimensional stability of flakeboard and particleboard. Forest Products Journal. 1980;30:23-29.

[9] Kamke FA, Casey LJ.: Fundamentals of flakeboard manufacture: internal-mat conditions. Forest Products Journal. 1988;38:38-44.

[10] Halligan AF.: A review of thickness swelling in particleboard. Wood Science and Technology. 1970;4: 301-312.

[11] Cai Z, Wu Q, Han G, Lee JN.: Tensile and thickness swelling properties of strands from Southern hardwoods and Southern pine: Effect of hot-pressing and resin application. Forest Products Journal. 2007;57:36-41.

[12] Van Houts JH, Wang S, Shi H, Pan H, Kabalka GW.: Moisture movement and thickness swelling in oriented strandboard, Part 1. Analysis using nuclear magnetic resonance microimaging. Wood Sci Technol. 2004;38:617-628.

[13] Hoover W, Michael LO, Hunt R, Lattanzi C, Bateman JH, Youngquist JA.: Modeling mechanical properties of single-layer, aligned, mixed-hardwood strand panels. Forest Products Journal. 1992;42:12-18.

[14] Nishimura T, Ansell MP, Ando N.: The relationship between the arrangement of wood strands at the surface of OSB and the modulus of rupture determine by image analysis. Wood Science and Technology. 2001;35:555-562.

[15] Sugimori M, Lam F.: Macro-void distribution analysis in strand-based wood composites using an X-ray computer tomography technique. J Wood Sci, 1999;45:254-257.

[16] Murakami K, Ueda H, Mathuda E, Zhang M, Kawasaki T, Kawai S.: Manufacture and properties of three-layered OSB. In: Abstracts of the 47th annual meeting of the JWRS; 1997; Kochi. Japan: p. 276

[17] Han G, Wu Q.: Oriented strand board from small-diameter Southern Pine trees. In: Proceeding of the 8th Pacific Rim Bio-Based Composites

Symposium; 2006; Kuala Lumpur: p. 110-120

[18] Steiner RP, Xu W.: Influence of flake characteristics on horizontal density distribution of flakeboard. Forest Products Journal. 1995;45:61-66.

[19] Xu W, Winistorfer PM, Moschler WW.: A procedure to determine water absorption distribution in wood composite panels. Wood and Fiber Science, Society of Wood Science and Technology. 1996;28:286-294.

[20] Maloney TM.: Resin distribution in layered particleboard. Forest Products Journal. 1970;20:43-52.

[21] Hill MD, Wilson JB.: Particleboard strength as affected by unequal resin distribution on different particle fractions. Forest Products Journal. 1978;28:44-48.

[22] Dunky M.: Particle size distribution and glue resin consumption: how to spare costs. In: Proceedings of the 2nd panel products symposium; 1998; Llandudno. Wales. UK

[23] Bekhta P, Hiziroglu S.: Theoretical approach on specific surface area of wood particles. Forest Products Journal. 2002;52:72-76.

[24] Ashraf MA, Abdalla AMA, Sekino N.: Veneer strand flanged I-beam with MDF or particleboard as web material II: Effect of resin type, application rate, strand dimension, and pressing time on the basic properties. J. Wood Sci. 2004;50:400-406.

[25] Brochmann J, Edwardson C, Shmulsky R.: Influence of resin type and flakeboard thickness on properties of OSB. Forest Products Journal. 2002;54:51-55.

Chapter 11

Effect of Materials Content on Dimensional Stability, Nano Roughness and Interfaced Morphology for Virgin or Recycled Polypropylene Based Wood Composites

Shamsul Haq

Abstract

The compositions of mango wood-polypropylene composites (WPCs) are formulated, with different compositions of virgin polypropylene (PP) or recycled PP, mango wood waste and a coupling agent. The compositions are fabricated via melt extrusion compounding pursued by injection hot molding. The tests of the prepared compositions are carried out for, water absorption, thickness swelling, surface properties at a nano-scale and interfaced morphology. Comparative study of WPCs composition has done on respective properties. All processing variable conditions are constant for different compositions. The recycled PP based wood composites with or without the coupling agent possessed superior properties in comparison to virgin PP based composites. FESEM images show that coupled composite is having the better bonding strength and smoothness along with a higher dimensional stability in comparison to none coupled composite. Future endeavor should be focused on optimizing the composition of reinforcement wood and recycled plastics matrix according to intended application. The quality of WPCs can also be improved with the co-ordination of latest development in technology and processing technique relevant to them. WPCs study supports "turning waste into something useful". This provides the mileage in price performance ratio and also the product's environmental footprints to be adjusted to suit the products application.

Keywords: polymer wood composites, interface morphology, nanoscale roughness, water absorption, sustainability, wood-based products

1. Introduction

Wooden materials are one of the firstly natural construction materials, as utilized by the human being. It is used in most of the cases. But it is applied in maximum cases, according to the traditional approach methodologies. This is showed by past experience and utilization tactics of wood. As a view of the above, in the field of engineering perspective and application, wood and the wood-based material is one of the low tech or second tier materials. But the fact is that the exploitation of wooden based material uninterrupted to steadily expand in construction and structural field. To reduce the consumption of neat wood or tree plant, many of the approaches are developed to create wood based material for sustained in different engineering applications. One of the approaches is based on the combination wood material with other material. Wood composites are based on a component of wood and thermoplastic are gaining an importance in the current century. Wood plastic composites are one of the newly developed materials for the different field of application. It is still a very new material in comparisons to the long history of natural lumber as a construction material and plastics.

The wooden article is readily biodegradable and renewable. On the other hand, thermoplastic has an option for the recyclability. For wood plastic composites (WPCs), thermoplastic and wood combinations have exerted enough pressure recently to develop the WPC products, due to their favorable properties, processing characteristics and eco-friendly advantages [1, 2]. Polypropylene (PP) belongs to the polyolefin family and most widely used plastics today. It is also light in weight. In WPCs, thermoplastics and wood flours are acting as a matrix and reinforcement materials respectively [3–5]. The filler wood flour has bearded the load across its interface with the matrix and procures the WPCs against break-down under the applied loads [6, 7]. Most of the physical and mechanical properties are broadly investigated and listed in the previous research works in this field [8–14]. Hygroscopic thickness swelling of composites was estimated for 50:50 wood plastic ratios [14]. Influence of artificial UV conditioning on the roughness of the composite was estimated for one composition of wood and PP [15]. Roughness characteristics of such like WPCs should be investigated for finding out its bonding and finishing properties [16, 17]. Also, WPCs are used in many interior and exterior applications. WPCs are gaining increasing potential in many construction applications including decking, fencing, (**Figure 1**) siding, paneling, furniture industries, windows, doors, siding, sign board, barrier, cabinet etc. In general, WPCs may be used in almost all application where wood or plastic already are utilized. Also, improving the performance of WPCs materials, additives may be used for the different function such like as Lubricant, pigments, colorant, an anti-microbial agent, antioxidants, UV stabilizer, fire retardants agent, coupling agent, stabilizers etc. Surface quality and dimensional stability of wood-based panels are one of the most

Figure 1.
Fencing and deck board made from WPCs.

important characteristics. It is stimulated by the manufacturing processes or their adhesive strength and also performs an important preface in the determination of the attributed product for an ultimate application. Therefore minute observation of dimensional stability, surface roughness and interfaced morphology of these WPCs materials is desirable. The objective of this work is to study various compositions of WPCs, fabricated by mango (scientific name-Mangifera indica) wood waste, virgin (v) PP, recycled (r) PP and maleic anhydride polypropylene (MAPP). Fabricated compositions of WPCs are compared and correlated in terms of water absorption, thickness swelling, nano-roughness and interface morphology. Also, estimated the effect of mango wood flour, vPP and rPP quantities are on the abovementioned properties. Effect of coupling agent on mango wood-PP composites is also studied in rPP matrix based compositions. Surface roughness parameters, roughness profile and 3D surface variations have analyzed the roughness of compositions.

The developments in the quality life of peoples are provided through better uses and exercises on available unutilized or utilized resources. From ancient times humans have fabricated wood and wood-based products as the most available materials in the surrounding. However, led to environmental consciousnesses are in exclusively on recycling of utilized materials, degeneration in forest resources and warming in global condition. These factors have forced the peoples to make efforts to form bio-composites from waste wood resources or non-wood residues [11]. Similarly, the plastic waste which is a major part of municipal solid waste (MSW) should also be utilized. In past research work, efforts have been made to utilize these waste thermoplastics and wood in sequence to decrease the eco-system imbalance and production of the neat plastics and wood. The important factor for promoting the recycling of a waste plastic product is to achieve an appropriate composition with the intended application. Worldwide environmental degradation motivated the researchers to focus on fabrication of the materials using several recycled and waste stream [2, 18]. Also, WPCs product can be re-molded under the application of heat many times after the useful service [19]. Correlation between nano-roughness and dimensional stability of above like mango-PP composites are not admitted as much as other engineered materials. So this study concerned for the surface, stability and interface between mango wood flour and polypropylene.

2. Materials, fabrication, and experimentation

Wood flour is biodegradable and cheap in cost, with low density and abrasion in comparison to synthetic reinforcement [20]. Mango wood sawdust waste is procured from the local Kanpur (U.P. India) saw mill. It is heated at 100°C for 2 hours in order to remove the moisture content up to less than 3%. The Sieve analysis is performed with 50 and 30 mesh sieve and the corresponding size of the wood particles are found between 297 μm < D <595 μm. Thermal degradation temperature of the wood flour is 200° C [21, 22], hence there is no mass loss at the processing and molding temperature in this study. Lubricant Zinc stearate ($C36H70O4Zn$) helps to improve the flow of melt in mold and also worked as a mold releasing agent. Polypropylene is obtained from the RIL, India. Recycled and virgin PP is used in this study. PP granules are heated in an oven at 50°C for 8 hours before melt compounding with reinforcing flours. It is a homopolymer, h10 ma grades with a density of 0.9 g/cm^3. MFI of PP is 11.9 ± 0.29 g/10 minute according to ASTM D1238. MAPP worked as a compatibilizer or coupling agent to WPCs. It is procured from Defense Material & Stores Research and Development Establishment Kanpur. MAPP (with Mw – 9, 100 by GPC, Mn – 3, 800 by GPC and maleated anhydride

1.5 wt.%) is used in this study. Twin screw micro conical co-rotating extruder is used for the melt compounding of the flours of wood with v or r PP and MAPP. Haake mini CTW5 lab performed the melt compounding and produced WPCs blend. Composites are formulated by the weight ratio in percentage. The different compositions and their codes are given in **Table 1** [23]. The twin screw with speed of 65 revolutions per minutes in the co-rotating extruder, compounding in the house is for two minutes and temperature range (start to end) in the chamber is 175 to 180°C. Same machine parameters are applied for the compounding of both wood flour with rPP or vPP and MAPP. The conical co-rotating twin screw, intercepted the agglomeration from start to end at the various stages of compounding due to its design characteristic. Haake mini jet micro injection molding is used to fabricate WPC samples as per ASTM standard by the molding of WPC blend or pellets. The temperature of the cylinder is about 188 to 190°C and of a mold is 55°C. The pressure of 450 bars is maintained for 7 sec. at the cylinder formfull filling the cavity of the mold.

Dimensional-stability (water absorption and thickness swelling) test is followed the ASTMD570–98 [24]. In this test, samples are immersed in a container of distilled water for 2 hours and then for 24 hours, at a temperature about 23 ± 1°C. All surface waters are wiped off with a dry cloth before all measurements. Surface topography measured by the attractive forces between the probe and the sample in a non-contact mode by AFM (Agilent Technologies, USA) at nano-level. AFM provides three-dimensional image and highresolution information. AFM cantilevers are fabricated with an integrated sharp tip of silicon nitride. The Agilent AFM, raster scanning in the X and Y axes, tip in the Z axis and produces the 2D profile and 3D surface structure of the composites. Surface roughness parameters data followed by ISO 4287 standards. It scans about 2 × 2 μm surface area of the sample. AFM produces the result with high accuracy and calibration but used with a limitation of little region scanning. AFM has collected a roughness parameter (Ra, Rz, and Rmax), 3D surface structure and roughness profile at a nanoscale [25]. The fracture interface of the samples arem characterized by FESEM, DMSRDE Kanpur. The scanning data between the interfaces of cracked composites are analyzed at 500× magnifications.

S. No.	Code for samples	WPCs composition	Polypropylene form	Polypropylene content	Mango flour content	MAPP (CA) content
Mango wood flour – polypropylene composites						
1.	I	vPP100	virgin	100	0	0
2.	II	rPP100	recycle	100	0	0
3.	III	vPP70W30	virgin	70	30	0
4.	IV	rPP70W30	recycle	70	30	0
5.	V	vPP60W40	virgin	60	40	0
6.	VI	rPP60W40	recycle	60	40	0
7.	VII	vPP50W50	virgin	50	50	0
8.	VIII	rPP50W50	recycle	50	50	0
9.	IX	rPP47W50CA3	recycle	47	50	3
10.	X	rPP45W50CA5	recycle	45	50	5

Table 1.
WPCs formulations of sample given below in table (percentage by weight).

3. Results and observation

3.1 Dimensional stability

Water absorption and thickness swelling of the compositions is examined for both rPP and vPP based WPCs. The dimensional stability test is also done in coupled composites. Data from the test is presented in **Tables 2** and **3**. It is found that the water absorption, increased with the higher quantity of mango flour in the WPCs composition. This trend is exactly right for both 2 and 24 hours water immersion for all samples. It is also calculated that the water absorption for 2 hours immersion increases from 0.02 to 6.51%, and after 24 hours water immersion, the water absorption rises from 0.056 to 9.08% dependent on the wood-plastic composite compositions. Sample V has a value of 3.72% and 7.46% and VI has a value of 3.38% and 7.04% for 2 hours and 24 hours respectively. It is established that rPP have lesser water absorption in comparisons to those fabricate of vPP for same plastic to a wood composition. The same behavior is pursued by sample III, IV, VII and VIII. Dispersion of wood flour and interface bonding, due to the existence of oxidized impurities are improved for rPP based composites [26]. The stability is increased by the presence of oxidized impurities in rPP by the development of ester bond with wood flour. So that micro gaps or crack at across the interface between the reinforcing flour and rPP matrix are reduced. It has provided a lower surface wetting for the same. That is why the dimensional stability of rPP based WPCs are higher than vPP based. It is pointed that the MAPP can significantly reduce the water absorption. This trend is followed by the samples IX and X for rPP based composites. In addition, the sample VIII exhibited more water absorption than sample VI. This shows that the composites made from higher plastic content have less water absorption.

It is also true for vPP based composites. This is due to the hydrophobic properties of polypropylene because it is covered the functional polar hydroxyls groups of wood flours and made chemically in active.

The composition of WPCs with high water absorption displayed higher thickness swelling for all sample compositions. The result of thickness swelling test is

Composite sample code	Water absorption (%)	
	(2 hours)	(24 hours)
I	0.024 (0.026)	0.078 (0.039)
II	0.02 (0.0245)	0.056 (0.0167)
III	2.88 (0.335)	5.94 (0.152)
IV	2.48 (0.192)	5.46 (0.114)
V	3.72 (0.165)	7.46 (0.207)
VI	3.38 (0.134)	7.04 (0.294)
VII	6.51 (0.222)	9.08 (0.228)
VIII	6.13 (0.12)	8.52 (0.13)
IX	3.38 (0.148)	6.6 (0.424)
X	2.7 (0.158)	5.63 (0.403)

The numbers in parenthesis are standard deviation value.

Table 2.
Result from water absorption test of the WPCs.

Composite sample code	Thickness swelling (%)	
	(2 hours)	(24 hours)
I	0.0125 (0.017)	0.0625 (0.022)
II	0.0063 (0.014)	0.05 (0.028)
III	2.07 (0.198)	5.16 (0.247)
IV	1.69 (0.175)	4.7 (0.214)
V	2.85 (0.164)	6.68 (0.376)
VI	2.57 (0.194)	6.26 (0.408)
VII	5.46 (0.199)	8.44 (0.247)
VIII	4.78 (0.219)	8.01 (0.215)
IX	2.49 (0.218)	6.1 (0.313)
X	2.08 (0.186)	5.19 (0.204)

The numbers in parenthesis are standard deviation value.

Table 3.
Result from thickness swelling test of the WPCs.

presented in **Table 3**. The existence of other polar groups and hydroxyl in several constituent elements of wood produces the poor compatibility between hydrophobic polypropylene and hydrophilic wood. So that, these wood-plastic based composites have the ability to take up water under humid environment due to the availability of various hydroxyl groups present for interaction via hydrogen bonding with water molecules. MAPP treated composition shows, lower thickness swell than the composition with the absence of MAPP at the same wood content. The anhydride moieties of MAPP enters into with hydrophilic groups of mango wood through an esterification reaction [27]. It forms monoester and diester bond between a wood particle and thermoplastic, consequential improve the compatibility. This is reduced water absorption sites in MAPP coupled compositions. Above properties of rPP based composites are improved by adding 3 to 5% coupling agent (MAPP).

3.2 Surface characterization

The roughness properties of composition at a nano-scale are estimated by the Average roughness (Ra), maximum roughness (Rmax) and mean peak-to-valley height (Rz) using AFM. The roughness parameters of mean values of are given in **Table 4**. The sample V and VII have reinforcing flour quantity 40 and 50 percentage respectively. The Ra values of sample V and VII are 1.47 nm and 2.38 nm respectively. It is clear from the above Ra values that with the decrease of plastic quantity from 60–50%, Ra value increases for the WPCs samples. A similar trend is also followed by rPP based wood composition VI and VIII. Roughnesses of the WPCs composition are significantly higher by the increase in the wood flour quantity for all composition of WPCs. On the other side, the Ra value of the sample VII and VIII are found to be 2.38 nm and 1.63 nm respectively. From these Ra values of sample VII and VIII, it can be concluded that vPP based composite has a more roughness in comparison to rPP based WPCs. The same trend is followed by the sample V and VI. It is because of the well dispersed wood flour in the rPP matrix [28]. Resultantly, rPP based WPCs obtained a good interfacial bonding between wood reinforcement and rPP matrices. Due to which, Ra value is lower for rPP

Composites code	Roughness parameters (mean value) (nm)		
	Ra	Rz	Rmax
V	1.47	4.97	9.95
VI	0.952	2.42	4.15
VII	2.38	9.12	16.4
VIII	1.63	3.92	8.18
X	0.74	2.59	3.64

Table 4.
Mean value of roughness parameters.

based composition. The roughness parameters (Rz and Rmax) also pursued the same trend of Ra result for samples.

The influence of coupling agent upon the roughness characteristics of WPCs composition VIII is also analyzed. With the addition of 5% MAPP, a roughness of the surface is decreased. Sample X has surface parameters, Ra, Rz, and Rmax values 0.74 nm, 2.59 nm, and 3.64 nm while VIII has values of 1.63 nm, 3.92 nm, and 8.18 nm. This proves that sample X has a smoother surface than sample VIII. This result shows the effect of MAPP and confirms the improvement of the interface bonding between the flours of wood and the rPP matrices for the sample [27, 29].

Particular roughness profile and the 3D surface variation of WPCs sample V, VI, VII, VIII and X also supported the result for the roughness parameters. 3D roughness variation is shown in **Figure 2**. Also, it must be noted that profile of a tip probe upon the surface does evaluate the variation in the roughness profile. 3D roughness image shows an actual 3D morphology of the zigzag structure for the respective composition at a nano-scale.

3.3 FESEM analysis

Mango wood flour of a weight ratio of 50% with vPP and rPP matrix is shown by FESEM morphology in **Figure 3** (VII) and (VIII) respectively. Outcomes of this morphology that interface bonding between vPP mango wood flour and rPP mango wood flour are showing a distinct lumen, cavity and gaps for sample VII and VIII. The pattern of **Figure 3** (VII) and (VIII) shows that the wood flours are so inadequately connected to the PP matrix. So it is recommended that the interface across the reinforcing flour and PP matrices are poorer due to the bad compatibility and dispersion. Mango wood flour shows a better and uniform dispersion with rPP matrix (sample VIII) in comparison to vPP matrix (sample VII). Because the mango reinforcement is in filling the small cavity and micro-gaps in the rPP matrix. Mechanical interlocking is improved due to better-dispersed flours filler in the rPP matrices [28]. So it possesses good entanglement in connections at the interface for composition VIII.

Image of coupled composite sample X in **Figure 3** (X) displayed no apparent cavity between reinforcement and matrix materials which indicates, the strong bond at the interface. It is observed from the morphology of sample X that there is a much little magnitude of ragged matrices, informing that the composite X is stable than those VII and VIII composites which have no coupling agent. The meshes of MAPP (coupling agent) are developed at the interface and fill the gaps. There is a lowest aggregation in the MAPP coupled composites. While in non-coupled composite, wood flour set out the aggregates for sample VIII. The presence of the -OH groups

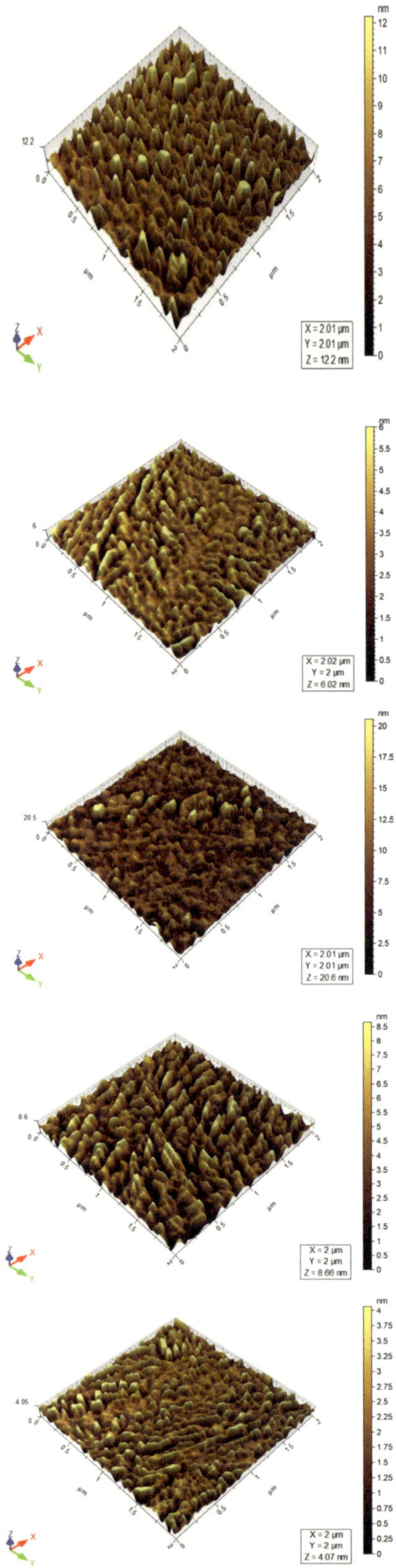

Figure 2.
3D surface structures for sample V, VI, VII, VIII, and X respectively.

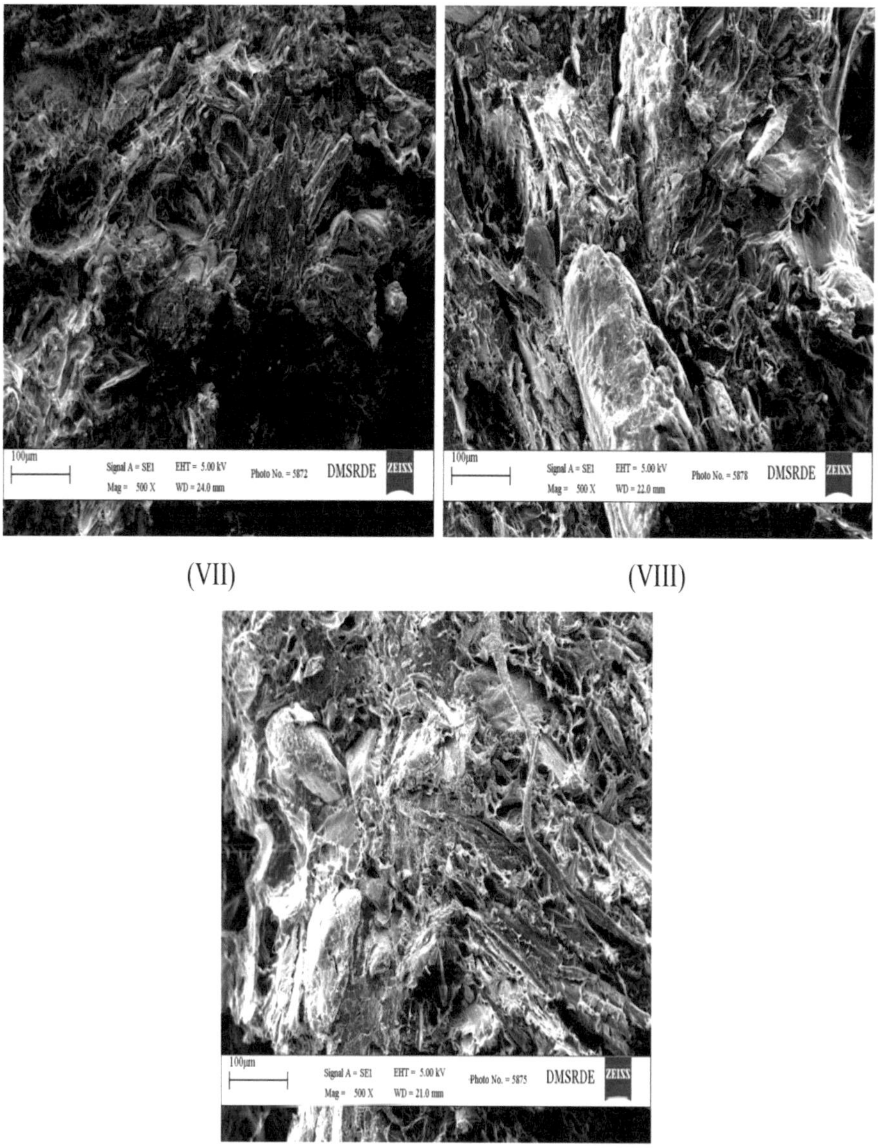

(VII) (VIII) (X)

Figure 3.
FESEM morphology of interface for sample VII, VIII and X respectively.

in the reinforcement flour, create hydrogen bonds and tempted to each other [19]. The addition of MAPP has certainly altered the chemical character of the wood flour surface [27] and improved crystallinity of the structure [23]. It is favorable for densification and reduces the roughness of WPCs [16, 26]. This enhances in turned the dimensional stability as well as surface smoothness.

4. Conclusions

Water absorption of the composites based on rPP is lower for those based on vPP in all set. Also, rPP based compositions show better smoothness in comparison to vPP based compositions. The coupling agent is improving the surface characteristics, highest in the same class and decreasing the water absorption in rPP based composites. This is reduced water absorption [(6.13% to 3.38% and 8.52% to 6.6%) & (6.13% to 2.7% and 8.53% to 5.63%)] for 2 h and 24 h and with addition of 3 and 5% MAPP respectively. The performance of surface roughness and water absorption of composites improved with reducing plastic content. The rPP based compositions displayed a better outcome in the same class either with MAPP or without MAPP, observed from the experimentations. FESEM interface images morphology is also validating the results. The Present research work demonstrates that all WPCs compositions which have lower water absorption or better dimensional stability shows higher smoothness and dense interface morphology and vice versa. The intended application of WPCs is suitable, where dimensional stability and smoothness condition merits are important especially in humid condition. Formulation contents and raw materials characteristics directly affect the quality of above properties. The present study will be further helpful for Wood based products development and construction applications.

Acknowledgements

I thanks to Shri S. B. Yadaw, Scientist "G", DMSRDE Kanpur, for his constant encouragement and suggestion for this study. MNNIT Allahabad and DMSRDE Kanpur resources are utilized for the grant no. is 2349/med/f-pf/2015.

Author details

Shamsul Haq
Department of Mechanical Engineering, Madhyanchal Professional University, Bhopal, India

*Address all correspondence to: shamsulhaq198615@gmail.com

References

[1] Nanthaya Kengkhetkit, Taweechai Amornsakchai. A new approach to "greening" plastic composites using pineapple leaf waste for performance and cost effectiveness. Mater. Des. 2014; 55:292-299.

[2] N Dwivedi, AP Khare, S Haq, Wood Plastic Composite: Emerging Material for an Environmental Safety—A Review Advances in Clean Energy Technologies 2021, 85-99.

[3] Sebastien Migneault, Ahmed Koubaa, Fouad Erchiqui, Abdelkader Chaala, Karl Englund, Michael P. Wolcott. Effects of processing method and fiber size on the structure and properties of wood–plastic composites. Composites: Part A 2009; 40 80-85.

[4] Tran LQN, Fuentes CA, Dupont Gillain C, Van Vuure AW, Verpoest I. Understanding the interfacial compatibility and adhesion of natural coir fibre thermoplastic composites. Compos Sci Technol 2013; (81):23-30.

[5] Jitendra Bhaskar, S. Haq and S. B. Yadaw "evaluation and testing of mechanical properties of wood plastic composite". Journal of Thermoplastic Composite Materials. 25 (4) (2012) 391-401.

[6] Bledzki A K, Faruk O. Wood fibre reinforced polypropylene composites: Effect of fibre geometry and coupling agent on physico-mechanical properties. Applied Composite Materials 2003; 10:365-379.

[7] Pelaez-Samaniego MR, Yadama V, Lowell E, Amidon TE, Chaffee TL. Hot water extracted wood fiber for the production of wood plastic composites (WPCs). Holzforschung. 2013; 67(2):193-200.

[8] Akiko Isa, Jun Minamino, Yoichi Kojima, Shigchiko Suzuki, Hirokazu Ito, RieMakise, Masaki Okamoto and Takashi Endo. The influence of dry-milled wood flour on the physical properties of wood flour/polypropylene composites. Journal of Wood Chemistry and Technology 2016; 36:2, 105-113.

[9] Diene Ndiaye, Laurent M. Matuana, Sandrine Morlat-Therias, Loıc Vidal, Adams Tidjani, Jean-Luc Gardette. Thermal and mechanical properties of polypropylene/wood-flour composites. Journal of Applied Polymer Science, 2011; Vol. 119:3321-3328.

[10] D.G. Dikobe, A.S. Luyt. Morphology and properties of polypropylene/ethylene vinyl acetate copolymer/wood powder blend composites. Express Polym Lett. 2009; 3 (3):190-199.

[11] A. Mohammed, A. Binhussain, Maher M. El-Tonsy. Palm leave and plastic waste wood composite for out-door structures. Constr. Build. Mater. 2013; 47:1431-1435.

[12] Shamsul Haq, Rajeev Srivastava (2015) investigations of wood thermoplastic composites for sustainable product applications. Discovery, 2015, 39(176), 8-14.

[13] Denni Kurniawan, Byung Sun Kim, Ho Yong Lee and Joong Yeon Lim. Effects of repetitive processing, wood content, and coupling agent on the mechanical, thermal, and water absorption properties of wood/polypropylene green composite Journal of Adhesion Science and Technology, 2013; Vol. 27, No. 12, 1301-1312.

[14] Najafi SK, Kiaeifar A, Tajvidi M, Hamidinia E. Hygroscopic thickness swelling rate of composites from sawdust and recycled plastics. Wood Sci Technol 2008; 42:161-168.

[15] Yao Peng, Ru Liu, Jinzhen Cao. Characterization of surface chemistry

and crystallization behavior of polypropylene composites reinforced with wood flour, cellulose, and lignin during accelerated weathering. Applied Surface Science 2015; 332: 253-259.

[16] Nadir Ayrilmis, Jan T. Benthien, Heiko Thoemen. Effects of formulation variables on surface properties of wood plastic composites. Composites: Part B 2012; 43: 325-331.

[17] Arusombuti S, Ayrilmis N. Surface characteristics and overlaying properties of flatpressed wood plastic composites. Eur J Wood Prod. 2011; 69: 375-382.

[18] Ashori A. Wood–plastic composites as promising green-composites for automotive industries!. Bioresour Technol 2008; 99:4661-4667.

[19] Shu-Kai Yeh, Rakesh K. Gupta. Improved wood–plastic composites through better processing. Composites: Part A 2008; 39: 1694-1699.

[20] Li X, Tabil LG, Panigrahi S. Chemical treatments of natural fiber for use in natural fiber reinforced composites: A review. J Polym Environ. 2007; 15 (1):25-33.

[21] Manuel Raul Pelaez-Samaniego, Vikram Yadama, Eini Lowell and Raul Espinoza Herrera (2013) A review of wood thermal pretreatments to improve wood composite properties. Wood Science and Technology 2013;(47):1285-1319.

[22] Christoph Burgstaller (2007) Processing of thermal sensitive materials – a case study for wood plastic composites. Monatshefte fur Chemie 2007;(138): 343-346.

[23] Shamsul Haq, Rajeev Srivastava (2016) Wood polypropylene (PP) composites manufactured by mango wood waste with virgin or recycled PP: Mechanical, morphology, Melt Flow Index and Crystalline Behaviour. J Polym Environ DOI 10.1007/s10924-016-0845-9.

[24] ASTM International (ASTM D 570-98) standard test method for water absorption of plastics. In annual book of ASTM standards 2005; vol. 08(01)35e7. West Conshohocken, PA.

[25] E.S. Gadelmawla, M.M. Koura, T.M.A. Maksoud, I.M. Elewa, H.H. Soliman roughness parameters. Journal of Materials Processing Technology 2002; 123, Issue 1, 133-145.

[26] Shamsul Haq, Rajeev Srivastava. Measuring the influence of materials composition on nano scale roughness for wood plastic composites by AFM. Measurement 2016; 91:541-547.

[27] M. Kazayawoko, J. J. Balatinecz, L. M. Matuana. Surface modification and adhesion mechanisms in woodfiber-polypropylene composites. Journal of Materials Science 1999; 34:6189-6199.

[28] Pei-Yu Kuo, Song-Yung Wang, Jin-Hau Chen, Huei-Chin Hsueh, Ming-Jer Tsai. Effects of material compositions on the mechanical properties of wood–plastic composites manufactured by injection molding. Materials and Design 2009; 30 (9):3489-3496.

[29] Qiu WL, Endo T, Hirotsu T. Interfacial interactions of a novel mechano chemical composite of cellulose with maleated polypropylene. J Appl Polym Sci 2004; 94(3):1326.

Chapter 12

Wood Plastic Composites: Their Properties and Applications

Kaimeng Xu, Guanben Du and Siqun Wang

Abstract

Wood plastic composites (WPCs) is one of crucial and potential engineering wood products that has been extensively employed in the fields of landscape, transportation, municipal engineering and building construction. It has gradually been used to replace the conventional wood-based composites. This chapter aims to introduce the properties and development of WPCs and illustrate how defects in their mechanical properties, biological and aging resistance, and flame retardance affect their global development. Herein, the effects on the biological durability of WPCs against algae, mold, fungi, and termites made with various wood species with different chemical extractive compositions, the natural weathering performance of WPCs and the mechanisms of protection against ultraviolet light and moisture, the effectiveness and mechanism of reinforcement of WPCs by novel alloy modification of linear and aromatic polyamides are reviewed. Additionally, the flame retardance properties, common testing methods as well as the performances of novel flame retardants for WPCs, are comparatively described. Lastly, the limitations and prospects of WPCs in future construction applications are also discussed.

Keywords: wood plastic composites, natural durability, reinforcement, flame retardance

1. Introduction

Under the global background of shortage crisis of forestry and fossil oil resources, it is estimated that there is an expectation for the application of natural fiber-reinforced composites from 12% in 2010 to 18% and 25% by 2020 and 2030, respectively [1]. With the fast-growing demands of engineering wood products for various applications, wood plastic composites (commonly abbreviated as WPC), as the most important one of natural fiber-reinforced composites, mainly produced by the thermoplastic polymers (polyethylene (PE), polypropylene (PP), polyvinyl chloride (PVC) and polystyrene (PS)) and the biomass particles and fibers from forestry and agricultural wastes (wood, bamboo, straw, stalk, husk and bast), which have been well considered as a crucial and novel candidate for residential and industrial constructions [2, 3]. WPC with a unique combination of advantages from wood and plastic has been extensively employed in landscaping facility, transportation, packaging, municipal engineering as well as the inner-outer decoration and structure of building [4–6], as seen in **Figures 1–3**. In recent years, WPC begins to partially replace some conventional and novel engineered wood products like wood

Figure 1.
The applications of WPC in various building construction.

Figure 2.
The applications of WPC in the indoor or outdoor ceilings and sidings.

veneer (WV), particle board (PB), medium-density fiber board (MDF), plywood (PW), oriented strand board (OSB) and laminated veneer lumber (LVL) [7].

In the initial developing period of WPC, manufacturers claim that there are many advantages for WPC including high strength and modulus, excellent dimensional stability, outstanding biological (fungi and termite) resistance and low maintenance requirement [8–10]. Hydrophilic biomass particles or fibers can be effectively encapsulated by hydrophobic thermoplastic resin to provide well protection against biological invasion. Although it had to be admitted that there was a superior weathering and biological durability for WPC than solid wood, natural aging and biological degradation of WPC by common microorganisms (mold, algae and decaying fungi) and harmful termites indeed can be observed with the damage

Figure 3.
The applications of WPC in the furniture, kitchen, and bathroom products.

of its external layer and interfacial bonding under the appropriate circumstance like mild ambient temperature, high humidity and oxygen content as well as strong ultraviolet (UV) radiation [11]. Meanwhile, the insufficient mechanical properties and poor flame retardance due to weak interfacial compatibility, high fire sensitivity of polyolefin resins and the strength limitation of conventional resins were gradually exposed as the service time of WPC extended, which may lead to potential failures and serious accidents in engineering construction.

Recently, a growing interest for WPC used in the field of construction becomes more popular with pervasive consciousness of environmental protection and sustainability. To further improve the mechanical properties, durability and flame retardance, selection of wood species with various extractives, alloy modification with linear and aromatic polyamides as well as the addition of flame retardants with single and synergistic effectiveness have been reported. This provides a significant technological support for WPC's future use in the construction.

2. Natural durability of WPC

2.1 Biological durability of WPC

The ever-expanding commercial market for WPC products in exterior construction engineering applications has placed more spotlights on improving the biological durability and lifetime of WPCs. The wood component of WPCs is considered as common fiber fillers with similar properties, including renewability, low density, high specific strength, and hydrophilicity. In fact, the species of wood plays a critical role in the comprehensive properties of WPCs, in addition to other factors, such as the plastic type, particle size, additives, and processing technology. Termites and other microorganisms, such as mold and fungi, can decrease the esthetic quality and mechanical strength of WPCs through discoloration and degradation by chemically altering the structures of lignin, hemicellulose, and cellulose fibers in the wood [12, 13]. However, algae colonization only affects the outer appearance of the WPC instead of the inner interfacial bonding [14]. To promote the quality and prolong the service time of the entire WPC structure, more attention needs to be paid to improving the biological durability during the engineering of the composites in the future. The chemical composition of wood extractives, such as tannins, fatty acids, aldehydes, ketones, sugars, and starches, have been reported to have both a negative and positive effect on the susceptibility of WPC to microbial invasion [11, 15, 16].

Despite of continuing emergence of new publications in this field, there are few reports on how the identity of the wood species affects the biological durability of WPCs. Different wood species contain varying compositions of volatile chemical components. The high toxicities of some natural wood extractives have been shown to exhibit a positive effect on the resistance of WPCs to microorganisms. The susceptibility of mold in PVC-based WPCs containing maple wood was found to be higher than the analogous WPCs made with pine wood [17]. *Gloeophyllum trabeum* (brown rot) demonstrated significantly higher mass losses in PE-based WPCs compared to *Trametes versicolor* (white rot). The ranking for the white rot testing fungus for the WPCs made by five wood species was Douglas-fir > black locust > white oak > ponderosa pine > poplar, while the ranking for the brown rot testing fungus was black locust > white oak > ponderosa pine > Douglas-fir > poplar [18]. PP-based WPCs made from maple or oak were more susceptible to fungus than the PP-based WPCs made with pine [19]. PE-based WPCs containing *Parthenium argentatum*, *Parthenium incanum*, and *Parthenium tomentosum* fibers displayed better termite resistance compared to the WPCs containing pine fiber, with mass losses of 7%, 6%, and 5%, respectively. PVC-based WPCs filled with different wood species exhibited varying resistance levels to four common biological organisms, including algae, fungi, mold, and termites, as shown in **Figures 4–7**. The trend for the algal resistance was: *Liquidambar formosana* > *Cunninghamia lanceolata* and *Melaleuca leucadendra* > *Eucalyptus grandis* × *Eucalyptus urophylla* and *Pinus massoniana*. The trend for the fungal resistance to *Coriolus versioolor* was: *Cunninghamia lanceolata* > *Pinus massoniana* > *Melaleuca leucadendra* > *Liquidambar formosana* > *Eucalyptus grandis* × *Eucalyptus urophylla*; for the fungal species *Poria vaporaria*, the trend was: *Cunninghamia lanceolata* > *Eucalyptus grandis* × *Eucalyptus urophylla*, *Melaleuca leucadendra*, *Pinus massoniana* > *Liquidambar formosana*. The trend for the mold resistance was: *Cunninghamia lanceolata* and *Melaleuca leucadendra* > *Eucalyptus grandis* × *Eucalyptus urophylla* > *Pinus massoniana* > *Liquidambar formosana*. The trend for the termite resistance was: *Cunninghamia lanceolata* > *Melaleuca leucadendra* > *Eucalyptus grandis* × *Eucalyptus urophylla* > *Liquidambar formosana* > *Pinus massoniana* [11, 14, 20].

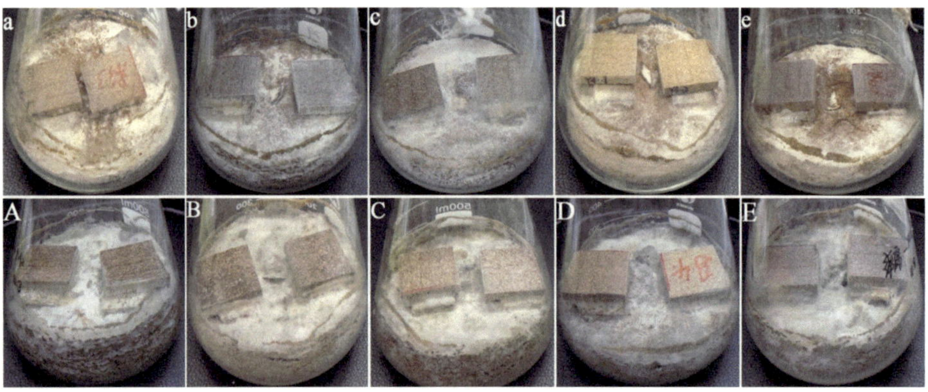

Figure 4.
The decay resistance to Coriolus versioolor *(in lowercase) and* Poria vaporaria *(in capital) testing of PVC-based WPC filled with different wood species: (a)* Cunninghamia lanceolata, *(b)* Melaleuca leucadendra, *(c)* Eucalyptus grandis × E. urophylla, *(d)* Pinus massoniana, *(e)* Liquidambar formosana *[20].*

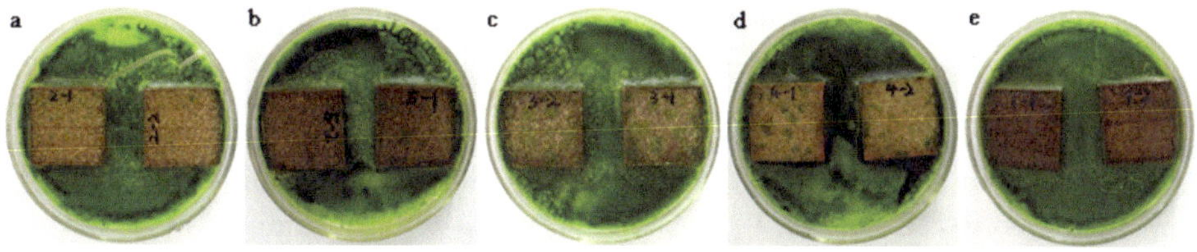

Figure 5.
The algae resistance of PVC-based WPC filled with different wood species: (a) Cunninghamia lanceolata, *(b)* Melaleuca leucadendra, *(c)* Eucalyptus grandis × E. urophylla, *(d)* Pinus massoniana, *(e)* Liquidambar formosana *[14].*

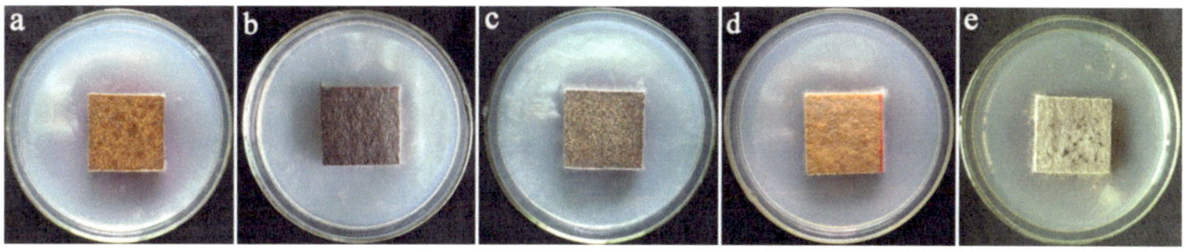

Figure 6.
The mound resistance of PVC-based WPC filled with different wood species: (a) Cunninghamia lanceolata, *(b)* Melaleuca leucadendra, *(c)* Eucalyptus grandis × E. urophylla, *(d) Pinus massoniana, (e)* Liquidambar formosana *[11].*

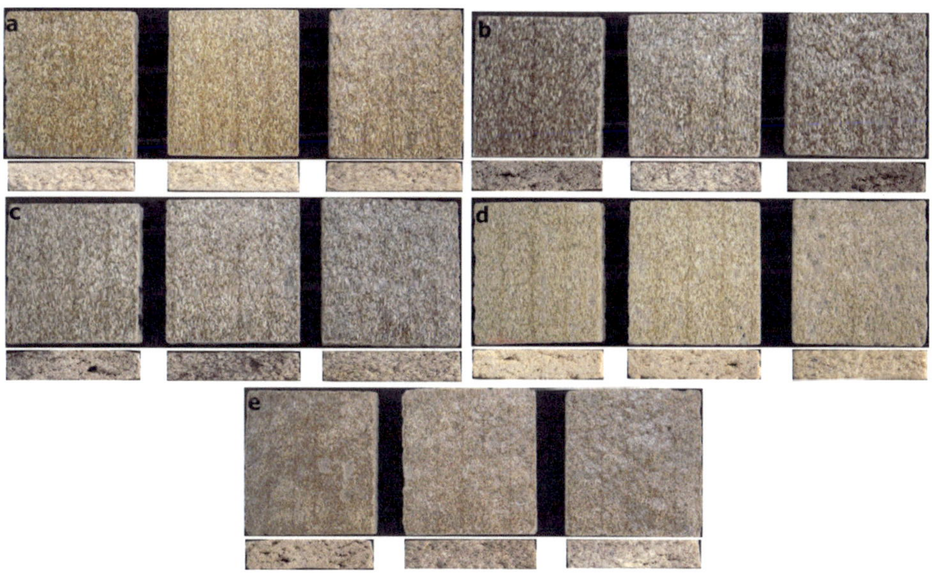

Figure 7.
The termite resistance of PVC-based WPC filled with different wood species: (a) Cunninghamia lanceolata, *(b)* Melaleuca leucadendra, *(c)* Eucalyptus grandis × E. urophylla, *(d)* Pinus massoniana, *(e)* Liquidambar formosana *[11].*

Several functional chemical extractives naturally present in wood varieties can either facilitate or inhibit the biodeterioration of WPCs. For example, 8-propoxycedrane, cedrol, α-cedrene, and β-cedrene in *Cunninghamia lanceolata*; 2,3-dihydro-2,2-dimethyl-3,7-benzofurandiol, 3-demethylcolchicine, and squalene in *Melaleuca leucadendra*; 2,3-dihydro-2,2-dimethyl-3,7-benzofurandiol and stigmast-4-en-3-one in *Eucalyptus grandis* × *Eucalyptus urophylla* were conducive to promoting the biological durability of WPCs. On the contrary, longifolene, caryophyllene, and α-pinene in *Pinus massoniana*; and 4-hydroxy-3,5-dimethoxybenzaldehyde, 3,5-dimethoxy-4-hydroxycinnamaldehyde, and cinnamyl cinnamate in *Liquidambar formosana* induced biodegradation [11]. Additionally, comparison to medium-density fiberboard, WPCs were found to demonstrate a better resistance to termites due to the presence of antifeedants, and there was a higher resistance level for PVC-based WPCs than for PE-based WPC made with the same wood particles. The biological durability of the WPC was potentially improved due to the intrinsic antimicrobial properties of PVC as well as the addition of inorganic fillers into the composites.

Similar results for the PP-based WPCs filled with poplar, MOSO bamboo, ramin, pine, gum, cedar, and rubberwood were reported, as observed in **Figure 8**. There was a poor mold resistance of the WPC made with MOSO bamboo due to the presence of 5-hydroxymethylfurfural in the bamboo species. The WPC made with ramin wood showed high resistance to various mold species, which could be attributed to the presence of γ-sitosterol, cis-9, cis-12-octadecadienoic acid, benzyl alcohol, and 8-propoxycedrane. However, 4-hydroxy-3-methoxycinnamaldehyde and dehydroabietic acid were determined to be the crucial chemical components that mediated the high resistance of WPCs made by pine. The resistance of WPCs made with gum fillers to molds can be explained by the synergistic effect of γ-sitosterol, stigmast-4-en-3-one, and cedrol. The relatively low fungal resistance of WPCs filled with rubberwood was attributed to the relatively high content of 5-hydroxymethylfurfural. In addition, hinokitiol, cedrol, α-cedrene, thymoquinone, totarol, β-cedrene, ferruginol, hinokiol, and benzyl alcohol were found to be vital chemical compounds in WPCs filled with red cedar for manifesting a high resistance of the composite to molds. The relatively high contents of cedrol, stigmast-4-en-3-one, and 8-propoxy-cedrane in Chinese fir provided the corresponding WPC with significant resistance to mold growth. Meanwhile, the relatively low contents of α-Cedrene, β-cedrene, α-cadinol, and γ-sitosterol were beneficial for mitigating the biodegradation of WPC [21].

2.2 Natural weathering of WPC

The exposure of WPCs to ultraviolet light and moisture represents two significant natural weathering sources of WPCs in outdoor applications, which affects

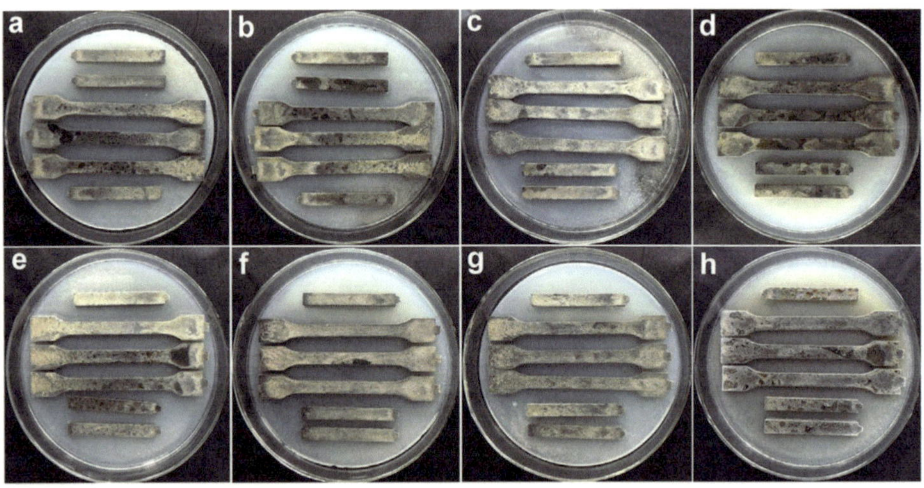

Figure 8.
Mold colonization of PP-based WPC with various wood species: (a) poplar, (b) MOSO-bamboo (c) Chinese fir (d) Ramin, (e) pine, (f) gum, (g) cedar, (h) rubberwood [21].

not only the visual appearance of the WPCs but also their mechanical properties by manifesting deleterious effects such as discoloration, furrows, and separated voids [22]. The photographs of unweathered and weathered WPCs are shown in **Figure 9**.

Chromophores naturally present in polymers and biomass generally accelerate the absorption of UV light, leading to the photodegradation of WPCs. Specifically, the photo-yellowing discoloration of weathered biomass can be attributed to the breakdown of lignin into smaller chromophoric monomers, including quinones, carboxylic acids, and hydroperoxyl radicals [23]. Polymer photochemical degradation is mediated by chain scission reactions, typically through Norrish Type I and II reactions (**Figure 10**). Norrish Type I reactions lead to the formation of free radicals by the cleavage of the ketones and aldehydes. On the other hand, carbonyl and vinylic groups are generated by Norrish Type II ring-opening reactions. Consequently, these photochemical degradation reactions lead to a reduction in the degree of chain entanglements in the amorphous zone, which results in the

Figure 9.
Photos of HDPE-based WPCs before and after natural weathering [23].

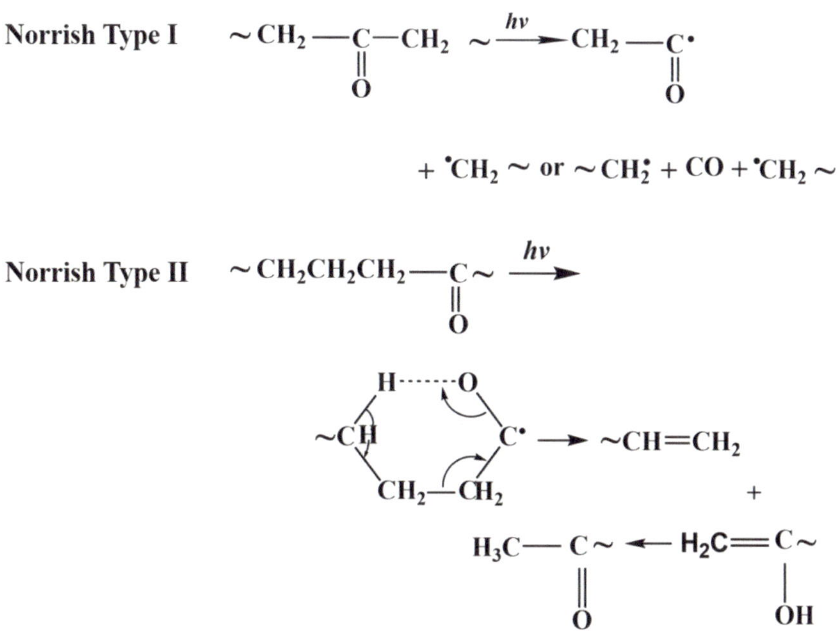

Figure 10.
The reaction mechanism for photochemical degradation of polymer in WPC [24].

crystallization of smaller polymeric units and individual monomers, reducing the mechanical properties of the WPCs [24]. Some chemical additives, such as photostabilizers, antioxidants, and pigments have been incorporated into WPC matrices to retard the discoloration of the composites.

Moisture in the air, especially during the rainy season, can be easily absorbed by the hydrophilic biomass particles in WPCs, during which the biomass particles begin to swell and create micro-voids or cracks in the polymer matrix, resulting in decreases in both the interfacial adhesion between biomass particles and plastics and the stress transfer efficiency from the matrix to the fibers (**Figure 11**). Therefore, the elastic modulus and flexural strength of the WPCs gradually decrease. In addition, the individual components of biomass (i.e., cellulose, hemicellulose, lignin, and extractives) have varying susceptibilities to photodegradation. In fact, UV light, warm temperatures, and aerial moisture typically have a synergistic effect on the natural weathering of WPCs. When the biomass particles in WPC undergo photodegradation, the surface chemical groups trend to become more hydrophilic due to the formation of water-soluble products from the breakdown of lignin. The surface wettability increases and appears more sensitive to moisture levels. It was also found that the anti-weathering discoloration of WPC was reinforced after the biomass was extracted or delignified [25]. The surface roughness of PP-based WPCs before and after weathering was measured by atomic force microscopy [26], which indicated that there was an increase in the roughness value after the photodegradation of PP, leading to the presence of some microcracks in the WPC and the loss of biomass particles, and a decrease in the interfacial bonding on the weathered surface.

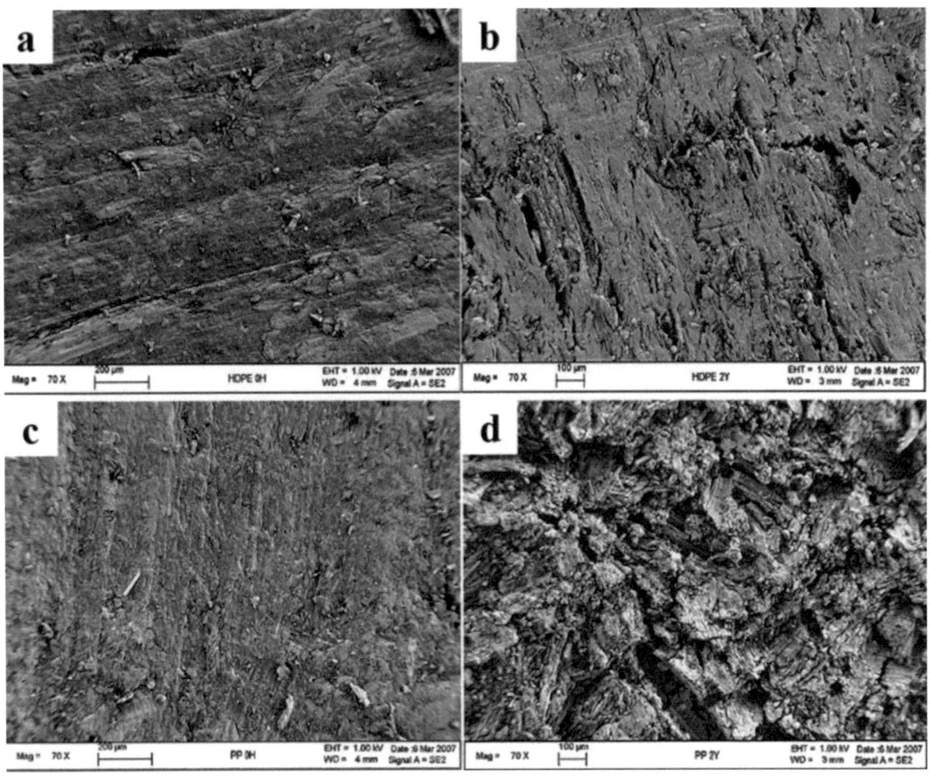

Figure 11.
Scanning electron micrographs of various WPCs before and after weathering: a, c. HDPE-based and PP-based WPC that was unweathered, respectively; b, d: HDPE-based and PP-based WPC that 2-yr exterior-weathered, respectively [23].

Many studies have evaluated the natural anti-weathering properties of WPCs in various applications. WPCs containing 40 wt.% of the polymer demonstrated better resistance to natural weathering a polymer loading of 50 wt.% [27]. High humidity also was determined by various researchers to be a principal determinant of the loss of flexural strength. In WPCs prepared with rubberwood fibers and several polymeric matrices, including HDPE, LDPE, PVC, PP, and PS, PP- and PS-based WPCs had the smallest reductions in tensile and flexural strengths of the WPCs by natural weathering when the mass fraction of polymer was 60 wt.%. These results demonstrated that PP and PS were the two ideal choices for WPC engineering applications in which WPCs are required to resist high stresses and natural weathering [28].

3. Reinforcement on mechanical properties of WPC

3.1 Linear polyamide reinforcement

The unfavorable mechanical properties of WPCs made with conventional PE, PP and PVC resins led to the placement of restrictions and safety risks on engineering construction products. Polyamides 6, 66, 1010, and 11 are typical engineering polymers that exhibit high strength and stiffness, excellent chemical stability, and outstanding heat resistance, and have the potential to promote the comprehensive properties of WPCs. PA6/biomass fibers composites have been directly fabricated by hot compression, extrusion, and injection at temperatures range from 230 to 250°C. However, severe thermal decomposition and poor dispersion of biomass fillers typically occurred during these manufacturing processes [29–31].

The melting and processing temperature of polyamide polymers can be decreased by the incorporation of lithium chloride (LiCl). LiCl and either maleic anhydride-grafted polyethylene (MA-g-PE) or maleic anhydride-grafted polypropylene (MA-g-PP) exhibited a synergistic effect on the mechanical strengths of PE- or PP-based WPCs with the reinforcement of the polyamide resin. Lithium cations (Li^+) can be easily embedded into the amorphous and crystalline phases of PA6 by the coordination of amide nitrogen or oxygen atoms, thereby breaking hydrogen bonds and disrupting three-dimensional arrangements. The formation of the copolymer PA6-g-PE by the condensation of amine functionalities in the PA6 polymer with the grafted anhydrides were found to improve the interfacial bonding ability between high-density polyethylene (HDPE) and PA6. Li^+ ions could also be complexed by the oxygen atoms of the anhydrides, which, too, promoted interfacial adhesion. On the other hand, biomass particles with rich hydroxyl groups synergistically reinforced the hydrogen bonds between PA6 and the wood flour (WF), underwent esterification reactions with MA-g-PE and WF, and complexed Li^+ ions [32]. The total reaction mechanism can be depicted as **Figure 12**. The addition of 2 wt%-2.5 wt% LiCl to the MA-g-PE polymer matrix led to a maximum flexural strength, flexural modulus, and impact strength of the PE-based WPC that exceeded 100 MPa, 6000 MPa, and 12 kJ/m^2, respectively. Similarly, the maximum flexural strength, flexural modulus, and impact strength of the PP-based WPC with 1 wt% LiCl were 115 MPa, 6500 MPa, and 10 kJ/m^2, respectively [33].

Increasing the molecular weight of PA6 is another effective way to improve its mechanical properties. The reaction of the chain extender 2,2′-(1,4-phenylene) bis(2-oxazoline) (PBO) with the terminal groups of PA6 was found to increase the molecular weight PA6 from 33000 to 72800 after the chain extender being incorporated into the melting polycondensation systems, the mechanism of which is shown in **Figure 13**. During the initial stage of chain elongation, the two

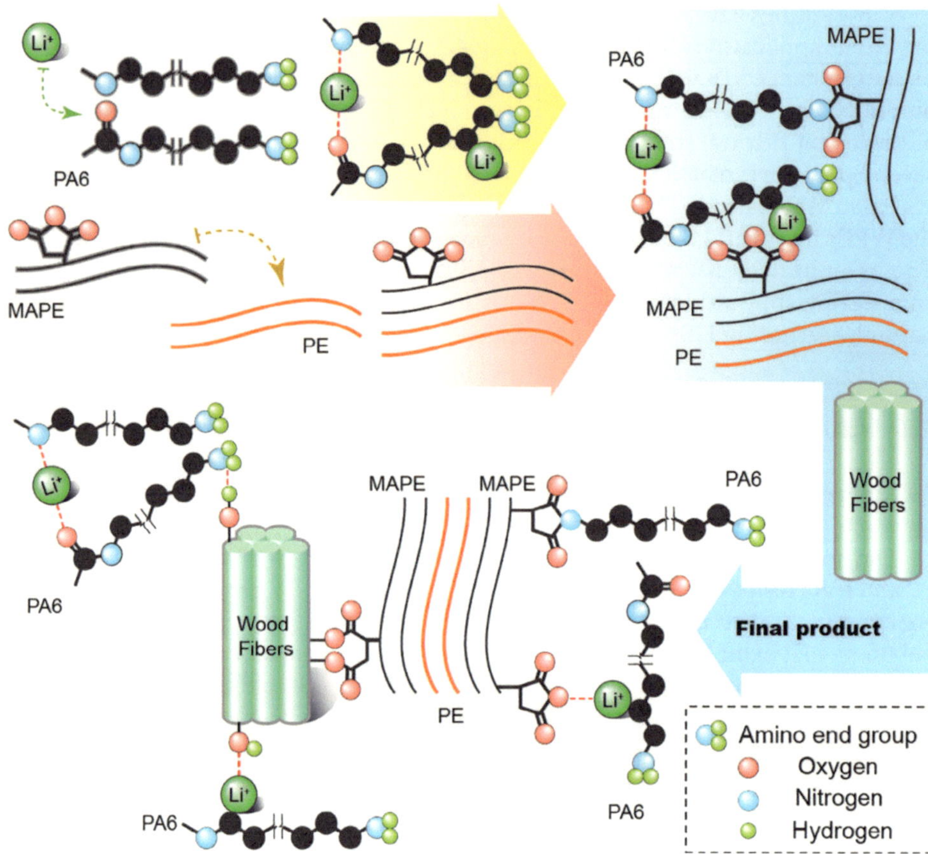

Figure 12.
Reaction mechanism diagram of LiCl and MA-g-PE in PE-based WPC [32].

oxazoline rings were opened, which then reacted with the two terminal carboxylic acid groups rather than the amine functionalities of the PA6 polymer, leading to the extension of the PA6 molecular chains. In addition, hydrogen bonding between the amide groups in PA6 and the hydroxyl groups in WF, as well as the noncovalent interactions between the oxygen and nitrogen atoms in PBO and the hydroxyl groups in WF, enhanced the interfacial bonding ability of PA6 and WF. PBO and LiCl were simultaneously used to modify the PA6-reinforced WPC, after which the melting point and processing temperature of PA6 decreased from 220–190°C and 240–210°C, respectively, which effectively mitigated the discoloration and thermal decomposition of the WPC. The modification of PA6 by PBO effectively improved the mechanical properties of the WPC by increasing the tensile strength, flexural strength, and modulus by 43%, 30%, and 37%, respectively [34].

Recently, the incorporation of novel, environmentally friendly polyamide copolymers matrices PA6/11 containing ε-caprolactam and 11-aminoundecanoic acid into WPC products was studied. The chain extension reactions of the PA6/11 copolymer were inhibited by the generating water from condensation reactions when the ε-caprolactam content was high. The melting temperature of the PA6/11 copolymers with the ratio of 40 to 60 can decreased to 120°C, which enabled the processing of the WPC below the thermal decomposition temperature of WF. In addition, when the ratio of PA6:PA11 was 70:30, the corresponding WPC exhibited outstanding ductility, with a strain at break of 15 to 20% [35].

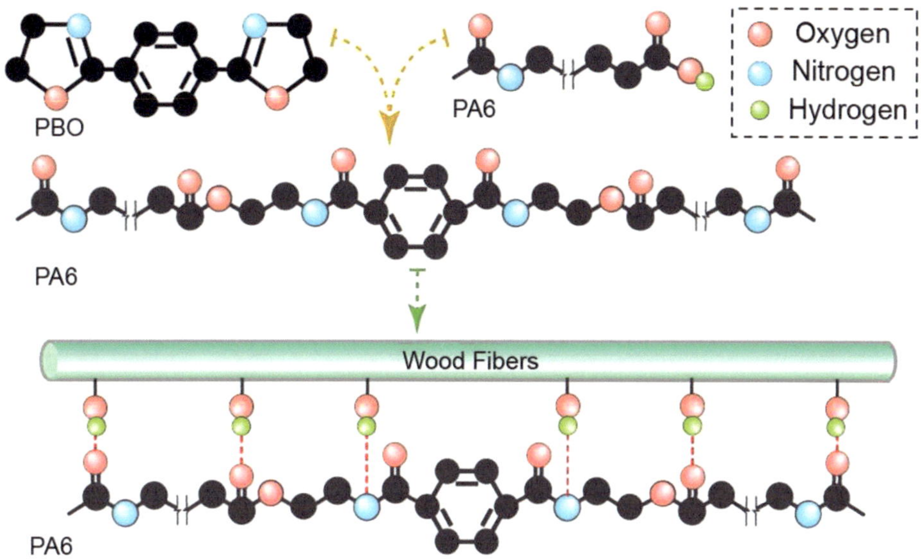

Figure 13.
Reaction mechanism diagram of the chain growth of PA6 [34].

The thermal modification of natural biomass fibers by superheated vapor has also proved to be a green, facile, and effective method for improving the mechanical and thermo-stable properties of WPCs. During this process, no chemical reactions ensue between the biomass fibers and exogenous reactants; instead, only chemical transformations within and between polymers, such as cellulose, hemicellulose and lignin, occur. Hemicelluloses are often hydrolyzed into oligomeric or monomeric structures at high temperatures. The hydroxyl groups of cellulose chains, as well as some chemical bonds that can be broken to form phenolic groups in lignin polymers, resulting in a reduction in the polarity of the biomass fibers and an improvement in their compatibility with non-polar HDPE resins [36]. The modification of rubber seed shell (RSS)-reinforced HDPE-based WPCs by superheated vapor at 200°C and 220°C is illustrated in **Figure 14**.

The superheated vapor treatment improved the mechanical strengths and thermal and dimensional stability of RSS, as well as its compatibility with HDPE. The fractured cross-section of the RSS/HDPE composites after RSS modification showed a rough morphology similar to a "thorn" due to strong interactions between the RSS and the HDPE matrix, such that the RSS fibers were pulled out during fracturing (**Figure 15**). The RSS particles modified at 200°C significantly enhanced the interfacial bonding ability of the RSS/HDPE composites and manifested outstanding mechanical properties. The flexural and tensile strengths of the RSS/HDPE composites were increased by 21.27% and 12.92% after modification of the RSS at 200°C and 220°C, respectively [37].

Superheated vapor modification mitigates the severe thermal degradation of biomass fibers during the heating processing, while the interfacial bonding can be promoted at the ideal ratio. The fractured surfaces of the PA6/HDPE/WF composite with HDPE:PA6 ratios of 8:2 and 7:3 both exhibited a rough appearance with many voids and separated structures, corresponding to the poor compatibility among WF, PA6, and HDPE [38]. These results were consistent with the weak interfacial bonding of PA6/HDPE/clay composites with when the weight ratio of HDPE:PA6 was 80:20 [39]. However, the HDPE:PA6 compatibility was markedly improved when the HDPE:PA6 ratio was 6:4, as the composite exhibited a coarse cross-section

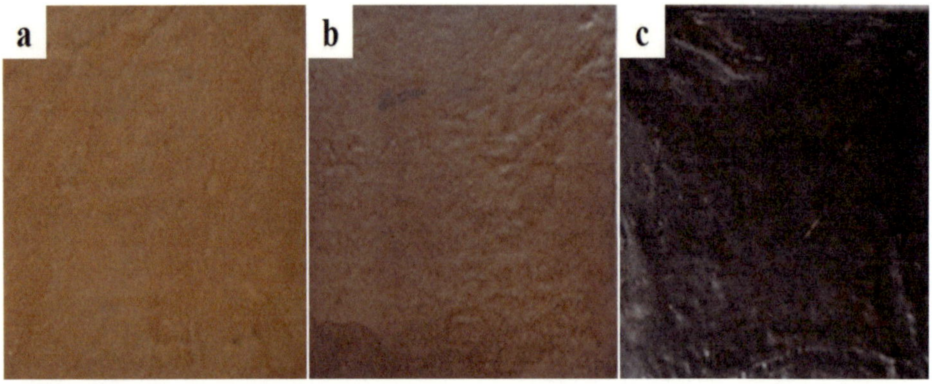

Figure 14.
The HEPE-based WPC with RSS modified by various superheated vapor temperatures: a. control, b. 200°C, c. 220°C [37].

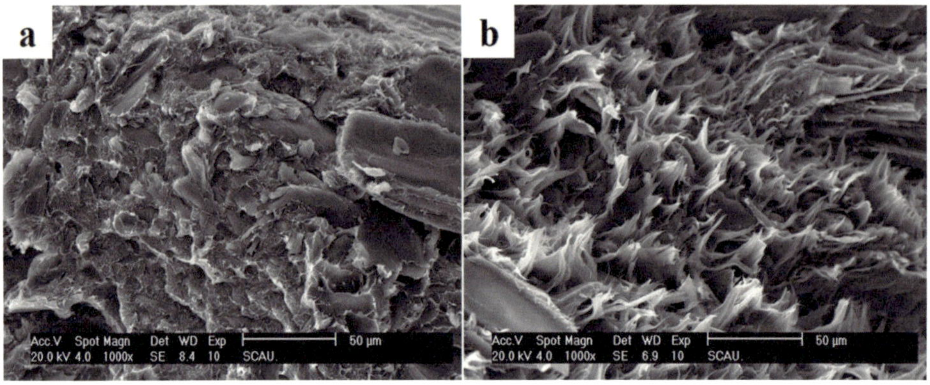

Figure 15.
Micro-morphology of the HDPE/RSS composites with unmodified and modified RSS by superheated vapor: a. unmodified, b. modified [37].

consistent with "thorns" similarly with **Figure 15**. The effect could be attributed to WF pulling out during fracturing with a strong interfacial bonding as mentioned above. The PA6 and modified wood fibers synergistically reinforced the interfacial adhesion only at a 6:4 ratio of HDPE:PA6, which was determined to be the optimal ratio. The flexural strength and flexural modulus of the PA6/HDPE/WF composite increased by 82.05% and 64.08%, respectively, compared to the control HDPE-based WPC. In addition, the tensile and impact strengths increased by 93.47% and 120.45%, respectively, and the maximum thermal degradation temperatures corresponding to the first and second decomposition stages increased by 7.17°C and 8.99°C, respectively. Moreover, the water absorption ratio of the PA6/HDPE/WF composite was controlled within 1.50%. Overall, these results demonstrated that the HDPE/PA6-modified WF composites have great potential in construction and building engineering applications as WPC products (**Figure 16**).

3.2 Aromatic polyamide reinforcement

Poly(p-phenylene terephthalamide), manufactured under the name Kevlar® by the Dupont corporation, is a linear polyamide featuring additional benzene rings in the polymer chain. Kevlar fibers (KFs) can be used as another novel reinforcement for WPC because they are resistant to combustion and have a high specific strength and modulus, superior toughness, and excellent chemical and thermal stability.

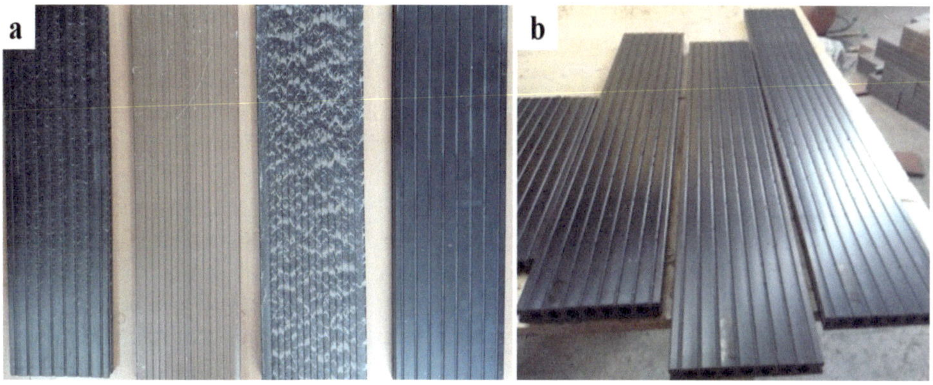

Figure 16.
The PA6/HDPE/modified WF composites: a. the samples with different ratios in the lab, b. the products in the factory.

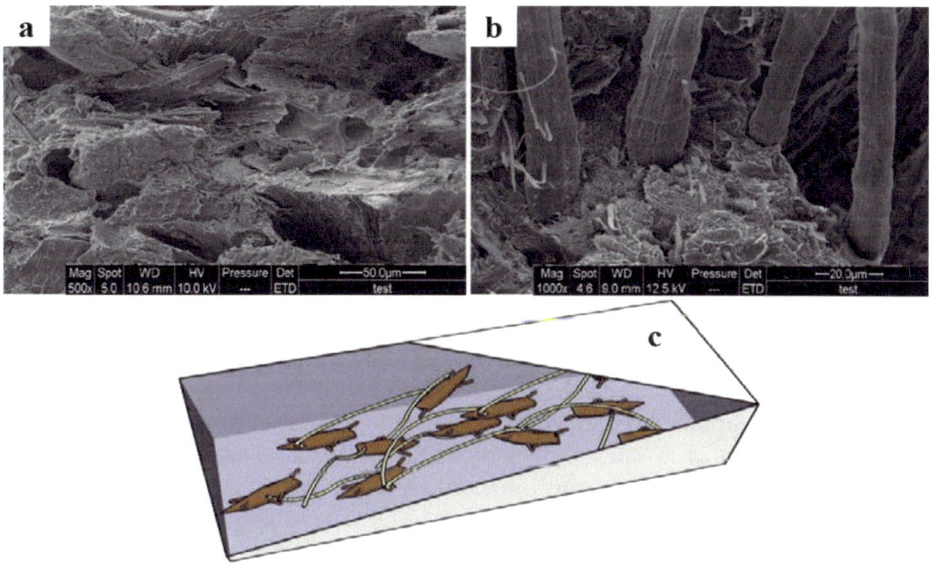

Figure 17.
The Kevlar reinforced HDPE-based WPC: a. HDPE/WF, b. HDPE/WF/KF, c. the assumed configuration of KF [40].

While KFs have been incorporated into WF/HDPE composites, their chemical inertness and low surface energy were found to be deleterious to the interfacial adhesion of the polymers within the composites. Prior to the addition of KFs, many voids and holes between the WF and the HDPE matrix on the fractured surface with the pullout of WF were attributed to poor interfacial adhesion (**Figure 17a**). The modification of KFs with allyl and 3-trimethoxysilylpropyl groups improved the compatibility between the WF and HDPE matrices, resulting in an increase in the mechanical properties of the composites (**Figure 17b**). The tensile strength and modulus of the KFs-reinforced WPC increased to 23.1% and 31.4%, respectively, while the notched and unnotched impact strengths of the reinforced WPC increased to 42.8% and 52.3%, respectively, when the KF loaded was below 3% compared to the unmodified WPC. In addition, the alkoxysilane groups on the KFs can be hydrolyzed to generate silanol groups or siloxane networks, which can engage in hydrogen bonding with the hydroxyl groups in WF. There was a physical anchor bonding among the melting of HDPE, KF, and WF. Meanwhile, free-radical reactions between HDPE chains and allyl groups may have occurred during the extrusion process. The flexible KFs with diverse S, J, C, and I configurations is shown in **Figure 17c**. The major structural failures included matrix, fiber, and interface fractures as well as fiber fibrillation and buckling. The various configurations of the KFs contributed to the mechanical interlocking among KF, WF, and HDPE components, which promoted an efficient stress transfer from the resin matrix to the KFs. Therefore, the reinforcement of WPCs with KFs can bestow the WFCs with the strength and toughness required for engineering applications [40]. However, until now, there has been very limited research conducted in the field of Kevlar and natural biomass fiber-reinforced composites. Nonetheless, the incorporation of KFs into WFCs can be considered a promising modification for the improvement of the mechanical properties of WPCs in various engineering applications in the future.

4. Flame retardance properties of WPC

The continued development of WPCs for a variety of different engineering applications, as well as the increased demand for more safety-consciousness products, has necessitated the development of flame-resistant materials as a substitute for some of the conventional engineering materials used in WPCs. Currently, PE- and PP-based WPCs are most often utilized in building exterior and outdoor products because PP and PE both have a high sensitivity to flame, which means they can quickly burn when ignited, leading to hazardous scenarios. Meanwhile, PVC-based WPCs are most often used in indoor products; although PVC has a relatively low flammability and inherent self-inhibited flame properties compared to PE and PP, PVC often does not comply with the safety requirements and regulations of governments because it releases a significant amount of toxic smoke while burning. Thus, it is necessary to improve the flame resistance of WPCs to expand their potential engineering applications. Many strategies that have been developed to enhance the flame resistance of WPCs can be divided into two general categories. The first is the incorporation of fire retardants into biomass particles and polymers during the compression, injection, and/or extrusion processes. The other strategy entails the pretreatment of the biomass particles with liquid flame retardants by impregnation, after which the particles can be combined with the polymers using the conventional processes.

The major flame retardants used in WPCs include halogenated and nonhalogenated chemicals. The halogenated flame retardants mainly include brominated and chlorinated molecules that function to reduce the heat release and inhibit combustion by quenching the burning of the materials. Nevertheless, they emit toxic or corrosive gases that harmful to the environment while burning. Hence, they have been gradually withdrawn from use because of added pressure from many governments around the world. In recent years, though, some novel halogen-free flame retardants like phosphorus-based flame retardants, boron-based flame retardants, metal hydroxide flame retardants and intumescent flame retardants have been developed.

4.1 The testing methods of flame retardancy for WPC

Many different evaluation methods of flame retardancy are available for various materials, among which the limited oxygen index (LOI), UL-94 V, and cone calorimeter burning testing are most frequently used for WPC products.

4.1.1 Limiting oxygen index

The LOI, which is also referred to as the critical oxygen index (COI) or oxygen index (OI), was first introduced by Fenimore and Martin in 1966. It is defined as the minimum concentration of oxygen in an atmosphere composed of a mixture of oxygen (O_2) and nitrogen (N_2) at 25°C required for supporting the continuous flaming combustion of a material sample. LOI has been included in several international and national standards, such as ISO 4589 and ASTM D2863 [41]. Specifically, the testing involves placing a WPC sample with a candle-like shape vertically in a glass column while feeding the sample an atmosphere of O_2 and N_2. After the WPC sample is ignited from the top, the oxygen concentration is decreased (i.e. the nitrogen concentration is increased) until the minimum oxygen concentration that allows for the sustained burning of the sample is determined. Prior to observation and evaluation, the sample should be at a stable burning stage. The testing can be considered as an effective result if one of two situations occurs: either the sample continues burning 3 min after the removal of the fire source or the consumed length of the sample is within 5 cm [42]; otherwise, a new sample should be retested at an adjustable O_2 concentration. The LOI test has been considered one of the most useful testing methods for determining the flame retardance of PE-based and PP-based WPCs since it can provide a precise rating on a numerical basis.

4.1.2 UL-94 V

The UL-94 V method was initially designed to evaluate the flammability of plastic materials for parts in devices and appliances and is also accepted throughout the world as a standardized method. Similar to the LOI, it also is used to assess the flame retardancy of PE-, PP- or PS-based WPCs. In this method, a bar-shaped sample is placed vertically and held from the top. The sample is ignited twice by a Bunsen burner, and the total sustained combustion time is recorded after each burning. A second burning will be conducted if the sample is self-extinguished after the first burning. Based on the results, the WPCs can typically be classified into V-0, V-1, and V-2 levels. The V-0 level means that the specimen is extinguished less than 10 s after it is ignited. The V-1 level corresponds to a maximum combustion time of a sample of less than 50 s without any combustible drips, or the combustion time for five samples is less than 25 s [43]. Often, standard cotton is positioned under the sample to determine whether the combustible drips will ignite the cotton. The sample is classified as V-2 when the combustion times comply with the requirements of V-1 level, but some of the flammable drips ignite the cotton.

4.1.3 Cone calorimeter

The LOI and UL-94 V methods generally are facile, fast, and reproducible. However, these methods typically only provide a single test parameter. The test results cannot be well-quantified and are difficult to correlate with the burning behavior of the material in a real fire scenario. In addition, the evaluation of identical materials using different experimental methods can generate conflicting results. Instead, cone calorimetry is the most effective way to measure and simulate the flammability behavior of a medium-sized WPC sample, as it provides abundant test parameters, including the time to ignition (TTI), heat release rate (HRR), mass loss rate (MLR), total heat release (THR), total smoke release (TSR), and effective heat of combustion (EHC) [42]. It also can be used to predict the large-scale testing results in fire-protection engineering, and it is incorporated into various international standards like ISO 5660 and ASTM E1354 because of its availability. The basic principle of cone calorimetry relies on the measurement of decreasing oxygen concentration in a mixture of gases during the combustion of a sample exposed to a constant heat flux.

4.2 Flame retardants for WPC

4.2.1 Phosphorus-based flame retardants

Phosphorus-based flame retardants can be traced back to the nineteenth century when ammonium phosphate was used to prevent the burning of theater curtains in France. The surface of wood materials has been coated with a mixture of phosphoric acid and other curable resins to modify the flammability of various materials in early Canada. Nowadays, it has become an important flame retardant for improving the flame resistance of WPCs. Phosphoric acid can facilitate the formation of char in the condensed phase and reduce the presence of flammable radicals in the gas phase, which leads to an enhancement of the flame retardance of WPCs [44]. Several kinds of phosphorus-based flame retardants, including ammonium polyphosphate (APP), aluminum hypophosphite (AHP), monoammonium phosphate (MAP), melamine phosphate (MP), melamine polyphosphate (MPP), and red phosphorus (RP), have been introduced into the formulations of polyolefin-based WPCs. **Table 1** lists the cone calorimetry testing results for WPCs containing several phosphorus-based flame retardants. It was concluded that there APP demonstrated a better flame resistance of WPC compared to DAP and MPP, while the effectiveness of DAP was lower than MPP.

4.2.2 Boron-based flame retardants

Boron-based flame retardants, which often contain additional functionalities, such as smoke suppressants, afterglow suppressants, and antitracking agents, are early used for cellulose-containing products [50]. However, the use of zinc borates (ZB) and a mixture of boric acid (BA) and borax (BX) as flame retardant in WPCs

WPC samples	Weight ratio	TTI (s)	pHHR (kW/m^2)	THR (MJ/m^2)	Reference
WF/PP	60/40	21	359	87	[45]
WF/PP/APP	60/40/20	22	208	68	[45]
WF/PP	40/60	19	535	319	[46]
WF/PP/MPP	36/56/8	20	470	300	[47]
WF/PP	40/60	19	535	319	[46]
WF/PP/DAP	36/56/8	19	495	299	[46]
WF/HDPE	30/70	28	194	71	[48]
WF/HDPE/APP	30/70/10	19	113	62	[48]
WF/HDPE	40/60	35	422	102	[49]
WF/HDPE/MPP	40/60/35	36	202	45	[49]
WF/HDPE	40/60	23	523	335	[46]
WF/HDPE/DAP	37/57/4	22	466	311	[46]

Table 1.
Results of cone calorimeter for polyolefin-based WPC with various phosphorus-based flame retardants.

has gained more attention. **Table 2** lists the cone calorimetry testing results for WPCs with two typical boron-based flame retardants. PP-based and PE-based WPCs demonstrated improved fire retardance due to increased TTI and decreased pHRR and THR values after incorporating ZB and BA-BX. The polyolefin WPC containing ZB demonstrated a superior effectiveness over BA-BX, but ZB has been determined to not be an ideal flame retardant for PVC-based WPCs because PVC has an intrinsic flame resistance given the presence of chlorine atoms in the polymer [54].

4.2.3 Metal hydroxide flame retardants

Metal hydroxides can consume a significant amount of heat energy from the surroundings as well as release water during thermal decomposition [55]. Magnesium hydroxide (MH) and aluminum trihydroxide (ATH) are two commonly used flame retardants in PP-based WPCs. **Table 3** lists the cone calorimetry testing results for WPC containing these metal hydroxides. The flame retardance of the WFC samples with varying weight ratios of WF, PP, and MH were different. Compared to the control samples, both WFC samples containing MH and AHH demonstrated a noticeable improvement in their flame resistance due to higher TTI and lower pHRR and THR values compared to the WFC samples without MH and AHH.

4.2.4 Expandable graphite (EG) flame retardants

Expandable graphite (EG) is a layered crystal consisting of sheets of carbon atoms tightly bound to each other. Some chemicals, such as sulfuric acid and potassium permanganate, can be inserted into the carbon layers of graphite. When subjected to heat, EG can expand up to 300 times compared to its original volume, which produces an insulating layer that can provide excellent flame resistance for WPCs [58]. EG appeared to be more effective for the polymer component of WPCs than the WF component when the weight ratios of the two varied from 60:40 to 40:60 (**Table 4**). Furthermore, EG exhibited the best flame retardance for PP-based WPC, with the lowest pHRR and THR values, compared to all the flame retardants mentioned above.

WPC samples	Weight ratio	TTI (s)	pHHR (kW/m^2)	THR (MJ/m^2)	Reference
WF/PP	40/58	19	505	253	[51]
WF/PP/ZB	40/58/10	22	402	237	[51]
WF/PP	40/60	19	535	319	[47]
WF/PP/BA-BX	37/57/4	20	465	318	[47]
WF/PE	50/45	25	505	373	[52]
WF/PE/ZB	50/35/10	26	320	255	[52]
WF/HDPE	40/60	26	516	319	[46]
WF/HDPE/BA-BX	37/57/4	25	510	320	[46]
WF/PVC	35/58	26	160	49	[53]
WF/PVC/ZB	35/58/6	27	209	43	[53]

Table 2.
Results of cone calorimeter for polyolefin-based WPC with various boron-based flame retardants.

WPC samples	Weight ratio	TTI (s)	pHHR (kW/m^2)	THR (MJ/m^2)	Reference
WF/PP	40/58	19	505	253	[51]
WF/PP/MH	40/48/10	25	407	249	[51]
WF/PP	50/46.7	21	563	93	[56]
WF/PP/ATH	50/36.7/10	25	467	99	[56]
WF/PP	50/43	28	395	162	[57]
WF/PP/ATH	40/43/10	30	336	152	[57]

Table 3.
Results of cone calorimeter for polyolefin-based WPC with various metal hydroxide flame retardants.

WPC samples	Weight ratio	TTI (s)	pHHR (kW/m^2)	THR (MJ/m^2)	Reference
WF/PP	60/40	21	358	87	[45]
WF/PP/EG	60/40/20	32	181	64	[45]
WF/PP	40/60	20	390	90	[59]
WF/PP/EG	40/60/25	24	99	39	[59]
WF/PP	60/40	28	358	102	[60]
WF/PP/EG	60/40/25	23	84	48	[60]

Table 4.
Results of cone calorimeter for PP-based WPC with expandable graphite flame retardants.

4.2.5 Prospect of flame retardants in WPC

While all WPC samples containing individual flame retardants displayed decreased pHRR and THR values, they still demonstrated limited effectiveness under the mandatory flame condition. In the future, the development of flame retardants for WPCs will be more focused on the synergism between multiple components to further improve the flame retardance and the interfacial compatibility in the compound system. Intumescent flame retardants (IRFs), which consist of a char-forming agent, dehydrating agent, and blowing agent, have attracted significant attention because of their environmental friendliness and high efficiency [61]. However, moisture absorption and incompatibility with polymers are potential drawbacks for blowing agents and carbonic sources. Therefore, novel flame retardants that are nontoxic, have excellent flame resistance, and are highly compatible with polymers should be developed in the future.

5. Conclusions

Wood plastic composites has a significant potential and brilliant prospect in the application of building construction. WPC products like sidings, ceilings, windows, and floorings with the special advantages in environmental protection and sustainable development can effectively promote the coordinated development of wood-based engineering products in the “next-generation” green construction. However, the main defects on low mechanical properties, weak biological and aging resistance and poor flame retardance limit its further development. The novel modification ways including the selection of wood species, alloy reinforcement and synergistic improvement on flame resistance can be used to effectively facilitate the promotion of comprehensive properties of WPC, which achieves to the related requirements in the field of building construction. Future works also should focus on these necessary problems. In addition, the coextrusion technology with solid woods, functional polymers or metals is also a promising modification method for engineered WPC for construction in the future.

Acknowledgements

This work is supported by the following grants and programs: 1. National Natural Science Foundation of China (32060381); 2. Yunnan Provincial Applied and Basic Research Grants (2019FB067); 3. Ten-thousand Program Youth Talent Support Program (2020); 4. "111" project.

Conflict of interest

There is no conflict of interest in this field.

Author details

Kaimeng Xu[1*], Guanben Du[1*] and Siqun Wang[2]

1 International Joint Research Center for Biomass Materials, Southwest Forestry University, Kunming, China

2 The University of Tennessee, Knoxville, TN, USA

*Address all correspondence to: xukm007@163.com and gongben9@hotmail.com

References

[1] Gurunathan T, Mohanty S, Nayak S K. A review of the recent developments in biocomposites based on natural fibres and their application perspectives. Composites Part A: Applied Science and Manufacturing. 2015;77:1-25. DOI: 10.1016/j.compositesa.2015.06.007

[2] Deka B K, Maji T K. Study on the properties of nanocomposite based on high density polyethylene, polypropylene, polyvinyl chloride and wood. Composites Part A: Applied Science and Manufacturing. 2011;42(6):686-693. DOI: 10.1016/j.compositesa.2011.02.009

[3] Adhikary KB, Park CB, Islam MR, Rizvi GM. Effects of lubricant content on extrusion processing and mechanical properties of wood flour-high-density polyethylene composites. Journal of Thermoplastic Composite Materials. 2011;24(2):155-171. DOI: 10.1177/0892705710388590

[4] Clemons C. Wood-plastic composites in the United States-the interfacing of two industries. Forest Products Journal. 2002;52:10-18.

[5] Ou RX, Zhao H, Sui SJ, Song YM, Wang QW. Reinforcing effects of Kevlar fiber on the mechanical properties of wood-flour/high-density-polyethylene composites. Composites Part A: Applied Science and Manufacturing. 2010; 41(9):1272-1278. DOI: 10.1016/j.compositesa.2010.05.011

[6] Ashori A. Wood-plastic composites as promising green-composites for automotive industries. Bioresources Technology, 2008;99(11):4661-4667. DOI: 10.1016/j.biortech.2007.09.043

[7] Xu K, Li K, Zhong T, Guan L, Xie C, Li S. Effects of chitosan as biopolymer coupling agent on the thermal and rheological properties of polyvinyl chloride/wood flour composites. Composites Part B: Engineering, 2014;58:392-399. DOI: 10.1016/j.compositesb.2013.10.056

[8] Klason C, Kubát J, Strömvall H E. The efficiency of cellulosic fillers in common thermoplastics. 1. Filling without processing aids or coupling agents. International Journal of Polymeric Materials. 1984;10(3):159-187. DOI: 10.1080/00914038408080268

[9] Mankowski M, Morrell J J. Patterns of fungal attack in wood-plastic composites following exposure in a soil block test. Wood and Fiber Science. 2000;32(3):340-345.

[10] Verhey S A, Laks P E, Richter D L. Laboratory decay resistance of woodfiber/thermoplastic composites. Forest Products Journal. 2001;51(9): 44-49.

[11] Xu K M, Feng J, Zhong T H, Zheng Z F, Chen T A. Effects of volatile chemical components of wood species on mould growth susceptibility and termite attack resistance of wood plastic composites. International Biodeterioration & Biodegradation. 2015;100:106-115. DOI: 10.1016/j.ibiod.2015.02.002

[12] Karimi A N, Tajvidi M, Pourabbasi S. Effect of compatibilizer on the natural durability of wood flour/high density polyethylene composites against rainbow fungus (Coriolus versicolor). Polymer Composites. 2007;28(3):273-277. DOI: 10.1002/pc.20305

[13] Dawson-Andoh B, Matuana L M, Harrison J. Mold susceptibility of rigid PVC/wood-flour composites. Journal of Vinyl and Additive Technology. 2004;10(4): 179-186. DOI: 10.1002/vnl.20027

[14] Feng J, Xu K M, Shi Q S, Huang X M, Li K F. Algal decay resistance of conventional and novel wood-based

composites. Bioresources. 2015;10(4):6321-6331. DOI: 10.15376/biores.10.4.6321-6331

[15] Morris P I, Cooper P. Recycled plastic/wood composite lumber attacked by fungi. Forest Products Journal. 1998;48:86-88. DOI: 10.1139/cjfr-28-1-154

[16] Pendleton D E, Hoffard T A, Adcock T, Woodward B, Wolcott M P. Durability of an extruded HDPE/wood composite. Forest Products Journal. 2002;52:21-27. DOI: 10.1023/A:1020158128506

[17] Dawson-Ahdoh B, Matuana L M, Harrison J. Mold susceptibility of rigid PVC/wood-flour composites. Journal of Vinyl & Additive Technology. 2004;10:179-186. DOI: 10.1002/vnl.20027

[18] Fabiyi J S, McDonald A G, Morrell J J, Freitag C. Effects of wood species on durability and chemical changes of fungal decayed wood plastic composites. Composites Part A: Applied Science and Manufacturing. 2011;42(5):501-510. DOI: 10.1016/j.compositesa.2011.01.009

[19] Lomeli-Ramirez M G, Ochoa-Ruiza H G, Fuentes-Talavera F J, Garcia-Enriquez S, Cerpa-Gallegos M A, Silva-Guzman J A. Evaluation of accelerated decay of wood plastic composites by *Xylophagus* fungi. International Biodeterioration & Biodegradation. 2009;63(8):1030-1035. DOI: 10.1016/j.ibiod.2009.08.002

[20] Xu K M, Feng J, Shi Q S, Li K F, Tu D Y. A comparative study on natural decay resistance for wood flour/PVC composites made by various wood species. China Forest Products Industry. 2016;43(2):13-17. DOI: 10.3969/j.issn.1001-5299.2016.02.003

[21] Feng J, Dong P, Li R M, Li C L, Xie X B, Shi Q S. Effects of wood fiber properties on mold resistance of wood polypropylene composites. International Biodeterioration & Biodegradation. 2019;140:152-159. DOI: 10.1016/j.ibiod.2019.04.005

[22] Friedrich D. Comparative study on artificial and natural weathering of wood-polymer compounds: A comprehensive literature review. Case Studies in Construction Materials. 2018;9:e00196. DOI: 10.1016/j.cscm.2018.e00196

[23] Fabiyi J S, McDonald A G, Wolcott M P, Griffiths P R. Wood plastic composites weathering: Visual appearance and chemical changes. Polymer Degradation and Stability. 2008;93(8):1405-1414. DOI: 10.1016/j.polymdegradstab. 2008.05.024

[24] Jabarin S A, Lofgren E A. Photooxidative effects on properties and structure of high density polyethylene. Journal of Applied Polymer Science. 1994;53(4):411-423. DOI: 10.1002/app.1994.070530404

[25] Chen Y, Stark N M, Tshabalala M A, Gao J M, Fan Y M. Weathering characteristics of wood plastic composites reinforced with extracted or delignified wood flour. Materials. 2016;9(8): 610. DOI: 10.3390/ma9080610

[26] Peng Y, Liu R, Cao J X. Characterization of surface chemistry and crystallization behaviour of polypropylene composites reinforced with wood flour, cellulose, and lignin during accelerated weathering. Applied Surface Science. 2015;332:253-259. DOI: 10.1016/j.apsusc.2015.01.147

[27] Te-Hsin Y, Tsu-Hsien Y, Wei-Cheng C, Shao-Yuan L. Characterization of the property changes of extruded wood–plastic composites during year-round subtropical weathering. Construction and Building Materials.

2015;88:159-168. DOI: 10.1016/j.conbuildmat.2015.04.019

[28] Ratanawilai T, Taneerat K. Alternative polymeric matrices for wood-plastic composites: Effects on mechanical properties and resistance to natural weathering. Construction and Building Materials. 2018;172:349-357. DOI: 10.1016/j.conbuildmat.2018.03.266

[29] de Arcaya P A, Retegi A, Arbelaiz A, Kenny J M, Mondragon I. Mechanical properties of natural fibers/polyamides composites. Polymer Composites. 2009; 30(3):257-264. DOI: 10.1002/pc.20558

[30] McHenry E, Stachurski Z H. Composite materials based on wood and nylon fibre. Composites Part A: Applied Science and Manufacturing. 2003;34(2):171-181. DOI: 10.1016/S1359-835X(02)00211-7

[31] Ozen E, Kiziltas A, Kiziltas E E, Gardner D J. Natural fiber blend-nylon 6 composites. Polymer Composites. 2013;34(4):544-553. DOI: 10.1002/pc.22463

[32] Xu S H, Fang Y Q, Chen Z Y, Yi S M, Zhai X L, Yi X, He J, Song Y M, Wang Q W. Impact of lithium chloride on the performance of wood fiber reinforced polyamide 6/high-density polyethylene blend composites. Polymer Composites. 2019;40(12):4608-4618. DOI: 10.1002/pc.25328

[33] Xu S H, Fang Y Q, Yi S M, He J, Song Y M, Wang Q W. Preparation and characterization of wood-fiber-reinforced polyamide 6-polypropylene blend composites. Journal of Applied Polymer Science. 2019;136(18):47413. DOI: 10.1002/app.47413

[34] Xu S H, Fang Y Q, Yi S M, He J, Zhai X L, Song Y M, Wang H G, Wang Q W. Effects of lithium chloride and chain extender on the properties of wood fiber reinforced polyamide 6 composites. Polymer Testing. 2018;72:132-139. DOI: 10.1016/j.polymertesting.2018.10.005

[35] Zierdt P, Mitzner E, Gomoll A, Theumer T, Lieske A. Synthesis of polyamide 6/11 copolymers and their use as matrix polymer in wood-plastic composites. Journal of Applied Polymer Science. 2016;133(46):44155. DOI: 10.1002/app.44155

[36] Kaboorani A, Faezipour M, Ebrahimi G. Feasibility of using heat treated wood in wood/thermoplastic composites. Journal of Reinforced Plastics and Composites. 2008;27 (16-17):1689-1699. DOI: 10.1177/0731684407084207

[37] Xu K M, Tu D Y, Chen T A, Zhong T H, Lu J. Effects of environmental-friendly modified rubber seed shell on the comprehensive properties of high density polyethylene/rubber seed shell composites. Industrial Crops and Products. 2016;91:132-141. DOI: 10.1016/j.indcrop.2016.06.036

[38] Xu K M, Zheng Z F, Du G B, Zhang Y L, Wang Z H, Zhong T H, Xie L K, Wang S Q. Effects of polyamide 6 reinforcement on the compatibility of high density polyethylene/environmental-friendly modified wood fiber composites. Journal of Applied Polymer Science. 2019;136(38):47984. DOI: 10.1002/app.47984

[39] Mallick S, Kar P, Khatua B B. Morphology and properties of nylon 6 and high density polyethylene blends in presence of nanoclay and PE-g-MA. Journal of Applied Polymer Science. 2012;123(3):1801-1811. DOI: 10.1002/app.34648

[40] Ou R X, Zhao H, Sui S J, Song Y M, Wang Q W. Reinforcing effects of Kevlar fiber on the mechanical properties of wood-flour/high-density-polyethylene composites. Composites Part A: Applied Science and

Manufacturing. 2010;41(9):1272-1278. DOI: 10.1016/j.compositesa.2010.05.011

[41] Van Krevelen D W, Te Nijenhuis K. Properties of polymers (fourth edition): Their correlation with chemical structure; their numerical estimation and prediction from additive group contributions. 2009.

[42] Yang X, Zhang W. Chapter11-Flame retardancy of wood-polymeric composites. Polymer-Based Multifunctional Nanocomposites and Their Applications. 2019: 285-317. DOI: 10.1016/B978-0-12-815067-2.00011-1

[43] UL-94. Standard for Tests for Flammability of Plastic Materials for Parts in Devices and Appliances. 1996.

[44] Schartel B. Phosphorus-based flame retardancy mechanisms-old hat or a starting point for future development. Materials. 2010;3 (10):4710-4745. DOI: 10.3390/ma3104710

[45] Yu F L, Xu F J, Song Y M, Fang Y Q, Zhang Z J, Wang Q W, Wang F Q. Expandable graphite's versatility and synergy with carbon black and ammonium polyphosphate in improving antistatic and fire-retardant properties of wood flour/polypropylene composites. Polymer Composites. 2017;38(4):767-773. DOI: 10.1002/pc.23636

[46] Ayrilmis N, Akbulut T, Dundar T, White R H, Mengeloglu F, Buyuksari U, Candan Z, Avci E. Effect of boron and phosphate compounds on physical, mechanical, and fire properties of wood–polypropylene composites. Construction and Building Materials. 2012;33:63-69. DOI: 10.1016/j.conbuildmat.2012.01.013

[47] Akbulut T, Ayrilmis N, Dundar T, Durmus A, White R H, Teker M. Effect of boron and phosphate compounds on thermal and fire properties of wood/HDPE composites. International Journal of Polymers and Technologies. 2011;3(2): 67-75.

[48] Pan M Z, Mei C T, Song Y X. A novel fire retardant affects fire performance and mechanical properties of wood flour-high density polyethylene composites. Bioresources. 2012;7 (2):1760-1770.

[49] Li L, Guo W, Guo C. Synergistic effect of melamine polyphosphate and aluminum hypophosphite on mechanical properties and flame retardancy of HDPE/wood flour composites. Wood Science and Technology. 2017;51(3):493-506. DOI: 10.1007/s00226-016-0877-2

[50] Shen K K. Chapter 11 - Review of recent advances on the use of boron-based flame retardants. Polymer Green Flame Retardants. 2014; 367-388. DOI: 10.1016/B978-0-444-53808-6.00011-1

[51] Ayrilmis N, Benthien J T, Thoemen H, White R H. Effects of fire retardants on physical, mechanical, and fire properties of flat-pressed WPCs. European Journal of Wood and Wood Products. 2012;70(1-3):215-224. DOI: 10.1007/s00107-011-0541-3

[52] Stark N M, White R H, Mueller S A, Osswaldb T A. Evaluation of various fire retardants for use in wood flour-polyethylene composites. Polymer Degradation and Stability. 2010;95(9):1903-1910. DOI: 10.1016/j.polymdegradstab. 2010.04.014

[53] Fang Y Q, Wang Q W, Guo C G, Song Y M, Cooper P A. Effect of zinc borate and wood flour on thermal degradation and fire retardancy of polyvinyl chloride (PVC) composites. Journal of Analytical and Applied Pyrolysis. 2013;100:230-236. DOI: 10.1016/j.jaap.2012.12.028

[54] Klyosov A A. Wood-plastic composites. John Wiley & Sons. 2007.

[55] Chapple S, Anandjiwala R. Flammability of natural fiber-reinforced composites and strategies for fire retardancy: a review. Journal of

Thermoplastic Composite Materials. 2010;23(6):871-893. DOI: 10.1177/0892705709356338

[56] Arao Y, Nakamura S, Tomita Y, Takakuwa K, Umemura T, Tanaka T. Improvement on fire retardancy of wood flour/polypropylene composites using various fire retardants. Polymer Degradation and Stability. 2014;100:79-85. DOI: 10.1016/j.polymdegradstab.2013.12.022

[57] Turku I, Nikolaeva M, Kärki T. The effect of fire retardants on the flammability, mechanical properties, and wettability of co-extruded PP-based wood-plastic composites. Bioresources. 2014;9(1):1539-1551. DOI: 10.15376/biores.9.1.1539-1551

[58] Nikolaeva M, Kärki T. A review of fire retardant processes and chemistry, with discussion of the case of wood-plastic composites. Baltic Forestry; 2011;17(2):314-326. DOI: 10.1016/j.foreco.2010.10.023

[59] Bai G, Guo C G, Li L P. Synergistic effect of intumescent flame retardant and expandable graphite on mechanical and flame-retardant properties of wood flour-polypropylene composites. Construction and Building Materials. 2014;50:148-153. DOI: 10.1016/j.conbuildmat.2013.09.028

[60] Guo C G, Zhou L, Lv J X. Effects of expandable graphite and modified ammonium polyphosphate on the flame-retardant and mechanical properties of wood flour-polypropylene composites. Polymer and Polymer Composites. 2013;21(7):449-456. DOI: 10.1177/096739111302100706

[61] Bourbigot S, Duquesne S. Fire retardant polymers: recent developments and opportunities. Journal of Materials Chemistry. 2007;17(22):2283-2300. DOI: 10.1039/B702511D

Chapter 13

Mineral-Bonded Wood Composites: An Alternative Building Materials

Halil Turgut Sahin and Yasemin Simsek

Abstract

The manufacturing of cost-efficient construction materials is at the center of attention these days. The development of engineeringly design products has occurred mostly over the past few decades. However, the term of mineral bonded wood composite is relatively new, covers many of the products, and is used to describe a material that is produced by bonding woody material with mineral-based substrates. At present, millions of tons of bio-based composite materials are now manufactured annually from many wood species. Woods are sustainable and engineeringly have enough performance properties in composite matrix systems for many end-use areas. Thus, their utilization processes and intended uses vary accordingly. But at manufacturing, many variables affect binder hydration in composite structure and the networking/bonding between wood and binder. The mineral bonded wood products are high in density and the appropriate strength in the construction industry, an important advantage to engineering applications appears to lie in their ability to absorb and dissipate mechanical energy. Despite their higher weight-to-strength ratio, especially cement and gypsum bonded wood composites have become popular, for use in many internal and external applications to meet increasingly stringent building design regulations for insulation, and failure in service due to deterioration.

Keywords: mineral binder, wood-cement composite, gypsum, magnesia cement

1. Introduction

Wood is one of the first raw materials for construction purposes. Its usage has continuously increased since human beings. However, as a result of excessive wood utilization, natural forests have become depleted at scarcity value. Therefore, consumers have become more aware of the destruction of the natural forests for wood supply. After technological developments and intensive studies, many valuable constructional elements have been developed from lignocellulosic in recent years [1–3]. In this context, numerous alternative biomass sources such as; agricultural and forest residues, low-value woody materials, annual plants have been considered to use as wood substitute alone or in combination with synthetic binders to manufacture construction materials [4–7].

One of the interesting materials has been invented which using inorganic minerals as the bonding agent, called mineral-bonded wood composites. These products were first produced by an Austrian carpenter using wood shavings and gypsum together in 1914. However, cement-bonded wood composites called Wood Wool Cement Board (WWCB) were also produced in Austria in the 1920s and several others in Europe followed. Moreover, Cement Bonded Wood Chip Boards called Durisol were invented and commercially produced in the 1930s. After that, the cement-bonded coarse wood particleboards called Velox boards were produced in the 1950s. The first Cement Bonded Particle Board (CBPB) called Duripanel was produced in 1970. Since the first invention of these products in Europe, these materials have developed further with calling different names in the market, spread to the rest of the world. These days, numerous mils have been built throughout the world, mostly in manufacturing panel form [8].

Due to variables and wide range of properties, mineral bonded wood composite materials could be broadly divided into two distinct groups;

1. Composite materials in which woody materials (i.e. fibers, sawdust, chips) are incorporated as an aggregate in the mineral matrix,

2. Composite materials in which the mineral binder acts purely as a binder, (i.e. wood wool cement board, particleboard, or fiberboard).

However, the three most common mineral-bonded composite products could be found in the market. These are;

- Cement-bonded composites,
- Gypsum-bonded composites,
- Magnesia cement-bonded composites,

All these mineral-based binders have been used to manufacture low-medium density (360 to 800 kg/m^3), and medium-high density (800 to 1.400 kg/m^3) products.

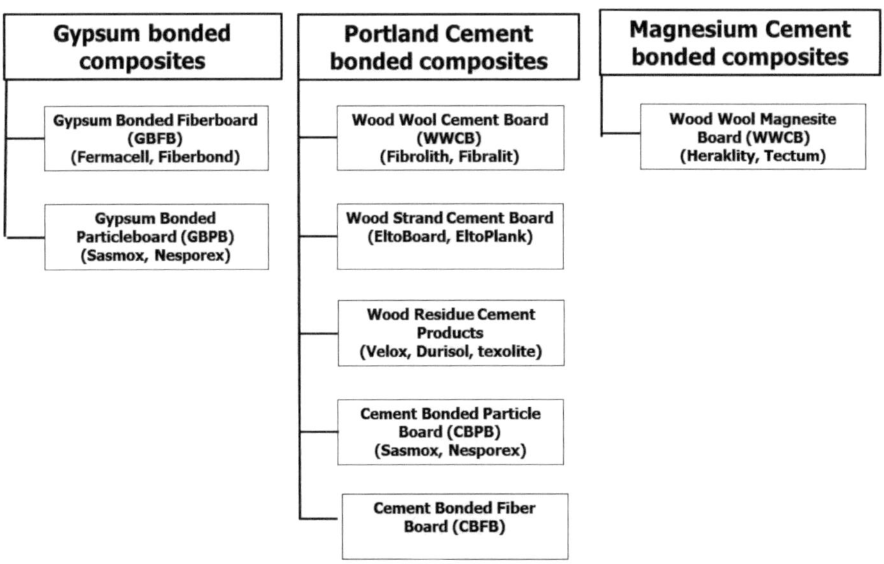

Figure 1.
The general classification of mineral-bonded composites [8].

Cement-bonded low-density products are usually called Excelsior and high-density products are called cement-bonded particleboard or fiberboard. However, Portland cement is the most common mineral binders while gypsum and magnesia cement are sensitive to moisture, and their useability is generally restricted to interior applications. Therefore, the panel materials bonded with Portland cement are considered to be more durable than others. These make cement-bonded products are useful material in both interior and exterior applications [2, 3, 9]. Although, synthetic resin-bonded panels (fiberboards and particle boards) are produced in much higher volumes due to the low cost and wider application areas, especially wood-cement bonded panels present several advantageous properties that make them more competitive for some special outdoor applications.

The general classification of mineral bonded wood composite materials and their commercialized names are briefly given in **Figure 1**. However, much valuable information and some excellent results have been provided on these products could be found elsewhere [2, 3, 8–12].

2. Compatibility of mineral binders with wood

The setting of inorganic binders is the result of a multi-complex chemical reaction causing a succession of crystallization stages. However, the hydration of mineral binder is an exothermic process which is possible to trace the compatibility by monitoring temperature changes. Typically, lignocellulosic have a cellular structure with various inhibitory substances (cellulose, hemicelluloses, lignin, extractives), and some of them dissolve in water and could be disturbed in the mineral binder crystallization [8–11]. Thereby, species compatibility varies with the type of binder and chemical constituents. But the term compatibility usually refers to 'the degree of binder setting after mixing with water and with a given wood in a fragmented form'. The inhibited reactions are generally characterized by;

- Hydration temperature,
- Hydration time,
- Retarding the cure of binders.

As briefly explain introduction section, the woody material can act both as an aggregate and as a reinforcing element in a mineral binder-based matrix system. In either case, the interaction between the binder and the element is very important. Because the particle-matrix interface is the diffusion zone, the matrix phases are connected either chemically or mechanically [8–12]. However, this diffusion region mostly influences the mechanical properties because the interfacial adhesion between particles and matrix characterizes composite materials. During hydration of cement, it created crystals in the contact layer with variable dimensions. But the crystals in the middle layer should be appropriate in the transition layer and resemble should be well bonded to each other. Because of inhibiting constituent presence, the form and dimension of the crystals could be modified. Thereby, some layers of altered crystals cannot be distinguished [9, 13].

However, cement is more sensitive to wood chemical constituents than either gypsum or magnesia cement, in most cases, the hydration times are the longest. Moreover, relative hydration times of gypsum, as also affected by selected wood species while sugars and extractives do not have as much effect on the curing and bonding of magnesia cement. The general comparative hydration properties of selected species with the inorganic binder are shown in **Table 1**.

The compatibility of wood with cement can be strongly influenced by

- Cutting and storage time of woods,
- Water to binder ratio of paste,
- Wood particle size of paste

But the content and type of sugars present in wood have been previously identified as the most critical compounds causing incompatibility, especially in softwoods [9, 13, 14]. Hence, it is very important to supply woody elements with homogeneous physical and chemical properties for the standardization of the manufacturing process.

However, the hydration of mineral binders can be improved by treating the particles or by using some additives. In the most cases, the pre-treatment is necessary, allowing to compatibility with mineral binders to obtain more suitable production [9, 14]. The aging or seasoning of wood and some degree use of chemical agents (i.e. CO_2, $CaCl_2$ and $MgCl_2$) found to be increase certain wood's compatibility with cement [9, 11–14]. The similar approaches could be useful for both gypsum and magnesia cement as well.

2.1 Cement-bonded composite materials

Portland cement is the most common type of binder in mineral based wood composite products. However, it reacts with water in a process called hydration and eventually solidifies into a hard mass. In general, the major ingredients of cement are three complex mixture of tricalcium silicate, dicalcium silicate, and tricalcium aluminate, which comprise more than 87% of the total weight [9, 13].

The advantages associated with wood elements in cement matrix system, include wide variety of species available, low density, high tensile strength, relatively low cost and well-developed technology to supply raw materials from renewable sources [9–11]. However, the use of cement in wood composites is faced with some limitations. One of the major drawback is the vulnerability of natural fibers to decompose in the alkaline environment of cement. In addition, some woods might exhibit incompatibility with cement due to specific chemical structures (sugars and extractives) that retard the cure of cement that impermeable hydrates are formed around unhydrated cement grains, which delay the setting and affect the final strength of the products [8–14]. Therefore, species selection can be important for the effective manufacturing process.

These drawbacks could be solved by several techniques which are effective in removing the detrimental components from wood. These are [9–14];

Wood or chemical	Gypsum	Magnesia cement	Portland cement
Inorganic binder	1.0	1.0	1.0
Glucose (1.0%)	1.20	1.20	Inhibited
Spruce wood	1.33	1.35	1.35
Beech wood	1.46	1.53	2.55

Table 1.
The comparative relative hydration properties (%) [13].

- Hot water extraction,
- Leaching in cold water,
- Chemical extractions,
- The use of cement curing accelerators like $CaCl_2$, $MgCl_2$ and $CaCO_3$,
- CO_2 treatment,
- Addition of pozzolans such as volcanic ash, fly ash, rice husk ash and condensed silica fume.

2.2 Gypsum-bonded composite materials

Gypsum is a sedimentary mineral and composed of calcium sulfate ($CaSO_4$) and water (H_2O), naturally in a crystalline form known as calcium sulfate dihydrate ($CaSO_4$•$2H_2O$). However, gypsum boards are typically made from a slurry of gypsum, water, and lignocellulosic fibers. Its structure is composed of interconnected needle-like calcium sulfate dihydrate crystals which entangle and rehydrated during the binder curing process (calcination), hardening to create a gypsum network [15]. Because hydrate crystals form in a gypsum-natural fiber network, wood chemical constituents especially sugars or some extractives may retard the hydration of the binder and alter crystalline structures. Typically, gypsum crystals are relatively long and have a hexagonal form, but the retarder chemicals could be influenced, the form and dimension of the crystals are altered [13, 14].

Manufacturing of wood-based gypsum boards required a higher binder (gypsum) than that needed in the bonding of composites with thermosetting resins. However, one of the main drawbacks of gypsum as a building material is its heaviness and brittleness. Hence, these boards do not have strong impact resistance for some building applications. These situations can be partly overcome by combining gypsum with various types of natural fibers (waste paper, agriculture waste fibers) to impinging better mechanical performance [15, 16]. Some advantages and disadvantages of gypsum-based composite materials are given in **Table 2**.

Advantages	Disadvantages
Ease of workability and adhesively attached to many substrates	Dry construction material that sensitive to water or moisture.
Acoustics properties that can be used partitions and floor/ceiling systems for control sound.	Very short period of setting times
Fire resistance material that gypsum could not support combustion	It has very brittle matrixes.
Cost effective manufacturing process and lower CO_2 emission compared to other construction materials (i.e. portland cement)	Water mold damage possible
Lightweight material	Environmentally not sound
The gypsum plaster supply chain has low energy consumption and	Harmful for health produce toxic SO_2 gas
Gypsum board is used to construct strong, high quality walls and ceilings	High wastage in use and manufacturing

Table 2.
Some advantages and drawbacks of gypsum bonded composites.

2.3 Magnesia-cement-bonded composite materials

Magnesia cement-based boards are formed by a chemical reaction between MgO and MgCl2, typically in a weight ratio of $MgO/MgCl_2$ (1.0/2.5–3.5 by weight). This product is quite similar to Sorel cement but has both organic additions (sawdust, wood flour) and inorganic fillers (sand, lime, or volcanic ash) [17]. However, the hydrated product is hard and strong, but the product decomposes over time by contact with water or air at high relative humidity (RH) [17, 18].

The first industrially made inorganic bonded wood composites were magnesia-bonded wood wool boards called the Heraklith boards in Europe and Tectum boards in the USA [13, 17, 18]. Recent studies show that half calcined dolomite can be partially substituted for magnesia. However, it was proposed that wood composites can also be made using a mixture of heavy magnesia and ground dolomite in combination with a solution of ammonium polyphosphate as the binder. The process is further simplified by using caustic calcined magnesia or half calcined dolomite in combination with a sparely soluble ammonium polyphosphate [13].

Fewer boards bonded with magnesia cement have been produced than portland-cement bonded panels, mainly because of cost. However, magnesia cement does offer some manufacturing advantages over portland cement. These are;

- The extractives and chemical constituents in lignocellulosic do not have as much effect on the curing and bonding,

- The magnesia cement is more tolerant of high water content during production.

These open up possibilities to use lignocellulosic not amenable to cement composites, without leaching or other modification, and to use alternative manufacturing processes and products.

In the production of this panel product, wood wool (excelsior) is laid out in a low-density mat. The mat is then sprayed with an aqueous solution of magnesia cement, pressed, and cut into panels. The cure of magnesia binders can be readily accelerated by the addition of heat. Wood-based boards made with this material are therefore compressed in a heated press. As with resin-bonded wood composites, total press time can be reduced by rapidly transferring heat to the center of the board. Steam injection pressing, a process whereby saturated steam is forced into a mat during pressing, is being successfully used to raise the center-line temperature of resin-bonded boards to curing temperatures in less than a second [13, 18, 19].

However, the addition of fluorine anhydrite has been caused by the modification of wood–magnesia bonding mechanisms that affect the stabilization of creeping deformations of the products. The fluorine anhydrite intensifies the processes of caustic magnesia solidification and causes the formation of thick structure in the wood–magnesia panel products [19].

3. Properties of mineral bonded composites

Typically, there are two types of water (free and chemically bound) in mineral bonded boards. This is important because it contributes to the fire resisting behavior. Hence, when exposed to fire, these materials undergo reactions in which the water is gradually driven off at temperatures above 100°C. However, considerable high-level heat energy is required to evaporate the free water and for the chemical reaction to release the water in the crystal structure. Moreover, those have also usually low heat transfer coefficients and are capable of quick release of the humidity.

However, wood-cement bonded composites have become environmentally benign sustainable materials for the constructions due to reducing material costs by combining a lower cost material [9, 20]. Hence, these products have very good dimensional stability, high fire resistance properties, and impart additional energy absorbing capacity to the matrix system. The wood cement composites typically show improved ductility, flexibility, and crack resistance when compared to neat cement concrete [9–12]. In addition, besides being high strength properties, these products could be provided well protection against decay and insect attacks as well [21, 22].

The mechanical properties of gypsum base panels are closely correlated with panel porosity, water/gypsum ratio, network structure, intercrystalline interaction, crystal sizes, and aging time. Although the hydrated gypsum and magnesia bonded boards are hard and strong, the product decomposes over time by atmospheric effects at high relative humidity (RH) [17]. Hence, the gypsum and magnesia bonded boards are not recommended as sheathing in exterior facades or any other application where the boards are in contact with a moist climate. In contrast, the cement-bonded boards showed excellent dimensional stability and only a slight reduction in mechanical properties after outdoor exposure for years, greatly outperforming other wood-based panels.

The modified magnesium polyphosphate-bonded particle boards are some similar properties to those obtained with magnesium sulfate while shrinkage decreases due to carbonization of wood–magnesia matrix, and the density, strength, and water resistance increases [13, 19].

There are numerous studies for determining the suitable board configurations on the end-use applications. But many properties such as; strengths, fire resistance, sound absorption, and insulation behaviors of panels, are primarily influenced by the density of the product and the binder/wood ratio. **Tables 3** and **4** show the general and physical comparative properties of mineral-based wood composite materials.

The cement-cellulosic substrate matrix is a complex system that can be given different properties and the resulting products can be used for a broad variety of applications. However, cellulosic fibers are well bonding ability to each other. Especially well-fibrillated fibers are more flexible and have a higher area available for bonding. This is possible by using refiners which are breaking the primary wall and the fibrils from the secondary wall will stick out. This will increase the surface area for bonding and therefore increase composite strength. The results presented in **Table 4** support this hypothesis.

	Gypsum boards	Wood-cement boards	Magnesia boards
Water resistance	+	+++	+
Fire resistance	+++	+++	+++
Fungal/mold/termite resistance	++	+++	++
Acoustic insulation	+++	++	++
Lightweight	+++	+	+
Nail holding capacity	+	+++	+++
Workability	+++	++	++
Durability	+	+++	++

Table 3.
General Properties of mineral bonded boards composites [9–12, 19, 23].

Wood composites	Density (kg/m^3)	Binder/wood ratio	MOR (MPa)	IB (MPa)
Cement-bonded particleboards	1000–1350	2.9	6.0–15	0.4–0.6
Cement-bonded fiberboard	1000–1350	10	12–20	0.8
Gypsum- bonded particleboards	1000–1200	4.0	6.0–9.0	0.3–0.6
Gypsum- bonded fiberboards	900–1000	5.0 to 6.0	4.0–7.0	0.3–0.5
Magnesia-bonded particleboard	900–1250	1.5	7.0–14	0.4–0.6
Magnesia-bonded fiberboard	700 to 1100	5.0	8.0 to 10	—

Table 4.
Comparative physical Propoerties of mineral bonded boards boards [23].

4. Uses of mineral bonded composites

An acceptable property from mineral bonded panels is dependent on both the type of the binder and wood properties. All these materials are considered non-toxic yet and commonly referred to as being virtually incombustible. Due to very high dimensional stability and physical properties, the cement-bonded products could be useful for many external applications including; exterior siding, agricultural buildings, pre-fabricated structures, mobile buildings, roofing, flooring, industrial and exterior domestic cladding, tunnel linings, highway sound-barriers, fire-barriers and paving tiles. However, the low-density cement-bonded boards (Excelsior) could be used for high-performance applications and improved acoustic and damping properties such as; fire-resistant, sound-absorbing walls, ceilings, and thermal insulation panels [24]. But it is important to note that the utilization of cement-bonded composites is highly dependent on construction techniques, esthetics, safety and energy regulations, and all the other underlying factors which determine public acceptance of a product. The vast literature on cement bonded-wood-based composites, their properties, and manufacturing variables could be found elsewhere [9, 20]. **Table 5** shows comparative use ability and **Figure 2** shows some examples of mineral bonded composites.

The gypsum-based composites are a well-known low-cost material and frequently used to finish interior wall and ceiling surfaces that are often called drywall, wallboard, or plasterboard [20, 25]. Thereby, these materials could be useful for both residential and non-residential construction applications. However, the paper-faced gypsum boards have been widely used since the 1950s for the interior lining of walls and ceilings which are appropriate to fire ratings [20]. The paper-faced gypsum boards also find use areas as exterior wall sheathings. The facings of drywall and gypsum sheathing panels are adhered to the gypsum core, providing the panels with impact resistance, and bending strength, and stiffness. An alternative to adhered facings is to incorporate lignocellulosic fiber (typically recycled paper fiber) in the gypsum core to make what is termed fiber-reinforced gypsum panels. Moreover, the gypsum sheathing panels are primarily used in commercial construction, usually over steel studding, and are distinguished from gypsum drywall by their water repellent additives in the paper facings and gypsum core [8, 15, 20]. It has already well established that natural fiber-reinforced gypsum panels (wood fibers) are typically stronger and more resistant to abrasion and indentation than paper-faced drywall panels and also have a moderate fastener-holding capability [8, 20, 25]. Although gypsum-based boards have usually been marketed for use as interior finish panels (drywall), some hydrophobic additives can provide a certain level of water resistance, for use as sheathing panels, floor, or roof

	Cement bonded composites	Gypsum bonded composites	Magnesia bonded composite
Exterior and partition walls	+++	+	+
Coating of the wall	+++	+++	++
Acoustic and thermal insulation	++	+++	++
Decoration	++	+++	++
Flooring	+++	+	++
Large size prefabricated elements	+++	+	++
Roofing, shingles and shade	++	++	++
Ceilings and architraves	++	++	++
Fire resistant construction	+++	+++	+++

Table 5.
General uses of mineral bonded composites (+: low level; ++: medium level; +++: high level).

Figure 2.
The uses of mineral-bonded composites in some applications.

underlayment, or tile-backer boards. In addition, the gypsum-based panels have a low thermal coefficient and low solid contents that these properties are good for insulating against heat and sound while the mechanical strength of the gypsum-based composites is still retained [8, 9, 16, 20].

Although composites bonded with magnesia cement are considered water sensitive, they are much less so than gypsum-bonded composites. One successful application of magnesia cement is a low-density panel made for interior ceiling and wall applications. However, the gypsum and magnesia-based wood composites have also presented high dimensional stability and resistance against biodegradation while well fire resistance and some level sound insulation properties. This is attributed to the lower content of organic matter and the crystal water in the binder. These special properties make these products could be useful for a wide variety of purposes in construction applications. Thereby, the low-density products could be useful as interior ceiling and wall panels while high-density panels could be used as complete wall and roof decking systems. Moreover, exterior-type panels are coated with stucco, and the interior is a gypsum board. These are also useful for decorative and sound barrier purposes in constructions [22, 26].

However, high-density magnesia and/or Portland cement-bonded boards can be used as flooring, roof sheathing, fire doors, and load-bearing walls. But complex shapes, such as decorative roofing tiles or non-pressure pipes, can be molded or extruded as well. The largest volume of cement-bonded wood-based composite materials manufactured in North America is fiber (pulp)-cement siding [27]. Moreover, cement-bonded panels can be used as low-cost housing systems in developing countries such as; rural prefabricated structures, mobile homes, structural insulation panels so on [9].

The Magnesia and gypsum boards might also be used outdoors but must be protected from direct exposure to the weather because of sensitivity to moisture.

5. Conclusions

The markets for inorganic-bonded wood composites vary throughout the world. However, there is a great potential for the use of wood species to make mineral-bonded composites. Substantial markets for these panels have been developed for various construction end-uses (i.e. sheathing and siding) with insulation partitions. These products may provide an option for using lignocellulosic residues for improved properties like fire and sound insulation characteristics and hence can be used as wall covering and filling material in the constructions. Moreover, inorganic-bonded boards could be adapted to the wood frame construction techniques used for residential housing.

The use of mineral-bonded composites is highly dependent on building codes, safety and fire regulations, construction techniques, esthetics, availability of materials, and all the other underlying factors which determine public acceptance of a product. However, cement-bonded wood composites (WBC) have been taken as very stable dimensionally when subjected to outdoor applications. However, gypsum boards are commonly used as a lining material in walls, ceilings, and wall partitions.

Author details

Halil Turgut Sahin[1]* and Yasemin Simsek[2]

1 Faculty of Forestry, Depertment of Forest Products Engineering, Isparta University of Applied Sciences, Isparta, Turkiye

2 Faculty of Engineering, Depertment of Civil Engineering, Suleyman Demirel University, Isparta, Turkiye

*Address all correspondence to: halilsahin@isparta.edu.tr

References

[1] Bowyer JL, Shmulsky R, Haygreen. Forest Products and Wood Science: An Introduction, 4nd ed. Iowa State University, Ames, IA. 2003. 553p.

[2] FPL. Wood Handbook: Wood as an Engineering Material. USDA Forest Service, FPL-GTR-190, Madison, WI. 2010. 509p.

[3] Maloney TM: The family of wood composite materials, Forest Products Journal. 1996. 46 (2):18.

[4] Rowell RM. Opportunities for composites fron agro-based resources, In: Rowell RM, Young RA, Rowell JK, editors. Paper and Composites from Agro-Based Resources, CRC Press, Inc, Boca Raton, FL; 1997. p.249-300.

[5] Youngquist JA, English BE, Scharmer RC, Chow P, Shook SR. Literature Review on Use of Nonwood Plant Fibers for Building Materials and Panels, USDA Forest Service, FPL-GTR 80, Madison, WI.1994.

[6] Sahin CK, Onay B. Alternative wood species for playgrounds wood from fruit trees, Wood Research. 2020. 65(1): 149-160.

[7] Sahin CK, Topay M, Var AA. A study on suitability of some wood species for landscape applications: surface color, hardness and roughness changes at outdoor conditions, Wood Research. 2020: 65(3):395-404.

[8] Van Elten GJ. Innovation in the production of cement-bonded particleboard and wood-wool cement board, 5th Int. Inorganic Bonded Wood and Fiber Comp. Mat. Conf. 1996. Spokane, Washington, USA.

[9] Sahin HT, Kaya AI, Yalçın OU, Kılınçarslan Ş, Şimşek Y, Mantanis GI. A study on the production process and properties of cement-based wood composite materials. The Journal of Graduate School of Natural and Applied Sciences of Mehmet Akif Ersoy University. 2019: 10(2): 219-228.

[10] Frybort S, Mauritz R, Teischinger A, Muller U. Cement bonded composites-a mechanical review. BioResources 2008: 3(2): 602-626.

[11] Jorge FC, Pereira C, Ferreira JMF. Wood–cement composites: A review. Holz als Roh– und Werkstoff. 2004: 62(5):370-377.

[12] Moslemi AA. Inorganically bonded wood composites. Chemtech. 1988: 18, 504-510.

[13] Simatupang MH, Geimer RL. Inorganic binder for wood composites: feasibility and limitations, Wood adhesive symposium proceedings, May 16-18, 1990. Madison, WI. p. 169-176.

[14] Hachmi M, Moslemi, AA, Campbell AG, Idaho M. A new technique to classify the compatibility of wood with cement. Wood Science and Technology. 1990. 354 (608): 345-354.

[15] Herhández OF, Bollatti MR, Rio M, Landa BP. Development of cork-gypsum composites for building applications, Construction and Building Materials. 1999. 13, 179-186.

[16] Shiroma L, Camarini G, Beraldo AL. Effect of wood particle treatment on the properties of gypsum plaster pastes and composites. Matéria (Rio de Janeiro). 2016. 21(4): 1032-1044.

[17] Hansen KK, Bunch-Nielsen T, Grelk B, Rode, C. Magnesium-Oxide boards cause moisture damage inside facades ın new danish buildings. Int. RILEM Conf. on Mat., Systems and Struc. in Civil Eng. Conf. 22-24 August 2016. Technical University of Denmark, Lyngby, Denmark.

[18] JamesHardie. Information on Magnesium Oxide (MgO) and Magnesium Chloride (MgCl) Building Products [Internet].2010. Availabe from: https://www.jameshardiepros.com/getattachment/58e0a4f4-9f6c-4220-9a57-ce9afe6474d3/information-on-magnesium-oxide-and-magnesium-chloride.pdf [Accessed 2021-04-11].

[19] Plekhanova TA, Keriene J, Gailius A, Yakovlev GI. Structural, physical and mechanical properties of modified wood–magnesia composite. Construction and Building Materials. 2007. 21, 1833-1838.

[20] Semple KE, Evans PD. Wood-cement composites-suitability of Western Australian mallee eucalypt, blue gum and melaleucas, Rural Industries Research and Development Corporation. 2004. Publication No. 04/102. Kingston, AU.

[21] Goodell B, Daniel G, Liu J, Mott FR. Decay resistance and microscopic analysis of wood-cement composites. Forest Prod J. 1997. 47(11/12):75-80.

[22] Kozlowski R, Mieleniak B, Helwig M, Przepiera A. Flame resistant lignocellulosic-mineral composite particleboards. Polymer Degradation and Stability. 1999. 64: 523-528.

[23] Moslemi, AA, Begum S. Wood Composites: Mineral-Bonded, Reference Module in Materials Science and Materials Engineering, 2017. Elsevier

[24] Berkenkamp R. Wood fiber-cement product, process and properties. In: Moslemi, AA. Editör, Inorganic-Bonded Wood and Fiber Composite Materials. Madison, WI, 2017. Forest Products Society. p.8-13.

[25] Youngquist JA. Wood-based Composites and Panel Products, In: Wood handbook: wood as an engineering material, USDA Forest Service, FPL; GTR-113. 1999. p. 10.1 10.31.

[26] Ashori A, Tabarsa T, Azizi K, Mirzabeygi R. Wood–wool cement board using mixture of eucalypt and poplar. Ind. Crop. Prod. 2011. 34 (1):1146-1149.

[27] Stark NM, Cai Z, Carll C. Wood-based composite materials: Panel products, glued-laminated timber, structural composite lumber, and wood-nonwood composite materials. In: Wood handbook: wood as an engineering material, USDA Forest Service, FPL; GTR-190. 2010. p. 11.1-11.28.

Chapter 14

Timber-Concrete Composite Structural Elements

Anita Ogrin and Tomaž Hozjan

Abstract

Timber-concrete composites are interesting engineered wood products usually used for structural elements, which are mainly subjected to bending load; from simple floor systems to long-span bridges. This way, the advantage can be taken of timber tensile strength and concrete compression strength. The chapter begins with an introduction of various types of timber-concrete composite structural elements regarding type of the element, connection type and types of timber and concrete. Next, specific characteristics and advantages of timber-concrete composite structural elements are thoroughly discussed from viewpoints of engineering, architecture, builders and ecology. Furthermore, basic mechanical principles of timber-concrete composite structural elements are presented and some design methods are briefly described. Finally, worldwide inclusion of timber-concrete composite structures in currently applicable standards is discussed.

Keywords: timber-concrete composite, floor systems, mechanical principles, advantages, design methods, standards

1. Introduction

Timber-concrete composite (TCC) structural elements are usually horizontal elements, which carry the load in one direction (so called one-way spanning elements) subjected to uniaxial bending. Timber and concrete part of TCC element are connected with one of several types of connectors in order to achieve composite action [1]. Since concrete has almost negligible tensile strength, timber is commonly on the lower side of the element, where tensile stresses are expected, and concrete is on the upper side, where compressive stresses occur. While there are also some reversed TCC structural elements with concrete on the bottom side [2] and even some TCC wall systems [3], this chapter focuses on most typical application of TCC structural elements described above.

A development of TCC structural elements began around 100 years ago in Germany, aiming at renovation and strengthening of existing timber floors. Paul Müller patented TCC floor system, made of upright timber boards with concrete topping in 1922. Another patent was received by Otto Schaub in 1939 for TCC slab made of timber ribs and concrete slab, connected with Z- or H-shaped steel connectors [1]. While in Europe TCC elements were firstly used as floor systems only, TCC bridges of short and medium span were developed in 1930s in America, as a result of shortage of steel for concrete reinforcement. Construction of TCC bridges spread to Australia and New Zealand in 1950s, where they were probably built by US army. In Europe, TCC bridges did not appear until 1990s [4].

a)

b)

Figure 1.
Examples of structures with TCC structural elements. (a) Bridge over river Agger, Germany [5]. (b) LifeCycle Tower One, Austria: prefabricated TCC slab during construction (left) and finished building (right) [6].

Nowadays, however, TCC structural elements are getting more and more popular worldwide. They are used both in restoration and renovation of existing, often historical structures as well as for new buildings and bridges. **Figure 1** shows TCC bridge over river Agger in Germany, which was built in 2014 [5], and 8-story building called LifeCycle Tower One in Austria, which has TCC horizontal structural elements and was built in 2012 [6].

In this chapter, various types of TCC floor systems are presented and described regarding type of the element and connection type, as well as types of timber and concrete. Next, specific characteristics and advantages of TCC structural elements in comparison with fully concrete or fully timber floor systems are thoroughly discussed from viewpoints of engineering, architecture, construction process and ecology. Furthermore, basic mechanical principles of TCC structural elements are presented, together with most often used simplified methods for their design. The chapter concludes with discussion on worldwide inclusion of TCC structures in currently applicable standards.

2. Variety of designs

TCC structural elements are very diverse. While there are only two main types of geometry of TCC floors (beam and slab type), there are a lot of different types of connections between timber and concrete with considerably different behaviour.

Also, there are several possibilities for type of concrete and even more for type of timber.

2.1 Beam type and slab type TCC floor systems

TCC structural elements are usually slab-like elements (floors in buildings and decks of bridges) and can be either of uniform thickness or ribbed, depending on several factors such as aesthetics, height limit and availability of materials. At the same time, TCC structural elements are almost always one-way-spanning elements, and can be modelled as beams, loaded with uniaxial bending. Therefore, there can be some confusion with the terms: is it a TCC beam or is it a TCC slab? Hereinafter, the following terms will be used: beam type of TCC floors and slab type of TCC floors. Beam type cross-section consists of a timber web and a concrete flange, i.e. timber part is much narrower than the concrete part. Consequently, the neutral axis of the entire TCC cross-section is located in the web. On the other hand, concrete and timber parts of slab type have equal widths and neutral axis is often in the concrete part. Besides the difference in appearance of both types, presented in **Figure 2**, there is also the difference in location of the neutral axis, which influences the portion of the concrete where tensile cracking can occur, and consequently influences appropriate design methods.

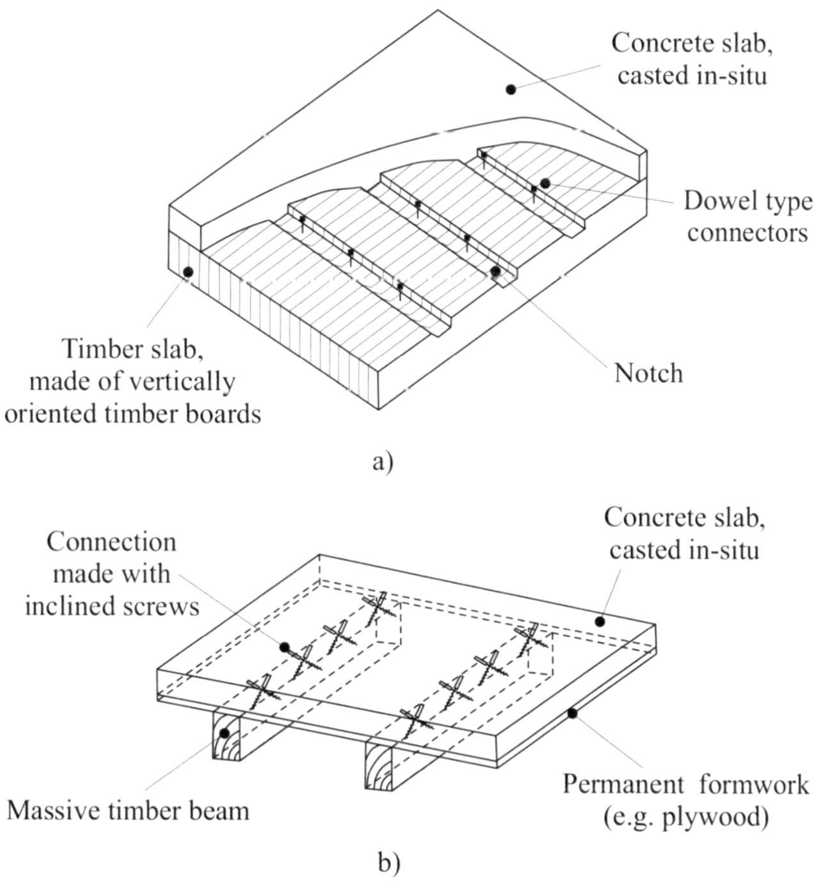

Figure 2.
Examples of slab type and beam type of TCC floor systems. (a) An example of slab type of TCC floors. (b) An example of beam type of TCC floors.

2.2 Materials

Timber part of TCC cross-section can be made from massive timber (beams or boards) or engineered wood products such as laminated veneer lumber (LVL), cross-laminated timber (CLT), glued laminated timber (Glulam), ... Especially with slab type of TCC floors there is also a matter of orientation of timber components; boards, plates and lamellas of LVL beams can be orientated either horizontally (placed one on top of another) or vertically (placed next to each other) [7]. Furthermore, while beam type of TCC floor usually has web made of one timber beam, there can also be two connected timber beams next to each other [8]. Such solution is necessary in case of certain type of connectors (e.g. nailplates), but can also be used in other cases.

Concrete in TCC structural elements is usually normal-weight concrete, reinforced only with small amount of steel reinforcement in order to prevent cracking of concrete due to its shrinkage. However, some research was conducted regarding the use of light-weight concrete with the aim of achieving smaller self-weight of TCC element (e.g., [9]). Furthermore, steel-fibre-reinforced concrete (SFRC) in TCC structural elements has proven to reduce possibility of explosive spalling of concrete in fire conditions [10].

The lower side of concrete part of TCC cross-section could be bare, or it can be covered with permanent formwork made of thin timber layer (e.g. plywood).

2.3 Timber-concrete connection

Timber-concrete connection has major influence on behaviour of TCC structural element. The perfect connection has enough strength for transfer of shear forces between the two materials, enough stiffness to allow only limited slip and enough ductility to avoid brittle failure of the connection. Additional, non-mechanical properties of the connection, such as cost and handiness, can also influence the choice on the connection type [1]. However, the perfect connection is nearly impossible to achieve in practice. Several connection systems have been developed, each with its own advantages:

- *Dowel type fasteners* include dowels, screws, inclined screws, nails and other metallic connectors. With exception of inclined screws, dowel type fasteners represent a connection system with the lowest stiffness and the largest ductility [11].

- *Notched connections*, where notches can be of various shapes (rectangular, circular, with vertical sides or with inclined sides, ...), are among the most brittle connections. Ductility of notched connection is greatly improved with addition of steel fastener (dowel or screw) into the indentation [11].

- When strength of *glued connections* is exceeded, they experience brittle failure. Their advantage is that they are very stiff and enable almost full composite action. Also, they ensure uniform distribution of shear forces and can be used for connection of timber with prefabricated concrete slab [1].

- *Connections with nailplates* have various designs of plates; one of the possibilities is a plate with lower part designed as a nailplate, positioned between two timber beams, and with the upper part designed as a perforated plate [8].

- *Friction based connections* are based on a system with upright positioned timber boards of altering heights, and are mainly used in Switzerland [1].

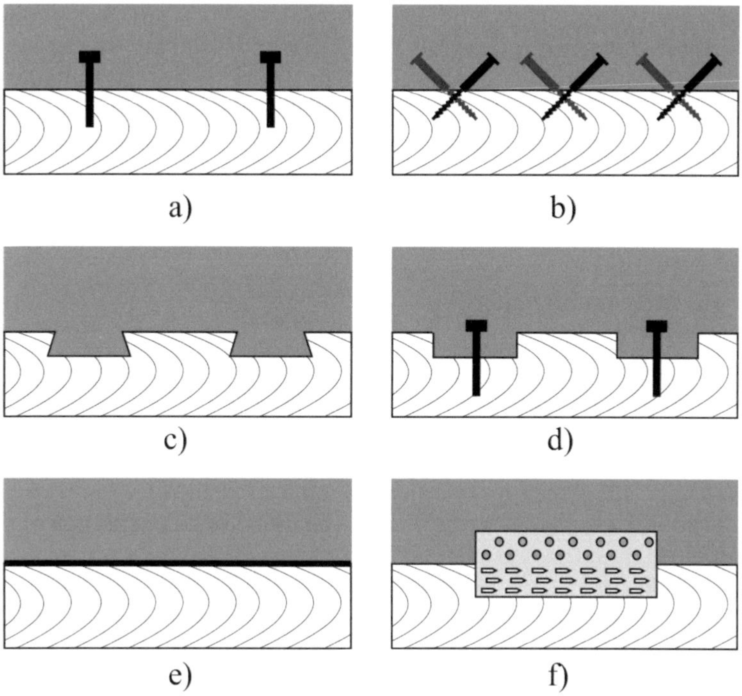

Figure 3.
Connection types. (a) Connection with dowels. (b) Connection with inclined screws at 45° in two rows. (c) Notched connection with inclined sides. (d) Notched connection with vertical sides and dowels. (e) Glued connection. (f) Connection with nailplates.

Some of the connection types are schematically presented in **Figure 3**. An interested reader can also find graphical comparison of stiffness and ductility of various connection systems (with respect to their normalised strength) in [11]. Additionally, more variations of connections can be found e.g. in [12].

In one-way-spanning simply supported floors, which is most often the case for TCC floors, the greatest shear force in the connection occur close to the supports, while it can be negligible in the middle of the span. The failure of one connector near the supports would lead to an additional load in the remaining connectors, which could result in more brittle failure of the TCC element [11]. In order to avoid that, the spacing between separate connectors near the supports is sometimes reduced, while at the midspan the spacing can be increased [13].

3. Characteristics of TCC structural elements

Due to their composite nature, TCC structural elements have pretty specific characteristics, some of them represent advantages over fully timber elements and other over fully concrete elements.

3.1 Aesthetics and convenience

From architectural point of view, timber, as a natural structural material, is often preferred over concrete. Therefore, timber parts of TCC structural elements are usually left to be visible due to aesthetic reasons. Furthermore, visible timber parts are desired when TCC structural elements are used in restoration of historical buildings.

In comparison with fully concrete structural elements of similar mechanical performance, TCC structural elements have lower self-weight. Among other benefits, this contributes to easier transport of materials to the construction site, as well as to smaller seismic forces on the finished structure.

Moreover, TCC structural elements can be prefabricated, either as a whole element or timber and concrete part separately. This can expedite the process of construction on site and the use of resources can be more controlled in a workshop, which result in smaller amount of waste.

3.2 Strength, stiffness and ductility

Strength of a structural element governs maximum load that the element can withstand without failure. In comparison with traditional timber-only floor systems, TCC floor systems (with properly connected timber and concrete parts) achieve up to three times higher load carrying capacity and up to six times greater bending stiffness [12]. Concrete on the upper side contributes to compressive load capacity of the TCC element, and timber on the lower side contributes to its tensile load capacity.

Bending stiffness of structural element is a resistance of an element against bending deformations, i.e. it determines how large deflections will occur under certain bending load. Greater bending stiffness of TCC elements, which means smaller deflection, is achieved due to elastic modulus of the concrete being several times higher than elastic modulus of the timber. In-plane rigidity of TCC floors is also improved when compared to timber floors. In fact, TCC floor systems can act as horizontal bracing of the structure and thus improve its seismic response, if both timber and concrete part of TCC element are connected to the walls [12].

Ductile structural elements can sustain large plastic deformations before failure. Properly designed and executed connection between timber and concrete can lead to improved ductility in comparison with fully concrete or fully timber floors. In order to achieve that, the connection must be sufficiently strong and rigid [1]. However, if the connection is too strong and stiff, a brittle tensile failure of the timber may occur prior to plastification of the connectors [12].

The bending stiffness of TCC structural element (which depends on the stiffness of the connection) influences distribution of deformation and stresses over TCC cross-section, and consequently importantly influences verification of ultimate limit states (ULS) as well as of serviceability limit states (SLS) [1]. As both timber and concrete are rheologically active materials, initial and long term response of TCC structural elements must be checked. Calculation of strength and stiffness of TCC elements is discussed in Section 5.

3.3 Sound insulation and vibrations

Normal-weight concrete has density around 2400 kg/m^3, which is three to five times higher than hardwood and around five to eight times higher than softwood (depending on the quality class of timber) [14]. As a result, TCC floor systems provide better insulation for air-transmitted sound than fully timber systems. Furthermore, in comparison with fully concrete systems, TCC floor systems provide improved insulation for impact noise, due to increased damping characteristics [12].

As already mentioned, TCC structural elements can achieve longer spans than fully timber floors, due to higher strength and bending stiffness. However, a research on one TCC slab type floor system has shown that an increase of height of concrete part with the aim of achieving higher load capacity and consequently longer span has its limit. When this limit span is exceeded, vibration performance becomes the controlling design parameter. The reason is in reduced natural frequency of TCC element as a result of increased self-weight of the element due to increased concrete height [15]. Because of low natural frequency, TCC floor systems could be categorised as susceptible to resonance [16]. Nevertheless, the viscous damping ratio of TCC systems is higher than that of fully timber systems. Consequently, "springiness" felt by the users when jumping or walking on the floors is reduced and users are less annoyed [12]. It was experimentally confirmed, that the achieved values of occupancy annoyance for TCC floors are way below the annoyance limit proposed in Eurocodes [16].

3.4 Fire conditions

Behaviour of TCC structural elements in fire conditions is an important factor in their design. First of all, there are certain phenomena, inherent to each of the two materials, which occur at exposure to high temperatures. In timber, being a combustible material, pyrolysis occurs at elevated temperatures, resulting in reduction of material properties and eventually (at approximately 300°C) in charring of material [17]. The charred layer of the timber part of TCC cross-section have negligible strength and stiffness and is for design purposes considered as mechanically completely ineffective layer. However, char works as a thermal insulation and therefore protects the remaining cross-section. On the other hand, mechanical properties of concrete are reduced when temperature exceeds 400°C [18]. Concrete part of TCC cross-section is usually quite thin slab, reinforced with small amount of steel reinforcement (aimed only at prevention of concrete cracking as a result of shrinkage). It has been shown that thin concrete slabs are more likely to experience explosive spalling than thicker ones [2]. Another research has shown that explosive spalling in beam type of TCC floor can be avoided with the use of concrete reinforced with steel fibres (SFRC) [10].

Geometry of the TCC cross-section changes during fire duration due to charring of timber and spalling of concrete. Additionally, height of the timber part can be reduced in case of laminated timber (LVL, CLT ...) if separate lamellas fall off [13]. This can occur either due to charring and result in lesser thermal insulation of the remaining cross-section or earlier due to failure of glued connection between lamellas, which causes additional reduction of otherwise effective cross-section.

Elevated temperatures influence behaviour of the connection between timber and concrete as well. Reduction factors for both strength and stiffness of the connection depend on: (i) type of the connection, (ii) initial timber cover i.e. distance between initial boundary of the cross-section exposed to fire and the connectors and (iii) development of elevated temperatures and related charring progress. It appears that lateral timber cover in beam type of TCC floors is more important for protection of the connection than bottom timber cover [19].

An obvious, yet important difference presents itself between beam type and slab type of TCC floor systems in fire conditions. With beam type, there are three sides of timber part (lower and both lateral sides) directly exposed to fire and charring progress. Additionally, almost entire lower side of the concrete slab is either immediately exposed or initially protected with thin layer of permanent formwork only. On the other hand, only lower side of timber slab in slab type of TCC floor system is directly exposed, therefore, heat transfer and charring process can be considered one-dimensional [7].

A different approach to TCC floor systems, so-called reverse TCC system, has concrete slab below timber beams and timber deck. An important function of concrete slab in such systems is fire protection of the timber part. The research has shown that very thin concrete slabs already provides efficient thermal insulation and thickness of the slab is actually governed by required cover depth of steel reinforcement and by prevention of explosive spalling of concrete [2].

Thus far, behaviour of TCC structural elements in fire has been extensively experimentally investigated [20] and both analytical simplified methods and numerical methods for fire safe design have been developed, see for example [19, 21, 22], respectively. However, the variety of designs and complexity of behaviour of TCC structural elements in fire conditions calls for further research.

3.5 Ecology aspect

Design of buildings with TCC structural elements can contribute to more ecological and sustainable built environment in several ways.

Firstly, TCC floor systems were developed and are still being used for strengthening of existing timber floors with added concrete layer or as their replacement. Consequently, the lifetime of renovated building is prolonged and thus its sustainability is improved.

Furthermore, in comparison with reinforced concrete structural elements with similar structural performance, TCC elements contain less concrete and, clearly, more timber. This brings the following ecological advantages:

- Unlike concrete, timber is a *renewable natural resource*.

- Carbon footprint is defined as sum of greenhouse gas emissions caused by organisation, event, product or individual expressed as carbon dioxide equivalent. Several studies have shown that timber-based buildings generally have *lower carbon footprint* and lower energy consumption during their life cycle (i.e. *lower embodied energy*) than concrete buildings with comparable heating and cooling requirements [23]. Research on cradle-to-gate environmental performance has also confirmed very low values of carbon footprint and embodied energy for prefabricated TCC wall element [3].

- Timber part of TCC element represents *storage for carbon*, which was captured there during tree growth. In order to avoid release of the captured carbon in the atmosphere and consequential increase of carbon footprint it is important to apply available carbon capture and storage (CCS) technologies for timber incineration at the end of its lifetime [23].

Since concrete, which is obviously less friendly to the environment than timber, importantly influences structural performance of TCC element, the question remains: "What geometry of TCC cross-section is optimal for satisfying both structural and environmental needs?" It was shown for a specific slab type of TCC floors, that for achieving longer spans than 7 m it is better to increase height of timber part, while keeping concrete thickness in specified range [15]. However, due to many different designs of TCC structural elements that is not always the case.

4. Mechanical principles

TCC structural elements are most often simply supported one-way-spanning floors (either slab type or beam type), subjected to vertical external load resulting in positive bending moment only. Each of the two parts of the TCC cross-section (i.e. timber part and concrete part) is usually rectangular. Mechanical principles in this section are presented accordingly.

Due to positive bending moment, tensile normal stresses occur on the lower half of each part of the TCC cross-section and compressive normal stresses occur on their upper halves. The connection between timber and concrete restricts horizontal slip on the contact and transfers shear force, which is in equilibrium with internal normal force in each part of the cross-section. The magnitude of the shear force and inherent magnitude of the constant normal stresses in each part of the cross-section are determined by the stiffness of the connection. Therefore, stiffness of the connection between timber and concrete has important influence on total stress distribution over cross-section. Theoretically, there are three possibilities: perfectly rigid connection, semi-rigid connection and without connection [22]. *Perfectly rigid connection* means that there is no horizontal slip between timber and concrete parts. Shear force is fully transferred, which enables full composite action of TCC cross-section. Such connection is desired because it results in the greatest bearing capacity of the TCC cross-section with the smallest possible total normal stresses as well as the smallest deflections of the element. However, perfectly rigid connections are practically impossible to achieve in reality, with exception of glued connections. On the other hand, timber and concrete parts *without connection* (i.e. with un-prevented slip between both parts) work completely separately. There is no shear force

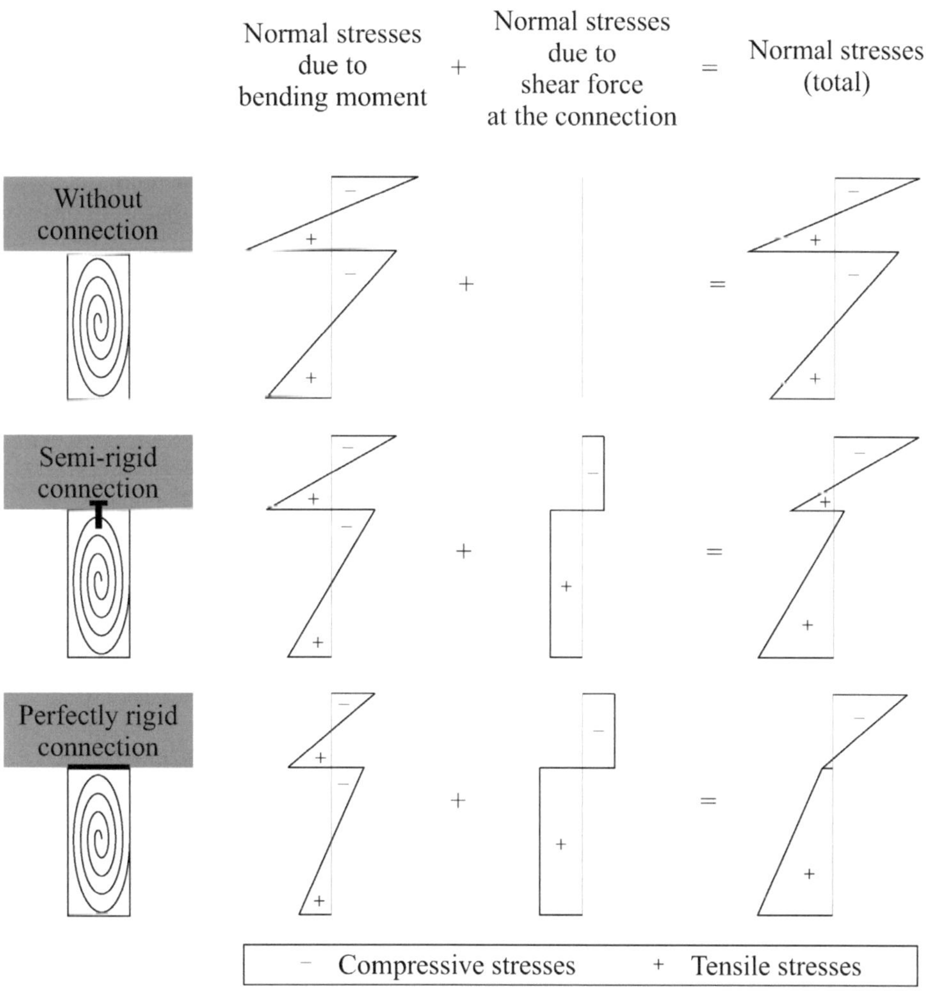

Figure 4.
Stress distributions over TCC cross-section for perfectly rigid connection, semi-rigid connection and without connection.

transferred and consequently there is no composite action between the two parts which causes the lowest bearing capacity of the TCC cross-section with the largest possible total normal stresses and deflections of the element. In reality, the connections between timber and concrete part of TCC element are *semi-rigid*; somewhere in-between the two aforementioned extreme options. Stress distributions over TCC cross-section for each of the three possibilities are depicted in **Figure 4**.

Failure of TCC structural element occur when any of its components fails; either the outermost fibres of timber part fail in tension, or the outermost fibres of concrete part fail in compression, or the connection fails due to exceeded shear capacity. Type of the connection determines its failure mode. For example, for notched type of the connection with dowels, shear failure of the dowel occur together with local compressive failure of timber (parallel to fibres) [22]. On the other hand, screwed connection with inclined screws (usually at 45°) in two rows work as a virtual truss with screws representing tensile and compressive diagonals; at the failure, tensily loaded screws fail due to withdrawal from timber part, while screws in compression experience buckling [19].

5. Currently available simplified design methods

Behaviour of TCC structural element is quite complex, mostly due to non-linear response of semi-rigid timber-concrete connection. Therefore, currently available analytical design methods introduce several simplifications.

Majority of simplified design methods for TCC structural elements (see e.g., [19, 24]) is based on so-called gamma method [25], developed in 1956, with one of its formulations being presented also in EN 1995-1-1, Appendix B [26]. The method introduces gamma coefficient (γ), which describes slip stiffness of the connection and depends on type of the connectors (their slip modulus) and on spacing of the connectors. Provided that geometry and material properties of the connected parts (i.e. timber and concrete part) are known, effective bending stiffness of TCC cross-section can be calculated. With that, stress distribution over cross-section can be determined.

Another approach includes a crude simplification and completely disregards slip in the connection. Thus, Bernoulli hypothesis can be applied and stress distribution over TCC cross-section in ULS (due to exceeded tensile or compressive bearing capacity) can be determined, see e.g., [21, 27]). The advantage of this method in comparison with the gamma method is in possibility to consider cracking of concrete as a result of exceeded tensile strength. Cracked concrete cannot contribute to effective stiffness of the cross-section; as the height of the cracked part depends on stress distribution, the effective bending stiffness according to gamma method could not be determined.

There are also some simplified methods available for fire safe design of TCC structural elements. These methods are often basically the same as the methods for design at normal temperatures, but upgraded to include charring of timber and reduction of material properties due to elevated temperatures. Usually (but not always), simplified methods for reduction of geometry and material properties, which are already established and described in standards for timber or for concrete, are applied [17, 18].

6. Recommendations for construction workers

While there are several proprietary TCC systems on the market [1], with detailed instructions for builders, Ceccotti [12] gives some general recommendations for construction of TCC structures. Mostly, he warns against water presence (use of wet timber should be avoided, timber should be protected from moisture when casting the concrete and connectors should be protected against corrosion). Regarding concrete, he advices to reinforce thicker layers of concrete, to use mixture with lower water/cement ratio and to leave propping in place longer than it is usual for fully concrete elements. Furthermore, he advices to avoid timber species, which could adversely react with cement (such as larch due to high-sugar-content extracts).

7. Inclusion of TCC structural elements in standards around the world

TCC structural elements are currently still very scarcely considered in design standards and guidelines worldwide. Most often, some information, which apply to timber-concrete connections, can be found in other standards (e.g. in standard for timber structures). There are also few specifications for timber-concrete bearing systems in some standards concerning bridges [1].

In *Europe*, where structural design is covered by Eurocode standards, some guidelines about general connections and even some more specific clauses regarding timber-concrete connections can currently be found in Eurocode 5, parts EN 1995-1-1 and EN 1995–2. A simplified method for calculation of moment resistance of mechanically jointed beams, known also as gamma method, is given in appendix B of EN 1995-1-1 and most of currently available methods for design of TCC structural elements are based on this method. However, as part of second generation of Eurocodes novel part of Eurocode 5, EN 1995-1-3, which will cover wholesome design of TCC structures, is in preparation and it is expected to be published soon [28].

According to information from European commission [29] there are several *Asian* and *African* countries that have already adopted Eurocodes or are currently in process of adoption and thus the same conclusions regarding inclusion of TCC structures in standards can be made for them as per European countries. Those countries are: Russia, Kazakhstan, Malaysia, Vietnam, Turkey, Madagascar, Angola, Ethiopia and Kenya. Additionally, China, India, South Africa and majority of Middle East countries have expressed interest in adopting Eurocodes [29].

Standards, which include some specifications for TCC design in *North America* and in *South America*, consider design of bridges [1]. In USA, there are AASHO and AASTHO codes, which date back to 1949 and 1983, respectively. In Canada, there is a Canadian Highway Bridge Design Code, which specifically allows two types of notched connections between timber and concrete. Guidelines for TCC floors can be found in recently published Design Guide for Timber-Concrete Composite Floors in Canada [30]. Greenland, on the other hand, has adopted Eurocode standards [29]. In Brazil, bridge design code has been published in 2006, which includes specifications for TCC bridges and analytical method based on shell theory. Otherwise, Eurocodes are being used in Brazil.

Oceania has its own design guide, namely Australia and New Zealand design guidelines, which is based on Eurocode 5. Some modifications have been made, though, in order to satisfy Australian and New Zealand rules for timber structures. Here, spans up to 8 m are allowed and two configurations of notched connections with screws are prescribed [1].

8. Conclusions

Increasing popularity of TCC structural elements over last decades is connected to many advantages of TCC systems over more established fully concrete and fully timber systems. In comparison with fully concrete elements, TCC elements have improved aesthetic and ecological component, reduced self-weight, as well as improved sound insulation for impact noise. Main advantages of TCC elements over fully timber elements are increased load-carrying capacity and stiffness, improved insulation for air-transmitted sound and decreased annoyance of users because of vibrations. Connection between timber and concrete is a very important component of TCC system; its strength and slip stiffness ensure composite action of timber and concrete part and, if the connection is properly designed, increased ductility of TCC structural element can be achieved.

TCC structural elements come in variety of designs; in particular, there is many different types of connectors, each with its own characteristic behaviour. If properties of the connection (shear strength, slip modulus, spacing, ...) are known, bending resistance of TCC element can be determined with one of existing simplified methods for design of TCC structures.

Design of TCC structural elements is scarcely represented in currently applicable standards. Lately, however, some design guidelines specifically for TCC floor systems were published in Canada. Furthermore, new part of Eurocodes, which will thoroughly consider design of TCC structural elements is already in preparation. Nevertheless, experimental research as well as development of more accurate design methods are still continuing, due to many different designs of TCC structural elements and their complex behaviour.

Acknowledgements

The research was financially supported by the Slovenian Research Agency (research core funding No. P2-0260). The support is gratefully acknowledged.

Conflict of interest

The authors declare no conflict of interest.

Abbreviations

CCS	carbon capture and storage
CLT	cross-laminated timber
Glulam	glued laminated timber
LVL	laminated veneer lumber
SLS	serviceability limit state
SFRC	steel-fibre-reinforced concrete
TCC	timber-concrete composite
ULS	ultimate limit state

Author details

Anita Ogrin*† and Tomaž Hozjan†
Faculty of Civil and Geodetic Engineering, University of Ljubljana, Ljubljana, Slovenia

*Address all correspondence to: anita.ogrin@fgg.uni-lj.si

† These authors contributed equally to this work.

References

[1] Dias A, Schänzlin J, Dietsch P. Design of Timber-Concrete Composite Structures: A State-of-the-Art Report by COST Action FP1402/WG 4. Aachen: Shaker Verlag; 2018.

[2] Meena R, Schollmayer M, Tannert T. Experimental and Numerical investigations of fire resistance of novel timber-concrete-composite decks. Journal of Performance of Constructed Facilities. 2014;28(6):1–8.

[3] Fortuna S, Dalla Mora T, Peron F, Romagnoni P. Environmental performances of a timber-concrete prefabricated composite wall system. Energy Procedia. 2017; 113: 90-97.

[4] Rodrigues JN, Dias AMPG, Providência P. Timber-concrete composite bridges: State-of-the-art review. BioResources, 2013;8(4): 6630-6649.

[5] Miebach Ingenieurbüro. Heavy Load Bridge Crossing the Agger River in Lohmar Schiffarth [Internet]. 2014. Available from: https://www.ib-miebach.de/en/projects/timber-bridges/timber-concrete-composite-bridges/timber-concrete-composite-bridge-schiffarth-de.html [Accessed: 2021-07-20]

[6] Rhomberg ventures. CREE by Rhomberg – LifeCycle Tower One [Internet]. 2012. Available from: https://www.rhombergventures.com/en/portfolio/cree_by_rhomberg/ [Accessed: 2021-07-20]

[7] Dagenais C, Ranger L, Cuerrier-Auclair S. Understanding fire performance of wood-concrete composite floor systems. In: Proceedings of WCTE 2016 – World Conference on Timber Engineering; 22–25 August 2016; Vienna, Austria; 2016. p. 5396-5404

[8] O'Neill J, Carradine D, Moss P, Fargiacomo M, Dhakal R, Buchanan A. Design of Timber-Concrete Composite Floors for fire resistance. Journal of Structural Fire Engineering, 2011;2(3): 231-242. DOI:10.1260/2040-2317.2.3.231

[9] Steinberg E, Selle R, Faust T. Connectors for timber–lightweight concrete composite structures. Journal of Structural Engineering, 2003;129(11). DOI:10.1061/(ASCE)0733-9445(2003)129:11(1538)

[10] Blesák L, Caldová E, Wald F. Unprotected timber-fibre reinforced concrete slab in fire. Wood research. 2015;60(4):605-615.

[11] Dias AMPG, Jorge LFC. The effect of ductile connectors on the behaviour of timber-concrete composite beams. Engineering structures. 2011; 33: 3033-3042.

[12] Ceccotti A. Composite concrete–timber structures. Progress in Structural Engineering and Materials. 2002; 4:264–275.

[13] Klippel M, Boccadoro L, Klingsch E, Frangi A. Fire tests on timber-concrete composite slabs using beech laminated veneer lumber. In: Proceedings of WCTE 2016 – World Conference on Timber Engineering; 22–25 August 2016; Vienna, Austria; 2016. p. 3910-3917

[14] EN 1991-1-1. Eurocode 1: Actions on structures. Part 1-1: General actions – Densities, self-weight, imposed loads for buildings. Brussels: CEN, European Committee for Standardization; 2000.

[15] Mirdad MAH, Daneshval H, Joyce T, Chui YH. Sustainability design considerations for timber-concrete composite floor systems. Advances in Civil Engineering. 2021; 6688076. DOI: 10.1155/2021/6688076

[16] Granello G, Palermo A. Vibration performance of timber-concrete-composite floors: A case study. Journal of the Structural Engineering Society of New Zealand, 2020;33(1):33-46.

[17] EN 1995-1-2. Eurocode 5: Design of timber structures. Part 1-2: General Structural fire design. Brussels: CEN, European Committee for Standardization; 2004.

[18] EN 1992-1-2. Eurocode 2: Design of concrete structures. Part 1-2: General actions. Actions on structures exposed to fire. Brussels: CEN, European Committee for Standardization; 2004.

[19] Frangi A, Knobloch M, Fontana M. Fire design of timber-concrete composite slabs with screwed connections. Journal of Structural Engineering. 2010; 136(2):219-228.

[20] Hozjan T, Bedon C, Ogrin A, Cvetkovska M, Klippel M. Literature review on timber-concrete composite structures in fire. Journal of Structural Engineering. 2019; 145(11):04019142. DOI:10.1061/(ASCE)ST.1943-541X.0002418

[21] Ogrin A, Hozjan T. Fire resistance of timber-concrete composite slabs – A simplified method. Materials and Structures. 2020;53:106.

[22] Frangi A, Fontana M. Elasto-plastic model for timber-concrete composite beams with ductile connection. Structural Engineering International, 2003;1: 47-57.

[23] Nässén J, Hedenus H, Karlsson S, Holmberg J. Concrete vs. wood in buildings – An energy system approach. Building and Environment. 2012; 51: 361-369.

[24] Yeoh D, Fragiacomo M, De Franceschi M, Heng Boon K. State of the art on timber-concrete composite structures: Literature review. Journal of Structural Engineering. 2011; 137(10): 1085-1095.

[25] Möhler K. On the Load Carrying Behavior of Beams and Columns of Compound Sections with Flexible Connections [Habilitation Thesis]. Germany: Technical Univ. of Karlsruhe; 1956. (in German)

[26] EN 1995-1-1. Eurocode 5: Design of timber structures. Part 1-1: General – Common rules and rules for buildings. Brussels: CEN, European Committee for Standardization; 2004.

[27] Van der Linden M. Timber-concrete composite floor systems [dissertation]. Delft: Delft university of Technology; 1999.

[28] Dias A. Timber-concrete composites – A new part in Eurocode 5. In: Proceedings of WCTE 2016 – World Conference on Timber Engineering; 22–25 August 2016; Vienna, Austria; 2016. p. 5650-5657

[29] European Commission. Use of EN Eurocodes outside EU-EFTA [Internet]. 2021. Available from: https://eurocodes.jrc.ec.europa.eu/images/worldmap-Eurocodes.png [Accessed: 2021-04-25]

[30] Shepard AB, Fisher EC, Barbosa AR, Sinha A. Fundamental behavior of timber-concrete composite floors in fire. Journal of structural engineering, 2021; 147(2):04020340.

Chapter 15

Cross-Laminated Timber: A Review on Its Characteristics and an Introduction to Chinese Practices

Zhiqiang Wang and Tianxiao Yin

Abstract

Cross-laminated timber (CLT) is a popular engineering wood product in recent years. It has some characteristics of configuration and mechanical properties, which makes it an excellent building material for floor, roof and other places. In Europe and North America, lots of middle and high-rise buildings have adopted CLT as their main structural component. CLT has recently been used to construct public buildings in China. As a building material, the lower rolling shear properties of CLT has always been a concern. To overcome this shortcoming of CLT, the structural composite lumber and bamboo have been employed to develop hybrid CLT. This chapter also presents the latest development and advances of CLT in China.

Keywords: Cross-laminated timber, Configuration characteristics, Rolling shear, Hybrid cross-laminated timber, Development in China

1. Introduction

1.1 Background

Cross-laminated timber (CLT) is an engineered wood product (EWP) that was originally developed in Europe in the 1990s, which has been gaining worldwide popularity in helping define a new class of structural timber products known as massive or "mass" timber. It is an engineered wood-based solution that complements the existing light frame and heavy timber options and is a suitable candidate for some applications that currently use concrete, masonry, and steel systems [1].

CLT is a kind of prefabricated engineering wood made of layers of lumbers stacked crosswise (typically at 90 degrees) and glued together on their wide faces and, sometimes, on their narrow faces as well, by structural adhesive. In some specific structural requirements, the lumbers of adjacent layers can be laminated in non-90 degrees groups. Usually, CLT consists of an odd number of layers, such as 3, 5, and 7 layers, and in some cases more. CLT products are usually 0.6 m, 1.2 m, 2.4 m and 3 m in width, up to 18 m in length and up to 508 mm in thickness.

At present, CLT is usually used in buildings as floor, wall and other structural components. CLT used for prefabricated wall and floor assemblies offers many advantages, and the 'reinforcement' effect provided by the cross-lamination in

CLT also considerably increase the splitting resistance of CLT for certain types of connection systems.

1.2 Development of CLT in China

The Chinese researchers and manufacturers began to develop and produce CLT materials and buildings around 2010. A number of research teams have carried out researches on the physical and mechanical properties and connection properties of CLT. Domestic fast-growing wood species such as poplar, eucalyptus and Japanese larch, and wood-based panels, such as construction OSB (COSB) boards, have been developed to produce CLT and hybrid CLT (HCLT) [2–4]. At the same time, the researchers also studied the embedment performance of CLT, the mechanical properties of the CLT joints connected by self-tapping screws (STS) and the mechanical properties of the new tongue-and-groove CLT joints [5–8]. In terms of the standard, the Standard for Design of Timber Structures (GB 50005–2017) and the Technical Standard for Multi-story and High-rise Timber Building (GB /T 51226–2017) have the corresponding provisions on CLT materials and CLT structure height. For example, for the structural system of pure timber structure, the maximum number of floors allowed for the CLT timber shear wall structure is 12. For the concrete core timber structure, the maximum number of floors allowed for the CLT shear wall structure is 18. The industry standard Cross-laminated Timber (LY/T 3039–2018) was officially implemented in May 2019.

China's domestic CLT manufacturers have also started from scratch, and now there are 4 CLT factories located in Hebei, Shandong, Zhejiang and Jiangsu provinces. Among them, Jiangsu Global CLT Co., Ltd. was established in 2017, with an annual capacity of 60,000 m^3 of CLT panels, as shown in **Figure 1**.

CLT buildings have been also developed in China. In 2019, Ningbo Sino-Canada Low-carbon Technology Research Institute Co., Ltd. built a 2-storey CLT public

Figure 1.
Jiangsu global CLT Co., ltd. and internal CLT production equipment (source: Photo obtained from Jiangsu global CLT Co., ltd).

Figure 2.
CLT public residential building in China (source: Picture obtained from https://www.sohu.com/a/391796450_100108650).

Figure 3.
6-story timber structure research and development center in China (source: Photo obtained from Shandong DENCHWOOD CLT Co., ltd).

Figure 4.
CLT office building in China (source: Photo obtained from Jiangsu global CLT Co., ltd).

residential building, with a total area of nearly 1,500 m^2, as shown in **Figure 2**. CLT are used as floor, wall and roof panels of this building, and the amount of CLT used is 215 m^3. In 2020, the first six-story pure timber structure building in China -- Shandong DENCHWOOD CLT Research and Development (R&D) Center project has been completed, as shown in **Figure 3**. The construction area of this building is 4771.96 m^2. The whole building employed glulam frame-shear wall structure as the main structure, and the elevator shaft and stairwell adopt 160 mm thick CLT as shear wall. In 2021, Jiangsu Global CLT Co., Ltd. completed a two-story, 1600 m^2 CLT office building, as shown in **Figure 4**. In this CLT building, CLT is used as floor panels (thickness 155 mm), wall panels (thickness 105 mm) and roof panels (thickness 105 mm), respectively, and the amount of CLT and glulam used are 406 m^3 and 82 m^3, respectively.

Figure 5.
CLT press (source: Photo obtained from Yantai Bohai woodworking machinery Co., ltd).

Besides, the development of CLT equipment is also under way in China. In 2015, Yantai Bohai Woodworking Machinery Co., Ltd. developed China's first automatic CLT assembly/glue-pouring/feeding/pressing production line. The press is shown in **Figure 5**, and it was put into use in Shandong Zhongyi Senke Wood Structure Co., Ltd. The maximum size of pressing CLT products is 24 m long and 3.5 m wide.

2. Characteristics of CLT

Generic CLT products consist of odd number of layers of lumber or structural composite panel stacked crosswise by applying structural adhesives, such as phenol-resorcinol formaldehyde (PRF), emulsion polymer isocyanate (EPI), melamine formaldehyde (MF), one-component polyurethane (PUR), etc. In some CLT product standards, lumber is required to be 6–45 mm thick, 40–300 mm wide, and width- thickness ratio greater than 4. For three-layer CLT, the thickness of the transverse layer ranges from 6 to 60 mm [9], as shown in **Figure 6**. Structural composite panels used as layer in CLT include laminated veneer lumber (LVL), laminated strand lumber (LSL), oriented strand board (OSB) and so on.

On some special cases, the adjacent layers of CLT can be assembled in the same direction. For example, the outermost layers of a 5-layer CLT can be successively set with two parallel layers [9], as shown in **Figure 7**. In addition, some CLT layers are glued at an angle of 45 degrees between adjacent layers [10], as shown in **Figure 8**. Some researchers have developed a box-based CLT system used in floor applications for more diverse structural performance [11], as shown in **Figure 9**.

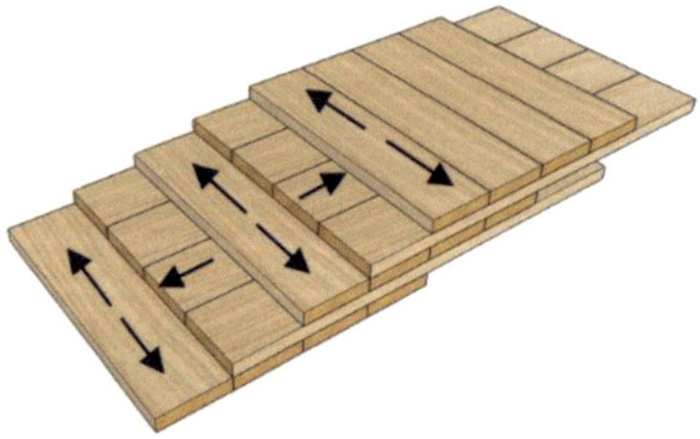

Figure 6.
Generic CLT configuration.

Figure 7.
CLT panel layup with two parallel layers in outermost layers.

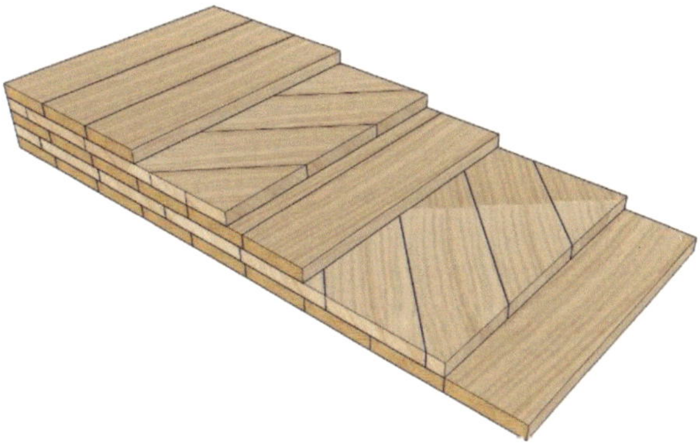

Figure 8.
CLT panel layup with ±45° alternating transverse layer configuration (adapted from Buck et al. [10]).

Figure 9.
CLT panel layup with box-based configuration (adapted from Chen [11]).

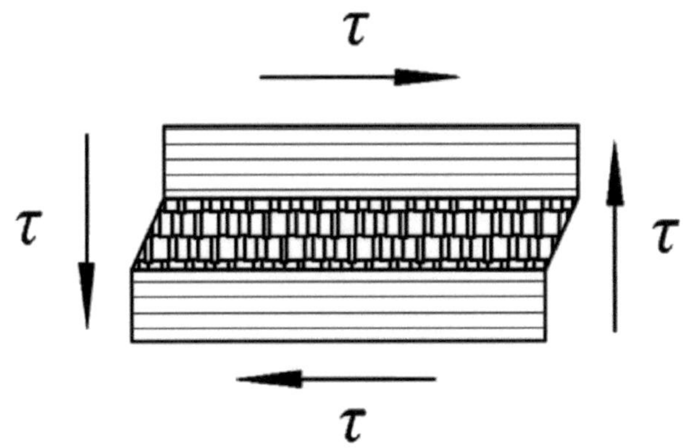

Figure 10.
Rolling shear of CLT.

In consideration of cost and the transverse deformation of the lumber, CLT may not apply adhesive between edges of the laminations at the same layer. If no adhesive is applied, the gap width between the laminations should not exceed 6 mm [9]. In addition, the vacuum pressurized CLT has stress relief to release the stress and reduce the warpage deformation of the CLT panel. However, the existence of gaps has adverse effects on CLT's mechanical properties (such as rolling shear properties), building physical properties (sound insulation, fire protection and thermal insulation properties), connection properties and product appearance, etc.

The orthogonality of CLT also makes the rolling shear failure of the transverse layers which is the key and main failure mode when CLT element is loaded with out-of-plane loads. Rolling shear, or called planar shear, usually refers to the behavior of shear strain occurring in the transvers layers, as shown in **Figure 10**. The cracks will initiate and propagate along the weak zones in the radial-tangential (RT) plane of wood transverse layer, resulting to low bending and shear strength properties. These weak zones include the earlywood/latewood boundary and wood rays. The rolling shear properties of wood are lower than those parallel to the grain of wood. For general wood or wood group used in CLT, such as spruce-pine-fir (SPF), Douglas Fir-L and Hem-Fir lumber, the assuming value of rolling shear modulus is only 50 MPa. The rolling shear properties is very important in the design of CLT products, which is the key factor in the design of CLT floor and roof panels [12].

3. Rolling shear properties of CLT

In recent years, researchers have mainly studied the rolling shear properties of CLT in two aspects: the evaluation method and influence factors of rolling shear properties of CLT. The employed testing methods for the rolling shear performance are mainly divided into two categories: compression shear and bending shear testing approaches. The influence factors of rolling shear properties of CLT include: the types of layer material (softwood, hardwood, and wood-based panels, etc.), macroscopic characteristics of lumber (growth ring orientation, and earlywood or latewood, etc.), processing technology (pressure and edge-gluing, etc.) and geometric characteristics of CLT, etc.

3.1 Factors influencing rolling shear properties

3.1.1 Types of layer materials

At present, CLT are mainly made of softwood, such as SPF, Norway spruce and other softwood. However, due to the low rolling shear modulus and strength of softwood, the development and utilization of hardwood and structural wood composite panels with high rolling shear properties to produce CLT has become one of the main research focus of CLT. Studies have shown that the rolling shear properties of some hardwood are higher than those of generic softwood. Aicher et al. [13, 14] studied the feasibility of using European beech wood (*Fagus sylvatica*) as transverse layer in CLT. They tested the rolling shear properties of European beech wood by compression shear method and found the rolling shear strength and modulus of European beech wood exceed the respective characteristic value for softwood by roughly factors of 5 and 7. In addition, a hybrid softwood-hardwood CLT build-up with outer layers of European spruce (*Picea abies*) and a center cross-layer of European beech (*Fagus sylvatica*) has been investigated with regard to out-of-plane bending. The novel investigations reveal the great potential of mixed softwood-hardwood CLT build-ups for structural elements in the building sector. Gong et al. [15] evaluated the rolling shear properties of cross hardwood lumber in HCLT. The tested wood species were spruce (*Picea mariana*), aspen (*Populus tremuloides*), white birch (*Betul papyrifera*) and yellow birch (*Betula alleghaniensis*). Based on their experimental results, it was found that the hardwoods (aspen and birches) exhibited a larger resistance to rolling shear stresses than that of softwoods (spruce). Ehrhart et al. [16] studied the rolling shear properties of some hardwoods and softwoods. They also found the hardwoods, such as birch (*Betula pendula Roth*), beech (*Fagus sylvatica L.*), poplar (*Populus spp.*), ash (*Fraxinus excelsior L.*), had higher rolling shear properties than that of softwood. In addition, some researchers have also studied the rolling shear properties of local wood, fast-growing wood and wood-based panels to evaluate the potential application of these materials in CLT. Wang et al. [17, 18] evaluated the rolling shear properties of normal and modified fast-growing poplar (*Populus tomentosa Carr*) modified by compression perpendicular to grain and impregnated with phenol-formaldehyde (PF) resin. Results showed that the characteristic value of rolling shear modulus and strength of normal fast growing poplar were 177 MPa and 2.24 MPa, respectively, which are much higher than the properties of SPF, indicating the fast-growing poplar can be used as transverse layer in CLT.

3.1.2 Macroscopic characteristics of lumber

The influences of macroscopic characteristics, such as growth ring orientation and pith, on the rolling shear properties of lumber were studied. Zhou et al. [19] investigated the influence of growth ring orientation and laminate thickness on the rolling shear properties of wood cross layer (WCL). They found that the growth ring orientation had a significant effect on the rolling shear modulus of WCL but did not on the rolling shear strength. And when the growth ring orientation was 45°, the maximum rolling shear modulus and strength could be obtained. Aicher et al. [13] tested the rolling shear properties of specimens with different growth ring orientation and pith board. It was found that the semi-quarter-sawn boards revealed the highest shear moduli whereas the quarter-sawn boards showed roughly 30% lower values. Wang et al. [2] used the compression shear test method of EN 408 to study the influence of macroscopic characteristics of fast-growing poplar sawn timber on the rolling shear properties. It were found the presence of pith had significant influence on the rolling shear properties of poplar board. Distance to pith and annual ring orientation both had effects on the rolling shear properties jointly. The rolling shear properties increase with the increase of distance to pith.

3.1.3 Manufacture technology of CLT

Yawalata et al. [20] studied the influence of different pressure on the rolling shear strength. When the pressure was 0.4 MPa and 0.1 MPa, the rolling shear strength of CLT was 2.22 MPa and 1.85 MPa respectively, and higher production pressure could improve the rolling shear performance of CLT. On the other hand, in order to prevent product deformation and save the cost of adhesives, the edge-gluing between laminations in the same layer are not applied in some manufacturers, resulting gaps between laminations. A study showed that the average width of the gaps in the transverse layers reaches 2 mm [21]. The maximum gap width allowed in the European CLT product standard EN 16351 is 6 mm. The existence of these gaps will have an impact on the rolling shear properties of CLT transverse layers. Gardner et al. [22] explored the effect of gaps between boards in transverse layers of CLT on shear strength. Five-ply specimens with gaps of 0, 6, 89 and 178 mm were subjected to short-span three-point bending tests. The digital imaging correlation (DIC) technique was used to quantify strains and displacements in transverse layers. It was found that panel shear capacity met the requirements of the PRG 320 standard for performance-rated CLT for the gap sizes tested, suggesting that small gaps did not reduce shear strength enough to warrant consideration in design. Wang et al. [23] evaluated the influence of edge-gluing and gap width (0 mm, 2 mm, 4 mm and 6 mm) between the transverse layers on the rolling shear properties. It was found that edge-gluing and gap size had a significant influence on measuring rolling shear strength rather than apparent rolling shear modulus by the modified rolling shear test method. With the gap size larger than 2 mm, its influence on measuring rolling shear strength became negligible.

3.1.4 Geometric characteristics of CLT

The research of effect of geometrical characteristics on rolling shear properties mainly focus on the thickness and the width-thickness ratio (γ)of layer. Sikora et al. [24] studied the influence of layer thickness on rolling shear properties. With the increase of the thickness of layers, the rolling shear properties tended to decrease, and the average rolling shear strength ranged from 1.0 MPa to 2.0 MPa. Li [25] used 35 mm and 20 mm thick layer to evaluate the influence of layer thickness on the rolling shear strength of CLT. It was found that the thickness had a significant effect on the rolling shear strength of CLT. Ehrhart et al. [16] tested CLT specimens with a constant thickness of 30 mm and different widths (60 mm, 120 mm and 180 mm), and found that the width-thickness ratio had a significant effect on the rolling shear strength and modulus. Gui et al. [26] studied the rolling shear properties of fast-growing eucalyptus lumbers with different width-thickness ratio (γ = 2,4 and 6), and used the DIC technique to record and evaluate the rolling shear strain distribution during rolling shear tests. The result showed that the mean values of rolling shear modulus and strength of eucalyptus layer were 260.3% and 88.2% higher than those of SPF layer with the same width-thickness ratio of 4, respectively. The rolling shear properties of eucalyptus layers increased as the width-thickness ratio increased. The high shear strain regions were primarily found around the gaps

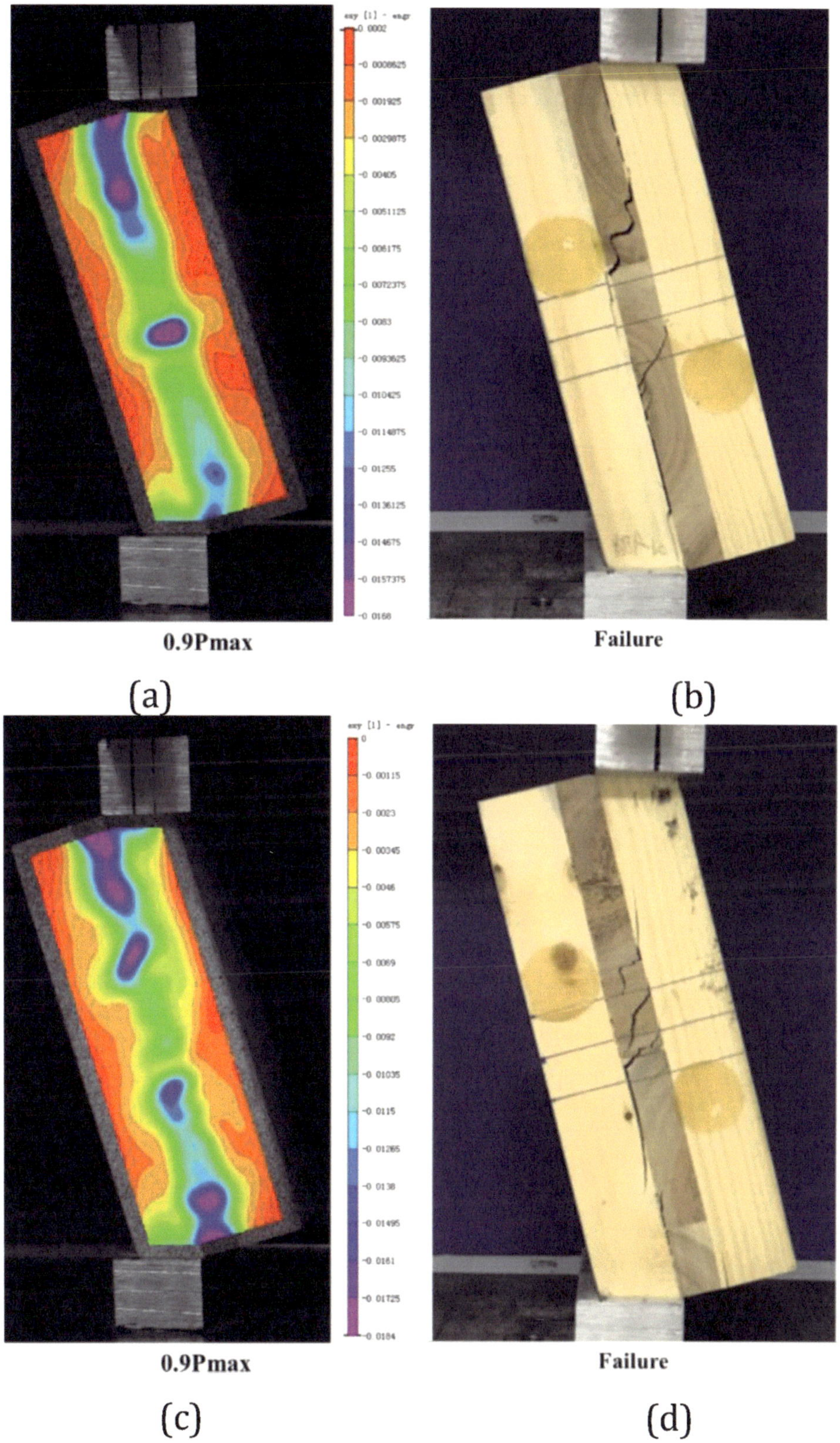

Figure 11.
Strain distribution and failure of eucalyptus CLT specimens under rolling shear: (a) and (b) width-thickness ratio (γ) = 4, and (c) and (d) γ = 6 (adapted from Gui et al. [26]).

between segments of transverse layer. The quantity of high shear strain regions increased as the width-thickness ratio of layer decreased, as shown in **Figure 11**.

3.2 Failure mechanism of CLT caused by rolling shear stress

The rolling shear failure mode of lumber is closely related to its macroscopic characteristics. Wang et al. [27] found that the rolling shear failure of SPF dimension lumber mainly happened at the weak macroscopic characteristics in the radial-tangential (RT)section of lumber, such as the junction of earlywood and latewood and wood rays, as shown in **Figure 12**. On the other hand, Yang [28] used the acoustic emission (AE) technique to monitor the changes of AE parameters during the rolling shear failure process of CLT. By clustering analysis of AE signals of different wood species, the relationship between AE signals and the mechanism of the rolling shear damage of CLT was established, as shown in **Figure 13**. There are two kinds of AE events in CLT specimens during rolling shear failure process: the main AE events with middle amplitude and the secondary AE signal with high amplitude.

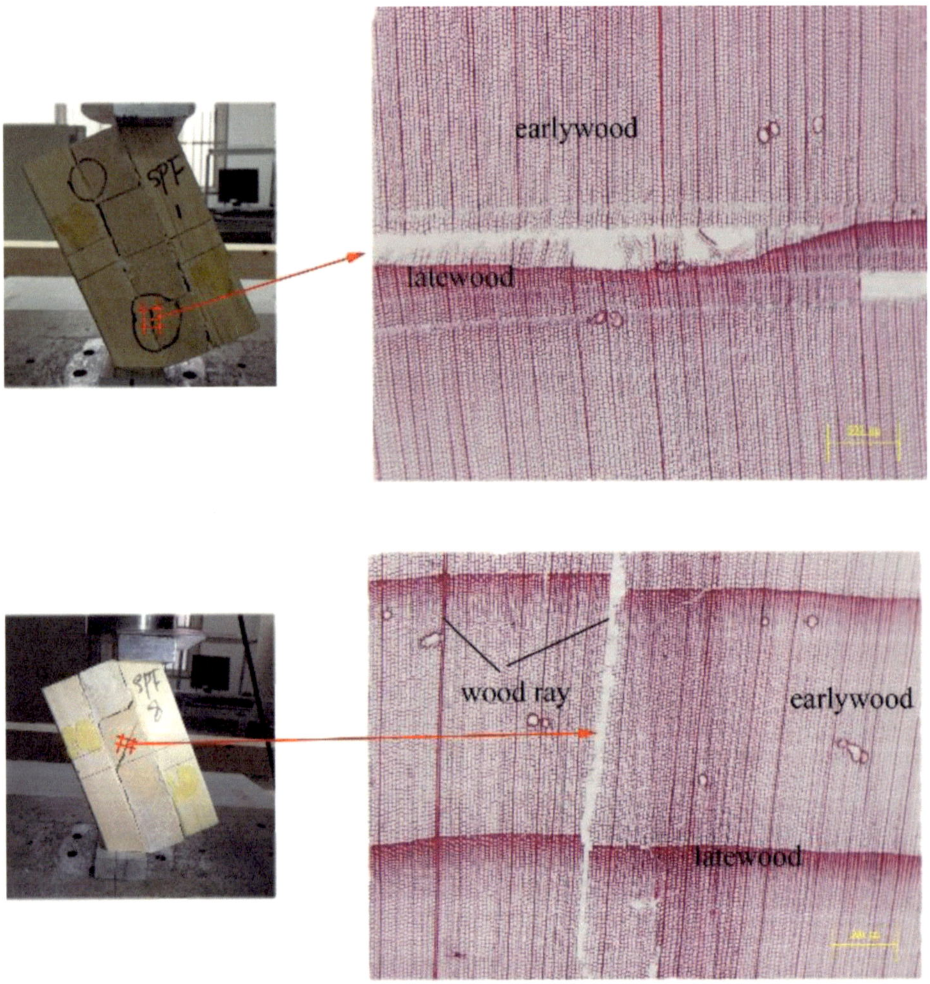

Figure 12.
Microcosmic images of failure modes of rolling shear of SPF CLT (adapted from Wang et al. [27]).

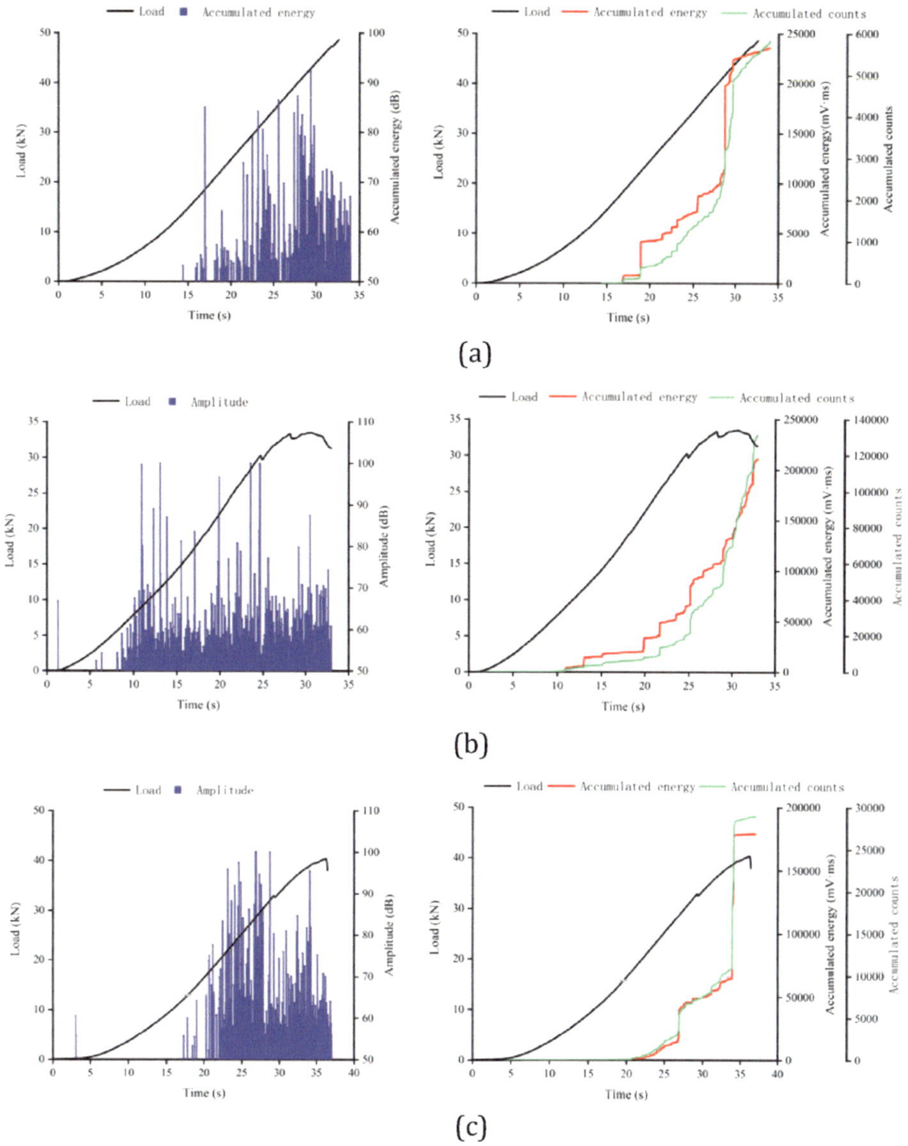

Figure 13.
AE activity for 3 groups of CLT specimens during rolling shear failure, (a) SSS-4, (b) SHS-2, and (c) SZS-3 (adapted from Yang et al. [28]).

Specimen	Main AE events			Secondary AE events		
	Percentage (%)	Amplitude (dB)	Duration (s)	Percentage (%)	Amplitude (dB)	Duration (s)
SSS-4	88	52–68	14–34	12	67–91	17–34
SHS-2	70	52–65	6–33	29	52–100	9–33
SZS-3	80	52–70	17–37	19	52–100	19–35

Table 1.
Statistical analysis of clustering features.

The duration of two kinds of AE events covered the whole process of rolling shear failure, while the duration of secondary AE events was slightly shorter than that of main AE events. The main AE events accounted for a high proportion, with AE amplitude within 70 dB, while the secondary AE events mainly distributed in the medium and high amplitude range, as shown in **Table 1**. Combined with rolling shear failure process of CLT, the main AE events might come from the accumulation of internal damage in the transverse layer. As the crack propagation, the accumulated energy was released continuously, thus producing the secondary AE events.

4. Research progress of hybrid CLT

In order to expand the source of CLT raw materials and improve the mechanical properties of CLT, the HCLT, i.e., CLT composed of laminations of different wood species, or composed of solid sawn wood and structural composite panels, has been developed. Compared with sawn timber, structural composite lumber (SCL) or structural wood-based panels have better mechanical properties and more sources of raw materials. Some studies have been conducted on the HCLT fabricated with sawn timber and SCL/ wood-based panels. The SCL/ wood-based panels employed in these studies included LVL, LSL, oriented strand lumber (OSL), COSB and plywood [27–35]. Wang et al. [29] fabricated the HCLT by mixing SPF and LSL, and when LSL was used as the outer layer (longitudinal layer), the modulus of elastic (MOE) and modulus of rupture (MOR) of HCLT increased by 19% and 36%, respectively, compared with the generic lumber CLT. When LSL was used as the cross layer (transverse layer), the MOE and MOR of HCLT were increased by 13% and 24%, respectively. Davids et al. [34] also obtained a similar research conclusion that the use of LSL hybrid structure could improve the bending performance of CLT, mainly because LSL had better mechanical properties (rolling shear properties and tensile properties) and more homogeneous mechanical properties than solid wood. Wang et al. [27] studied the mixing of SPF and LVL according to different layups and formed three kinds of HCLT. It was found that due to the low rolling shear properties of LVL, the bending mechanical properties of CLT could be greatly improved only when LVL was placed in the outer layer. Other researchers studied the bending properties of HCLT mixed with Korean larch plywood (*Larix kaempferi Carr.*) and North American Douglas fir (*Pseudotsuga menziesii Franco*). When larch plywood was placed in the transverse layer of CLT, the experimental results showed that the MOE and MOR of CLT were improved with the increase of the number and thickness of larch plywood [35].

Furthermore, layers with different wood species have been studied to fabricate HCLT. Wang et al. [18] studied the mechanical properties of CLT mixed with different wood species. The results showed that the MOE, MOR and shear strength of HCLT formed by placing poplar lumber in the cross layer and Douglas fir lumber in the outer layer could been improved to some degree. Ukyo et al. [36] investigated the out-of-plane shear strength of HCLT with outer layers of hinoki (hinoki cypress, *Chamaecyparis obtusa*) and inner layers of sugi (Japanese cedar, *Cryptomeria japonica*). The influence from rolling shear properties of transverse layers on the shear strength of CLT, stress analysis was conducted using the shear analogy method. Pang et al. [37] analyzed swelling (S_w) and shrinkage (S_h) behaviors of CLT made of different species and various layer thickness and combinations. Compared to S_w and S_h of CLT made of larch, S_w and S_h of CLT made of pine showed lower values.

Bamboo has been extensively applied in composite industries due to its faster growth, higher specific strength and rigidity, and a relatively lower water swelling ratio compared with wood. The potential of using bamboo as CLT lamination has

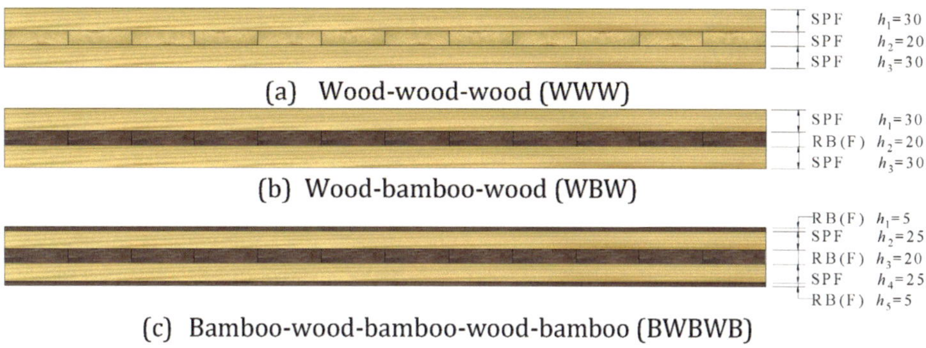

Figure 14.
Configuration of bending specimen of CLT and CLTB (unit: mm). (a) Wood-wood-wood (WWW), (b) wood-bamboo-wood (WBW), (c) bamboo-wood-bamboo-wood-bamboo (BWBWB) (adapted from dong. [46]).

got attention in recent years with the development of CLT. So far, there are not many studies on composite cross-laminated timber-bamboo (CLTB). Wei et al. [38] investigated the feasibility of manufacturing composite cross-laminated timber (CCLT) from bamboo parallel strand lumber and hem-fir lumber. They found the rolling shear strength of CCLT was first governed by the low strength of hem-fir lumber, particularly in the direction perpendicular-to-grain. Munis et al. [39] added strips of *Dendrocalamus asper* bamboo species on the outer layers of generic wood CLT as reinforcement, and found a significant increase in the parallel compression-to-grain in the configuration of the CLT reinforced with bamboo in the longitudinal direction. Barreto et al. [40] evaluated the structural performance of CLTB produced from wood (*Pinus spp.*) and bamboo strips (*Dendrocalamus giganteus*). Their results showed a good potential of this composite CLTB for structural uses.

Bamboo resources are abundant in China. To make good use of bamboo, several engineered bamboo products (EBPs), such as bamboo scrimber, bamboo glulam and bamboo plywood, have been developed and utilized [41, 42]. Bamboo scrimber is a new engineering material made of bamboo via defibring and compositing technology that utilizes up to 90% raw materials [43]. Many studies have been carried out on the mechanical properties, e. g. tension and bending properties, of EBPs [41–45], however there are few studies on the rolling shear properties of them.

Dong [46] evaluated the rolling shear properties of two bamboo scrimbers and one bamboo plywood and the bending properties of CLTB. The configuration of the CLTB is shown in **Figure 14**. The results indicated that the rolling shear modulus and strength of bamboo scrimber were 92.65% and 98.53%, 337.89% and 120.31% higher than those of bamboo plywood and SPF dimension lumber, respectively. The bending properties of CLT can be improved by using bamboo scrimber as transverse or longitudinal layers, or a combination of the two layers. Compared with the generic SPF CLT, the apparent bending modulus of CLTB specimen having bamboo scrimber as transverse layer, increased by 3.54%; furthermore, the apparent bending modulus and strength of CLTB specimen, having bamboo scrimber both as transverse layer and the outermost longitudinal layer, increased by 23.69% and 60.43%, respectively.

5. Conclusions

Compared with other engineered wood products, CLT has unique structural and mechanical properties, which makes CLT widely used in medium and high-rise timber construction. In addition to Europe, other countries and regions in the world are also paying attention to the development of CLT material and buildings. Hardwood, fast-growing wood, wood-based panels, bamboo, and local wood will be widely used in CLT materials.

In general, due to the orthogonal structure of CLT and the orthotropy of wood, the rolling shear properties have a significant effect on the mechanical properties of CLT. On the other hand, the layer material, the assembly structure, the fabricate process and the test method all affect the rolling shear properties of CLT. The mixing of different layer materials can effectively improve the rolling shear properties of CLT, and AE technology is helpful to evaluate the process and mechanism of rolling shear failure of CLT.

In China, product standards of CLT have been gradually established and improved, meanwhile, production and construction of CLT are in the stage of development. The development and production of HCLT using fast-growing wood, bamboo or wood-based panels, which are abundant in China, will greatly promote the development and application of CLT in China.

Acknowledgements

The authors wish to acknowledge the National Natural Science Foundation of China (Grant Nos. 32071700 and 31570559).

Author details

Zhiqiang Wang* and Tianxiao Yin
College of Materials Science and Engineering, Nanjing Forestry University, Nanjing, China

*Address all correspondence to: wangzhiqiang@njfu.edu.cn

References

[1] Gong M. Lumber-Based Mass Timber Products in Construction. IntechOpen; 2019. DOI: http://dx.doi.org/10.5772/intechopen.85808

[2] Wang Z, Dong W, Wang Z, et al. Effect of macro characteristics on rolling shear properties of east-growing poplar wood laminations. Wood Research. 2018; 63(2), 227-238

[3] Gong Y, Wu G, Ren H. Block shear strength and delamination of cross-laminated timber fabricated with Japanese larch. BioResources. 2016; 11(4), 10240-10250. DOI: 10.15376/biores.11.4.10240-10250

[4] Liao Y, Tu D, Zhou J, et al. Feasibility of manufacturing cross-laminated timber using fast-grown small diameter eucalyptus lumbers. Construction & Building Materials. 2017; 132, 508-515. DOI: 10.1016/j.conbuildmat.2016.12.027

[5] Dong W, Li Q, Wang Z, et al. Effects of embedment side and loading direction on embedment strength of cross laminated timber for smooth dowels. European Journal of Wood and Wood Products. 2020; 78:17-25. DOI: 10.1007/s00107-019-01490-z

[6] Dong W, Wang Z, Zhou J, et al. Embedment strength of smooth dowel-type fasteners in cross-laminated timber. Construction & Building Materials. 2020; 233: 117243. DOI: 10.1016/j.conbuildmat.2019.117243

[7] Zhang J. Research on shear performance of CLT with T&G joints [thesis]. Nanjing: Nanjing Forestry University; 2020 (In Chinese)

[8] Dong W, Yue Y, Song H, et al. Shear strength of cross-laminated timber connected with self-tapping screws. China Wood Industry. 2018; 32 (5): 1-5. DOI: 10.19455/j.mcgy.20180501 (In Chinese)

[9] European Committee for Standardization, EN 16351:2015, Timber Structures –Cross Laminated Timber - Requirements, 2015

[10] Dietrich B, Wang X, Olle H, et al. Bending properties of cross laminated timber (CLT) with a 45° alternating layer configuration. BioResources, 2016; 11(2), 4633-4644

[11] Chen Y, Structural performance of box based cross laminated timber system used in floor applications [thesis]. Vancouver: The University of British Columbia; 2011

[12] Gagnon S, Karacabeyl E. CLT Handbook: Cross-Laminated Timber, FPInnovations, Pointe-Claire, Canada, 2019

[13] Aicher S, Christian Z, Hirsch M. Rolling shear modulus and strength of beech wood laminations. Holzforschung. 2016; 70(8):773-781. DOI: 10.1515/hf-2015-0229

[14] Aicher S, Hirsch M, Christian Z. Hybrid cross-laminated timber plates with beech wood cross-layers. Construction & Building Materials. 2016, 124:1007-1018. DOI: 10.1016/j.conbuildmat.2016.08.051

[15] Gong M, Tu D, Li L. Planar shear properties of hardwood cross layer in hybrid cross laminated timber; September 2015; Quebec City. 5th International scientific conference on hardwood processing; 2015. p. 85-90

[16] Ehrhart T, Brandner R. Rolling shear: Test configurations and properties of some European soft- and hardwood species. Engineering Structures. 2018; 172:554-572. DOI: 10.1016/j.engstruct.2018.05.118

[17] Wang Z, Fu H, Luo H, et al. Planar shear properties of fast-growing poplar

in hybrid cross laminated lumber. Journal of Nanjing Tech University (Natural Science Edition). 2016; 38(5):116-120. DOI: 10.3969/j.issn.1671-7627.2016.05.019 (In Chinese)

[18] Wang Z, Fu H, Chui YH, et al. Feasibility of using poplar as cross layer to fabricate cross-laminated timber. In: World Conference on Timber Engineering; 10 August 2014; Quebec city, Canada, 2014

[19] Zhou Q. Development of evaluation methodology for rolling shear properties in cross laminated timber (CLT) [thesis]. New Brunswick: University of New Brunswick; 2013

[20] Yawalata D. Development of technology for cross laminated timber building systems [thesis]. Vancouver: University of British Columbia; 2011

[21] Uibel T, Blaß HJ. Load carrying capacity of joints with dowel type fasteners in solid wood panels. August 2006; Florence: Proceedings of CIB-W18; 2006. P. 1-10

[22] Gardner C, Davids W, Lopez-Anido R, et al. The effect of edge gaps on shear strength and rolling shear modulus of cross laminated timber panels. Construction & Building Materials. 2020; 259, 119710. DOI: 10.1016/j.conbuildmat.2020.119710

[23] Wang Z, Zhou J, Dong W, et al. Influence of technical characteristics on the rolling shear properties of cross laminated timber by modified planar shear tests. Maderas. Ciencia Y Tecnología. 2018; 20(3):469-478. DOI: 10.4067/S0718-221X2018005031601

[24] Sikora K, Mcpolin D, Harte A. Effects of the thickness of cross-laminated timber (CLT) panels made from Irish Sitka spruce on mechanical performance in bending and shear. Construction & Building Materials. 2016; 116:141-150. DOI: 10.1016/j.conbuildmat.2016.04.145

[25] Li M. Evaluating rolling shear strength properties of cross-laminated timber by short-span bending tests and modified planar shear tests. Journal of Wood Science. 2017; 63(4):331-337. DOI: 10.1007/s10086-017-1631-6

[26] Gui T, Cai S, Wang Z, et al. Influence of aspect ratio on rolling shear properties of fast-grown small diameter eucalyptus lumber. Journal of Renewable Materials. 2020; 8(9):1053-1066. DOI: 10.32604/jrm.2020.011645

[27] Wang Z, Fu H, Gong M, et al. Planar shear and bending properties of hybrid CLT fabricated with lumber and LVL. Construction & Building Materials. 2017; 151:172-177. DOI: 10.1016/j.conbuildmat.2017.04.205

[28] Yang Y. Research on rolling shear failure mechanism of cross-laminated timber based on acoustic emission [thesis]. Nanjing: Nanjing Forestry University; 2020 (In Chinese)

[29] Wang Z, Gong M, Chui YH. Mechanical properties of laminated strand lumber and hybrid cross-laminated timber, Construction & Building Materials. 2015; 101: 622-627. DOI: 10.1016/j.conbuildmat.2015.10.035

[30] Davids W, Willey N, Lopez-Anido R, et al. Structural performance of hybrid SPFs-LSL cross-laminated timber panel, Construction & Building Materials. 2017; 149: 156-163. DOI: 10.1016/j.conbuildmat.2017.05.131

[31] Zhou J, Niederwestberg J, Chui YH, et al. Bending properties of innovative multi-layer composite laminated panels; 20-23 August 2018; Seoul: World conference on timber engineering (WCTE); 2018

[32] Li Q, Wang Z, Liang Z, et al. Shear properties of hybrid CLT fabricated with lumber and OSB, Construction & Building Materials. 2020; 261: 120504.

DOI: 10.1016/j.conbuildmat.2020.120504

[33] Pang S, Ahn K, Kang S, et al. Prediction of withdrawal resistance for a screw in hybrid cross-laminated timber. Journal of Wood Science. 2020; 66 (1): 79. DOI: 10.1186/s10086-020-01926-8.

[34] Davids W, Willey N, Lopez-Anido R, et al. Structural performance of hybrid SPFs-LSL cross-laminated timber panels. Construction & Building Materials. 2017; 149:156-163. DOI: 10.1016/j.conbuildmat.2017.05.131

[35] Choi C K, Ojima E, Kim K J, et al. Analysis of mechanical properties of cross-laminated timber (CLT) with plywood using Korean Larch. BioResources. 2018; 13(2),2715-2726. DOI: 10.15376/biores.13.2.2715-2726

[36] Ukyo S, Miyatake A, Shindo K, et al. Shear strength properties of hybrid (hinoki cypress and Japanese cedar) cross-laminated timber. Journal of Wood Science. 2021; 67:23. DOI: 10.1186/s10086-021-01954-y

[37] Pang S, Jeong G. Swelling and shrinkage behaviors of cross-laminated timber made of different species with various lamina thickness and combinations. Construction & Building Materials. 2020; 240:117924. DOI: 10.1016/j.conbuildmat.2019.117924

[38] Wei P, Wang J, Wang L, et al. An exploratory study of composite cross-laminated timber (CCLT) made from bamboo and Hemlock-fir Mix. BioResources. 2019; 14 (1): 2160-2170. DOI: 10.15376/biores.14.1.2160-2170

[39] Munis R, Camargo D, De Almeida A, et al. Parallel compression to grain and stiffness of cross laminated timber panels with bamboo reinforcement. BioResources. 2018; 13 (2): 3809-3816. DOI: 10.15376/biores.13.2.3809-3816

[40] Barreto M, De Araujo V, Cortez-Barbosa J, et al. Structural performance analysis of cross-laminated timber-bamboo (CLTB). BioResources; 2019 14 (3): 5045-5058. DOI: 10.15376/biores.14.3.5045-5058

[41] Wei Y, Yan S, Zhao K, et al. Experimental and theoretical investigation of steel-reinforced bamboo scrimber beams. Engineering Structures. 2020; 223: 111179. DOI: 10.1016/j.engstruct.2020.111179

[42] Wang Z, Li H, Fei B, et al. Axial compressive performance of laminated bamboo column with aramid fiber reinforced polymer. Composite Structures. 2021; 258: 112298. Doi: 10.1016/j.compstruct.2020.113398

[43] Wei Y, Zhou M, Zhao K, et al. Stress–strain relationship model of glulam bamboo under axial loading. Advanced Composites Letters. 2020; 29: 1-11. Doi: 10.1177/2633366X20958726

[44] Wei Y, Ji X, Duan M, et al. Flexural performance of bamboo scrimber beams strengthened with fiber-reinforced polymer. Construction & Building Materials. 2017; 142: 66-82. DOI: 10.1016/j.conbuildmat.2017.03.054

[45] Li H, Su J, Xiong Z, et al. Evaluation on the ultimate bearing capacity for laminated bamboo lumber columns under eccentric compression. Structures. 2020; 28 (12): 1572-1579. DOI: 10.1016/j.istruc.2020.10.004

[46] Dong W. Evaluation of bonding technology and mechanical properties of bamboo-wood composite cross-laminated timber [thesis]. Nanjing: Nanjing Forestry University; 2019 (In Chinese)

Chapter 16

Various Geometric Configuration Proposals for Dovetail Wooden Horizontal Structural Members in Multistory Building Construction

Hüseyin Emre Ilgın, Markku Karjalainen and Olli-Paavo Koponen

Abstract

Adhesives and metal fasteners have an important place in the content of engineered wood products (EWPs). However, adhesives may cause toxic gas emissions due to their petroleum-based nature, while metal fasteners may adversely affect the reusability of these products. These issues also raise important questions about the sustainability and environmental friendliness of EWPs. Thus, there is still room for a solution that is solid and completely pure wood, adhesive- and metal-connectors-free dovetail wood board elements (DWBEs). There are many studies on the technological, ecological, and economic aspects of these products in the literature, but no studies have been conducted to assess the technical performance of DWBEs. This chapter focuses on DWBEs by proposing various geometric configurations for horizontal structural members in multistory building construction through architectural modeling programs. In this architectural design phase, which is one of the first but most important stages, the proposed configurations are based on a theoretical approach, considering contemporary construction practices rather than structural analysis or mechanical simulation. Further research, including technical performance tests, will be undertaken after this critical phase. It is believed that this chapter will contribute to the dissemination of DWBEs for innovative architectural and structural applications, especially in multistory wooden structures construction.

Keywords: timber/wood, dovetail wood board elements, engineered wood products, sustainability, multistory building construction, architectural modeling

1. Introduction

Climate change is dangerously close to spiraling out of control [1, 2]. The probability of this critical phenomenon attributed to human factors is over 90% and requires urgent management of our operations [3]. Buildings are the major contributors to the climate crisis, producing about 40% of annual global CO_2 emissions [4]. Additionally, building operations are responsible for 28% of these total annual emissions, while building materials and construction (often referred to as embedded carbon) are responsible for an additional 11% annually [5].

In this sense, wood as a renewable material is unquestionably ecological and environmentally friendly in terms of low carbon emissions during processing and carbon sequestration: one cubic meter of growing wood can bind about one ton of CO_2 from the atmosphere. If the dry mass of wood is 500 kg, about half of this mass is carbon, namely 250 kg. Thus, timber, which is at the forefront of addressing European climate policy, is considered one of our best allies in resolving climate change, especially due to its environmentally friendly features [6, 7].

Moreover, thanks to its numerous technological benefits such as dimensional stability, uniform strength, ecological properties such as low carbon emission, engineered wood products (e.g., cross-laminated timber, laminated veneer lumber) (**Figure 1**) are increasingly becoming a viable solution in high-rise structures [8–11] as in the cases of the 26 m and 8-story Carbon 12 (Portland, 2018) (**Figure 2**) [12] and the 85 m and 18-story Mjøstårnet (Brumunddal, 2019) (**Figure 3**) [13].

With the standardization of the building industry, adhesives and metal fasteners are often used in engineered wood products (EWPs), replacing traditional wood-to-wood assemblies [14]. It is worth noting here that engineered wood products (EWPs), also called mass timber, composite wood, artificial wood, or fabricated wood, include a range of derivative wood products manufactured by bonding or fastening strips, particles, fibers or wood veneers or boards of wood with adhesives or other fixing methods to generate composite material. Adhesives play a critical role in EWPs, especially by protecting the wood, making the structure light and robust, preventing shrinkage and expansion caused by natural humidity [15, 16], while metal fasteners ensure the overall integrity of the wooden structure [17, 18]. However, adhesives can cause problems in the sustainability and environmental friendliness of EWPs due to toxic gas emissions [19, 20], and similarly, metal fasteners can negatively affect the reusability and recyclability of EWPs [21, 22]. Therefore, there is still room for a solution consisting of solid and completely pure wood, dovetail wood board elements (DWBEs) [23]. Numerous studies have been conducted in the literature on the technological, ecological, and economic aspects of EWPs in construction with different building solutions [24] such as [25–30], and there is limited understating of DWBEs, which mostly includes structural analysis of connection details (e.g., [31–39]). Here, the dovetail wood board elements (DWBEs) can be defined as solid/massive and pure wood structural elements such as floor

(a) (b)

Figure 1.
Engineered wood products: (a) cross-laminated timber; (b) laminated veneer lumber (sources: Wikipedia—https://en.wikipedia.org/wiki/Cross-laminated_timber; https://en.wikipedia.org/wiki/Laminated_veneer_lumber).

Figure 2.
Carbon 12 (Portland, 2018) (source: Wikipedia—https://en.wikipedia.org/wiki/Carbon12).

Figure 3.
Mjøstårnet (Brumunddal, 2019) (source: Wikipedia–https://en.wikipedia.org/wiki/Mj%C3%B8st%C3%A5rnet).

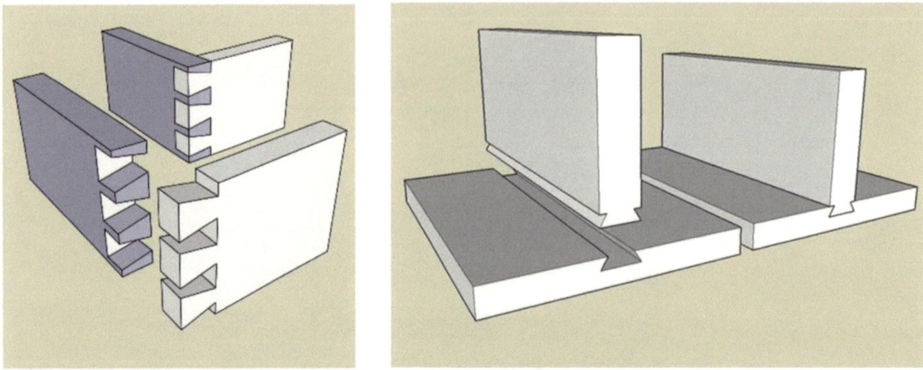

(a) (b)

Figure 4.
Dovetail joint as one of the oldest joining techniques in furniture design: (a) a through dovetail joint; (b) a sliding dovetail joint (sources: Wikipedia—https://en.wikipedia.org/wiki/Dovetail_joint).

slabs that use plug-in dovetail form in the joint detail and do not use adhesives and metal connections. More importantly, no studies have been conducted to assess the technical performance of DWBEs in multistory construction [40]. This chapter focuses on DWBEs—based on one of the oldest joining methods (**Figure 4**)—as sustainable material alternatives for ecological engineering solutions by suggesting various geometric configurations for flooring of multistory building construction through architectural modeling programs.

It is worth mentioning here that ecological engineering combines contemporary environmental engineering practices with ecological principles to achieve ecologically oriented goals [41–44]. Today, ecological engineering has become an essential tool, with the use of sustainable materials such as wood to tackle challenges resulting from the climate crisis. Given that buildings account for around 40% of annual global CO_2 emissions, the construction of wooden (multistory) buildings, especially with dovetail wooden elements as more environmentally friendly pure wood material, will contribute significantly to the fight against climate change in terms of ecological engineering approach.

It is believed that this chapter will contribute to the creation of higher value-added circular economy opportunities to support European climate policy as part of bio-economy and sustainable development through the dissemination of DWBEs for diverse and innovative structural applications in the construction sector as an ecological engineering-based solution.

In this chapter, wood or timber refers to engineered wood products (EWPs), e.g., cross-laminated timber (CLT—a prefabricated multi-layer EWP, manufactured from at least three layers of boards by gluing their surfaces together with an adhesive under pressure), laminated veneer lumber (LVL—made by bonding together thin vertical softwood veneers with their grain parallel to the longitudinal axis of the section, under heat and pressure), and glue-laminated timber (Glulam—made by gluing together several graded timber laminations with their grain parallel to the longitudinal axis of the section). Moreover, in this research, "multistory building" and "tall building" are defined as a building with over two-story and eight-story, respectively.

2. Research Methods

This study was conducted through an extensive literature survey mainly including international peer-reviewed journals and similar research projects.

Furthermore, this chapter was carried out with architectural modeling methods used in the solution of research and design problems in architectural activities. This method enables architects to think, write, discuss, and disseminate as a bridge from theory to practice [45]. It is widely utilized in architectural design research, where it is used by architects as a tool for research methodology [46, 47].

In addition, currently, there is no single approach to creating the object and subject of architectural activity, which inevitably leads to significant differences in research methods and architectural design of objects, especially at such important

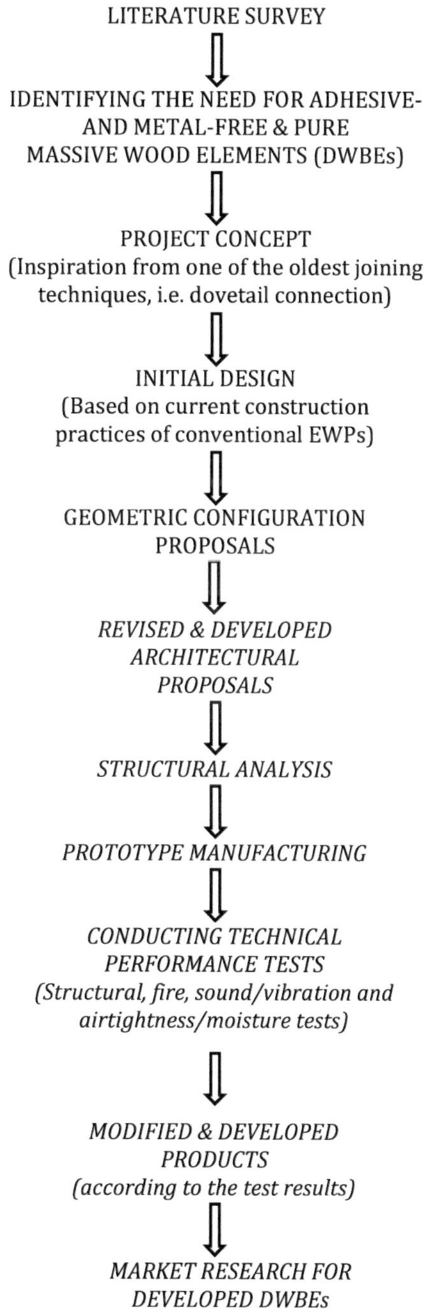

Figure 5.
The phases of the preliminary design used as a research method in this study and project's next phases (in italics).

levels of solving this problem [48]. On the other hand, the precise operation of text and project interaction in architectural design research remains a highly debated and relatively unformed topic [49–52].

Hence, main business applications such as AutoCAD, SketchUp, parametric modeling and information modeling methodology of buildings, and complex object modeling methods used in modern architectural design applications (e.g., [53, 54]) were utilized in this study. Here, creative proposals are realized through a mix of drawings and models as visual representations to encourage a fresh and lively approach to architectural research.

Figure 5 shows the phases of the preliminary design used as a research method in this study and the next phases of the project.

3. Findings

DWBE's innovation is based on a new way of combining the understanding of the properties and potential of wood, traditional woodworking skills, mechanical capability to mill large wood boards efficiently and precisely, digital machining control, and digital design. Thus, the architect, structural designer, and production

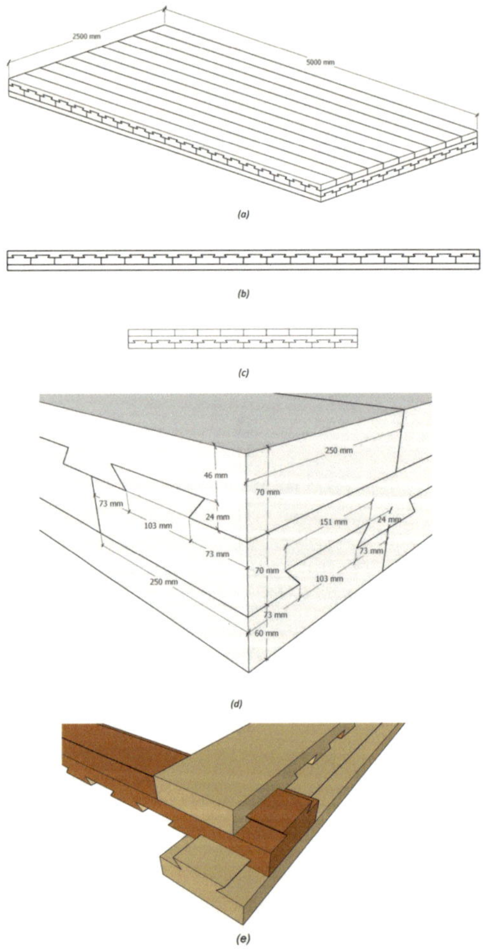

Figure 6.
Dovetail wood elements as floor slab alternative-1 (solid type): (a) isometric view; (b) front view; (c) side view; (d) with representative dimensions; (e) detail.

unit can work on the same file, and the result is the same as desired. The number of layers can be varied, and the width and thickness of the wood can also vary according to need, and the hardness of the board is completely formed without adhesives, nails, staples, or other materials, with no size limitation unlike traditional EWPs such as CLT and LVL.

According to current construction practices of conventional EWPs and critical discussions with various industry professionals as the first step in design, geometrically original and technically sound 2D and 3D horizontal (e.g., floor slab) frame models are presented below.

For comparison with the CLT of equivalent dimensions, the optimal test size of the dovetail wood board will usually be taken as about 200 mm thick (three-layer),

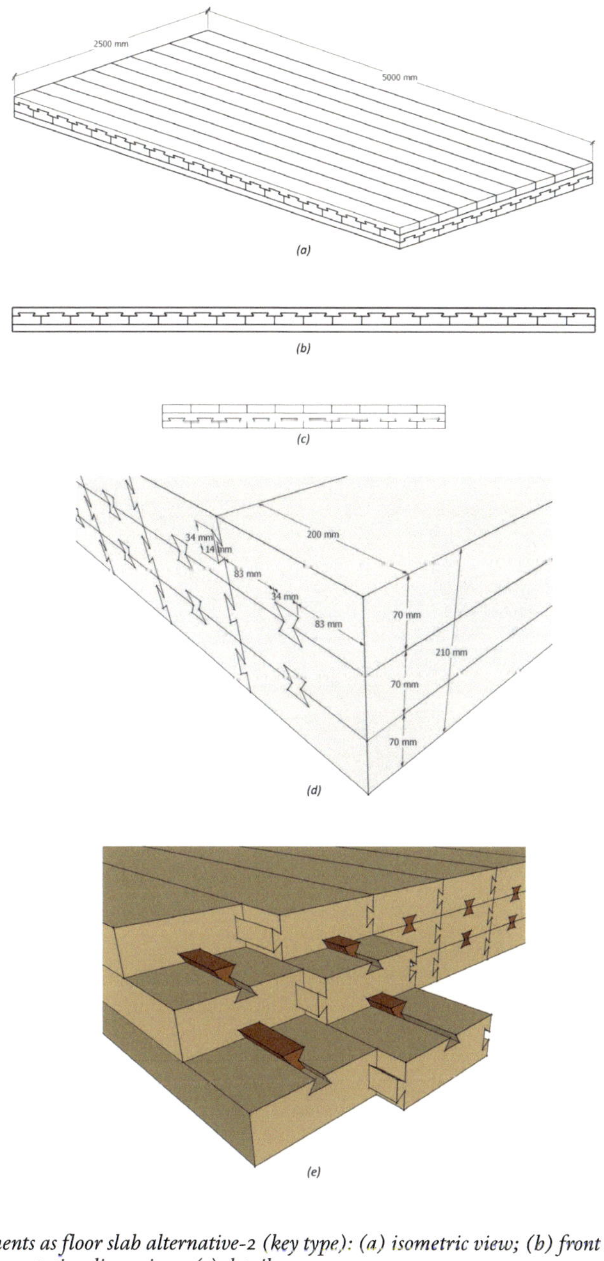

Figure 7.
Dovetail wood elements as floor slab alternative-2 (key type): (a) isometric view; (b) front view; (c) side view; (d) with representative dimensions; (e) detail.

2500 mm wide, and 5000 mm long. Additionally, in the light of the abovementioned practical knowledge and discussions, the design, which was initially considered as five-layer, has been revised to three-layer to minimize the amount of waste products. Moreover, considering structural tests and other performance tests such as fire safety and sound insulation, it is predicted that the dimensions of building components may change, especially after structural analysis.

As can be seen in **Figure 6**, the "solid type" can be used as dovetail wood elements as an alternative to the floor slab, inspired by conventional dovetail connection, one of the oldest joining methods used in ancient temples and churches.

Figure 7 highlights "key type" with a similar structural mechanism with key laminated timber beams. However, since in the current literature, there are few studies (e.g., [55, 56]) on key-laminated beams that show similarities to our key type proposal in **Figure 6**, advantages and disadvantages of our key type can be

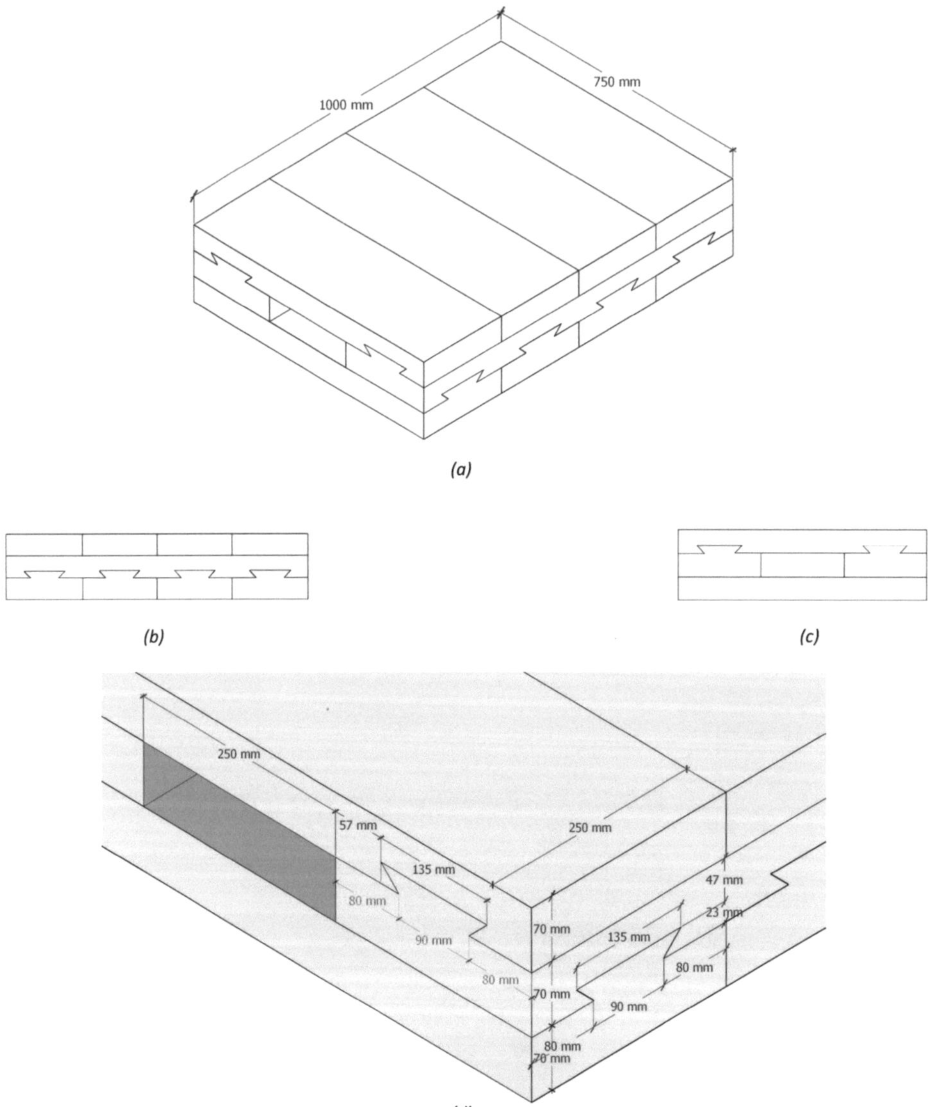

Figure 8.
Dovetail wood elements as floor slab alternative-3 (hollow type): (a) isometric view with dimensions; (b) front view; (c) side view; (d) with representative dimensions.

revealed as a result of structural analyses and subsequent technical performance tests in the laboratory.

As shown in **Figures 8** and **9**, hollow type is a viable alternative due to its many advantages including reducing the dead load, improving the weight-to-strength ratio, ease of plumbing or electrical work, thus saving construction cost as in the cases of hollow concrete slab [57, 58] and hollow-core cross-laminated timber [59, 60].

After the geometric configuration design phase, structural analyses will be made, and it is planned to proceed to the prototype manufacturing phase. In this phase, softwood from gymnosperm trees such as pines and spruces will be used, taking into consideration its advantages such as workability, sustainability, and cost.

Here, first, the solid type shown in **Figure 6** will be manufactured and technical performance tests will be carried out compared with equivalent CLT elements as mentioned above. In this sense, as a result of the first technical evaluations made

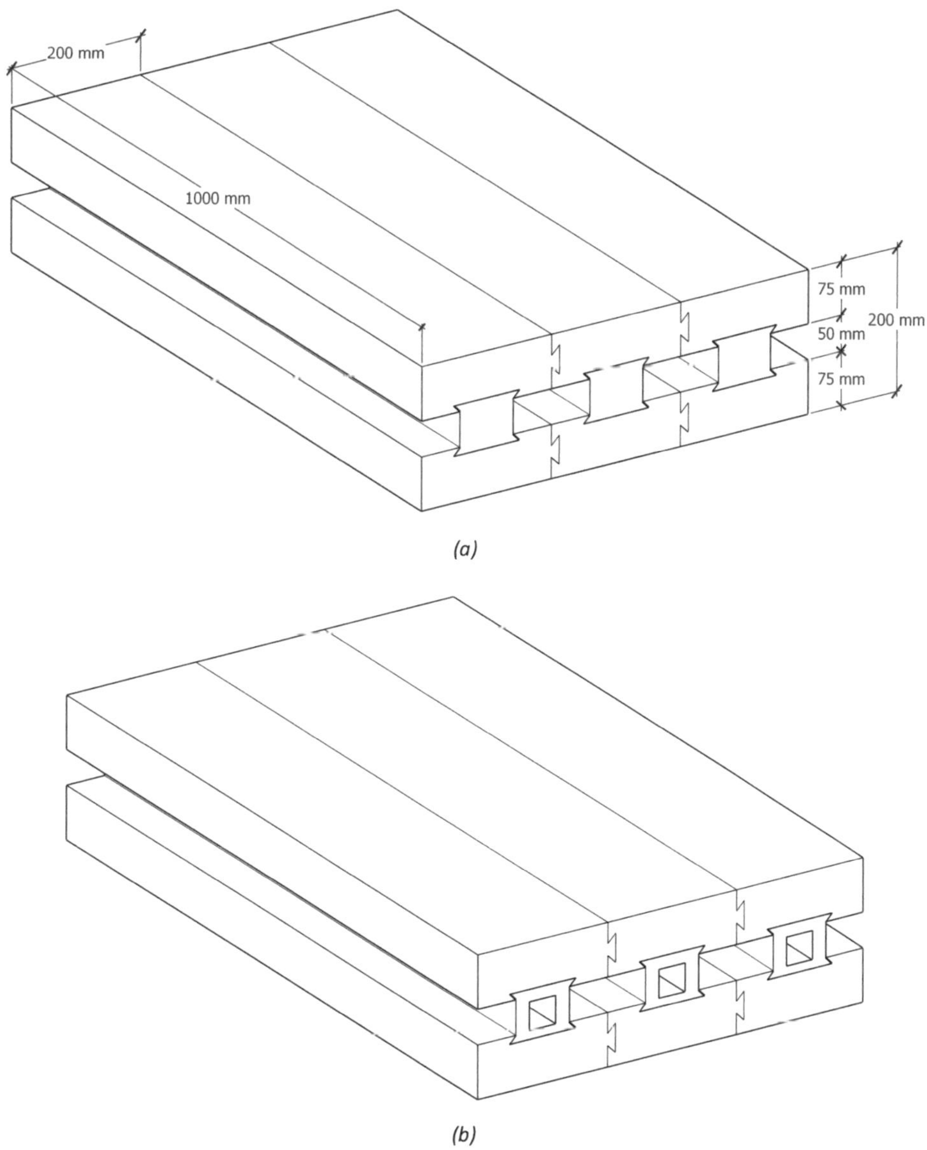

Figure 9.
Dovetail wood elements as floor slab alternatives-4 (hollow type-2).

with wood construction experts and structural designers, it is predicted that the structural performance tests, which represent the most critical phase of the comparison study to be conducted, will yield positive results considering DWBE's solid structure and joint details as seen in **Figure 6**. On the other hand, considering the future performance tests to be carried out as mentioned above, it is anticipated that as shown in **Figures 8** and **9**, hollow types' weight-strength advantage will make a great contribution to their structural performance, and the porous structure will make a big difference in their sound insulation performance.

4. Conclusion

Since there are no patented adhesive and nonmetallic dovetail wood panel solutions in the timber market [40], it has not been possible to carry out a comprehensive discussion on the similarities and differences of our proposals with other solutions. This study aimed to present various geometric configurations for dovetail wooden horizontal structural members in multistory building construction as ecologically sound engineering solutions through architectural modeling programs as the first step to developing DWBEs. The results are at the architectural design stage based on a theoretical approach taking into account current construction practices, but developed products will be finalized through market research after the technical performance tests (e.g., structural, fire, sound insulation, and airtightness tests) and necessary optimization processes. Therefore, currently, though DWBEs uptake for commercial applications is very limited, due to new studies such as DoMWoB project (*Dovetailed Massive Wood Board Elements for Multi-Story Buildings—see Acknowledgments*), the potential of the "innovative dovetail concept" will be further used in high-rise construction.

Acknowledgements

This project has received funding from the European Union's Horizon 2020 research and innovation program under the Marie Skłodowska-Curie grant agreement No [101024593].

Funding

This project has also received funding (60 000 EUR) from the Marjatta and Eino Kolli Foundation for funding the technical performance tests including fire safety, structural, moisture transfer resistance and air-tightness, and sound insulation.

Thanks

The authors thank Dr. Ozlem Nur Aslantamer (Atılım University) for her great contribution to both initial design ideas and drawings.

Author details

Hüseyin Emre Ilgın*, Markku Karjalainen and Olli-Paavo Koponen
Tampere University, Tampere, Finland

*Address all correspondence to: emre.ilgin@tuni.fi

References

[1] Alsheyab MAT. Recycling of construction and demolition waste and its impact on climate change and sustainable development. International journal of Environmental Science and Technology. 2021. DOI: 10.1007/s13762-021-03217-1. https://link.springer.com/article/10.1007/s13762-021-03217-1#citeas

[2] Li D, Huang G, Zhu S, Chen L, Wang J. How to peak carbon emissions of provincial construction industry? Scenario analysis of Jiangsu Province. In: Renewable and Sustainable Energy Reviews. Leicestershire, UK: Loughborough University; Vol. 44. 2021. DOI: 10.1016/j.rser.2021.110953

[3] IPCC. Climate Change 2013: The Physical Science Basis. Contribution of Working Group I to the Fifth Assessment Report of the Intergovernmental Panel on Climate Change. Cambridge and New York: Cambridge University Press; 2013

[4] UN Environment programme, building sector emissions hit record high, but low-carbon pandemic recovery can help transform sector—UN report. Available from: https://www.unep.org/news-and-stories/press-release/building-sector-emissions-hit-record-high-low-carbon-pandemic [Accessed: November 20, 2021]

[5] Architecture 2030. Why the Building Sector? Available from: https://architecture2030.org/why-the-building-sector/ [Accessed: November 20, 2021]

[6] Bosman R, Rotmans J. Transition governance towards a bioeconomy: A comparison of Finland and The Netherlands. Sustainability. 2016;**8**(10):1017. DOI: 10.3390/su8101017

[7] Wood Building Programme. Finnish Ministry of Environment. Available from: https://www.ym.fi/enUS/Land_use_and_building/Programmes_and_strategies/Wood_Building_Program/Wood_Building_Programme(47800) [Accessed: November 20, 2021]

[8] Harte AM. Mass timber-the emergence of a modern construction material. Journal of Structural Integration Maintenance. 2017;**2**(3):121-132

[9] Shahnewaz M, Tannert T, Alam MS, Popovski M. In-plane stiffness of cross-laminated timber panels with openings. Structural Engineering International. 2017;**27**:217-223

[10] Stepinac M, Šušteršic I, Gavric I, Rajcic V. Seismic design of timber buildings: Highlighted challenges and future trends. Applied Sciences. 2020;**10**:1380

[11] Ramage MH, Burridge H, Busse-Wicher M, Fereday G, Reynolds T, Shah DU, et al. The wood from the trees: The use of timber in construction. Renewable and Sustainable Energy Reviews. 2017;**68**:333-359

[12] Carbon 12. Available from: https://www.carbon12pdx.com/about [Accessed: November 20, 2021]

[13] Abrahamsen RB. Mjøstårnet-Construction of an 81 m Tall Timber Building. Garmisch-Partenkirchen, Germany: Internationales Holzbau-Forum IHF; 2017

[14] Gamerro J, Bocquet JF, Weinand Y. A calculation method for interconnected timber elements using wood-wood connections. Buildings. 2020;**10**(3):61. DOI: 10.3390/buildings10030061

[15] Frihart CR. Introduction to special issue, wood adhesives: Past, present, and future. Forest Products Journal. 2015;**65**:4-8

[16] Solta P, Konnerth J, Gindl-Altmutterb W, Kantnerc W, Moser J,

Mitterd R, et al. Technological performance of formaldehyde-free adhesive alternatives for particleboard industry. International Journal of Adhesion and Adhesives. 2019;**94**: 99-131

[17] Nguyen MN, Leicester RH, Wang C-H, Foliente GC. Corrosion effects in the structural design of metal fasteners for timber construction. Structure and Infrastructure Engineering. 2013;**9**(3): 275-284. DOI: 10.1080/15732479. 2010.546416

[18] Robert H, Falk PE, Baker AJ. Fasteners for exposed wood structures. Wood Design Focus. 1993;**4**(3):14-17

[19] Khoshnava SM, Rostami R, Zin RM, Štreimikienė D, Mardani A, Ismail M. The role of green building materials in reducing environmental and human health impacts. International Journal of Environmental Research and Public Health. 2020;**17**(7):2589

[20] Kozicki M, Guzik K. Comparison of VOC emissions produced by different types of adhesives based on test chambers. Materials. 2021;**14**(8):1924. DOI: 10.3390/ma14081924

[21] Chang WS, Nearchou N. Hot-pressed dowels in bonded-in rod timber connections. Wood and Fiber Science. 2015;**47**(2):199-208

[22] Sotayo A, Bradley D, Bather M, Sareh P, Oudjene M, El-Houjeyri I, et al. Review of state of the art of dowel laminated timber members and densified wood materials as sustainable engineered wood products for construction and building applications. Developments in the Built Environment. 2020;**1**:1-11

[23] Ilgın HE, Karjalainen M, Koponen O. Review of the Current State-of-the-Art of Dovetail Massive Wood Elements. London, UK: IntechOpen; 2021

[24] Toivonen R, Lähtinen K. Sustainability—A Literature Review on Concealed Opportunities for Global Market Diffusion for the Crosslaminated Timber (CLT) in the Urbanizing Society. Curitiba, Brazil: BioProducts Business; 2019

[25] Lu B, Lu W, Zhong M, Wu W, Zhou P. Experimental investigation and analytical model of cross-laminated timber wall with coupled U-shaped flexural plate connectors. Construction and Building Materials. 2021;**307**:124984. DOI: 10.1016/j.conbuildmat.2021.124984

[26] Pečnik JG, Gavrić I, Sebera V, Kržan M, Kwiecień A, Zając B, et al. Mechanical performance of timber connections made of thick flexible polyurethane adhesives. Engineering Structures. 2021;**247**:113125. DOI: 10.1016/j.engstruct.2021.113125

[27] Lukacs I, Björnfot A, Tomasi R. Strength and stiffness of cross-laminated timber (CLT) shear walls: State-of-the-art of analytical approaches. Engineering Structures. 2019;**178**:136-147

[28] Xing Z, Wang Y, Zhang J, Ma H. Comparative study on fire resistance and zero strength layer thickness of CLT floor under natural fire and standard fire. Construction and Building Materials. 2021;**302**:124368. DOI: 10.1016/j.conbuildmat.2021.124368

[29] Ronquillo G, Hopkin D, Spearpoint M. Review of large-scale fire tests on cross-laminated timber. Journal of Fire Sciences. 2021;**39**(5):327-369. DOI: 10.1177/07349041211034460

[30] Li M, Zhang S, Gong Tian Z, Ren H. Gluing techniques on bond performance and mechanical properties of cross-laminated timber (CLT) made from *Larix kaempferi*. Polymers. 2021;**13**:733

[31] Jasieńko J, Nowak T, Karolak A. Historical carpentry joints. Journal of Heritage Conservation. 2014;**40**:58-82

[32] Jeong GY, Park MJ, Park JS, Hwang KH. Predicting load-carrying capacity of dovetail connections using the stochastic finite element method. Wood and Fiber Science. 2012;**44**(4): 430-439

[33] Robeller C, Weinand Y. Interlocking folded plate—integral mechanical attachment for structural wood panels. International Journal of Space Structures. 2015;**30**:111-112. DOI: 10.1260/0266-3511.30.2.111

[34] Pozza L, Scotta R, Trutalli D, Pinna M, Polastri A, Bertoni P. Experimental and numerical analyses of new massive wooden shear-wall systems. Buildings. 2014;**4**:355-374

[35] Pozza L, Scotta R, Trutalli D, Polastri A. Behaviour factor for innovative massive timber shear walls. Bulletin of Earthquake Engineering. 2015;**13**:3449-3469

[36] Jeong GY, Song JK. Evaluation of structural properties of dovetail connections under tensile load using three methods of data analysis. Journal of Materials in Civil Engineering. 2017;**29**(10):6017011

[37] Sha B, Wang H, Li A. The influence of the damage of mortise-tenon joint on the cyclic performance of the traditional Chinese timber frame. Applied Sciences. 2019;**9**(16):3429. DOI: 10.3390/app9163429

[38] Xie Q, Zhang B, Zhang L, Guo L, Wu Y. Normal contact performance of mortise and tenon joint: Theoretical analysis and numerical simulation. Journal of Wood Science. 2021;**67**:31. DOI: 10.1186/s10086-021-01963-x

[39] Mashrah WAH, Chen Z, Liu H, Amer MA. Mechanical behaviour of a novel steel dovetail joint subjected to axial compression loading. Engineering Structures. 2021;**245**:112852. DOI: 10.1016/j.engstruct.2021.112852

[40] Ilgın HE, Karjalainen M. Preliminary design proposals for dovetail wood board elements in multi-story building construction. Architecture. 2021;**1**(1):56-68. DOI: 10.3390/architecture1010006

[41] Murdock JN. Stream restoration. In: Jørgensen SE, Fath BD, editors. Encyclopedia of Ecology. Amsterdam: Elsevier, Academic Press; 2008. pp. 3390-3397. DOI: 10.1016/B978-008045405-4.00077-X

[42] Mitsch WJ. Ecological Engineering: A New Paradigm for Engineers and Ecologists, National Academy of Engineering, Engineering Within Ecological Constraints. Washington, DC: The National Academies Press; 1996. DOI: 10.17226/4919

[43] Schönborn A, Junge R. Redefining ecological engineering in the context of circular economy and sustainable development. Circular Economy and Sustainability. 2021;**1**:375-394. DOI: 10.1007/s43615-021-00023-2

[44] Matlock MD, Morgan RA. Ecological Engineering Design Restoring and Conserving Ecosystem Services. Hoboken, NJ: Wiley; 2010

[45] Fraser M. Design Research in Architecture: An Overview. London: Ashgate; 2013

[46] Akšamija A. Research Methods for the Architectural Profession. New York, NY, USA: Routledge; 2021

[47] Short CA. What is 'architectural design research'? Building Research & Information. 2008;**36**(2):195-199. DOI: 10.1080/09613210701811015

[48] Vasilenko NA. General System Principles of Architectural Systems Formation. IOP Conference Series: Materials Science and Engineering. 2020;**753**:32047

[49] Luck R. Design research, architectural research, architectural design research: An argument on disciplinarity and identity. Design Studies. 2019;**65**:152-166. DOI: 10.1016/j.destud.2019.11.001

[50] Gao Y. Design research in architecture: An overview edited by Murray Fraser. The Design Journal. 2015;**18**(2):291-294. DOI: 10.2752/175630615X14212498964439

[51] Eilouti B. Scenario-based design: New applications in metamorphic architecture. Frontiers of Architectural Research. 2018;7(4):530-543. DOI: 10.1016/j.foar.2018.07.003

[52] Luck R. Participatory design in architectural practice: Changing practices in future making in uncertain times. Design Studies. 2018;**59**:139-157. DOI: 10.1016/j.destud.2018.10.003

[53] Bhooshan S. Parametric design thinking: A case-study of practice-embedded architectural research. Design Studies. 2017;**52**:115-143. DOI: 10.1016/j.destud.2017.05.003

[54] Fu F. Design and Analysis of Tall and Complex Structures. Oxford and Cambridge: Butterworth-Heinemann, Elsevier; 2018

[55] Miller JF, Bulleit WM. Analysis of mechanically laminated timber beams using shear keys. Journal of Structural Engineering (New York, N.Y.). 2011;**137**(1):124-132. DOI: 10.1061/(ASCE)ST.1943-541X.0000273

[56] Miller JF. Design and analysis of mechanically laminated timber beams using shear keys [PhD dissertation]. Michigan, United States: Civil Engineering, Michigan Technological University; 2009

[57] Ahmed IM, Tsavdaridis KD. The evolution of composite flooring systems: Applications, testing, modelling and Eurocode design approaches. Journal of Constructional Steel Research. 2019;**155**:286-300. DOI: 10.1016/j.jcsr.2019.01.007

[58] Al-Shaarbaf IA, Al-Azzawi AA, Abdulsattar R. A state of the art review on hollow core slabs. ARPN Journal of Engineering and Applied Sciences. 2018;**13**:9

[59] Huang H, Lin X, Zhang J, Wu Z, Wang C, Wang BJ. Performance of the hollow-core cross-laminated timber (HC-CLT) floor under human-induced vibration. Structure. 2021;**32**:1481-1491. DOI: 10.1016/j.istruc.2021.03.101

[60] The CLT Handbook. CLT Structures–Facts and Planning. Stockholm, Sweden: Swedish Wood; 2019. Available from https://www.svenskttra.se/siteassets/5-publikationer/pdfer/clt-handbook-2019-eng-m-svensk-standard-2019.pdf [Accessed: November 20, 2021]

Chapter 17

Review of the Current State-of-the-Art of Dovetail Massive Wood Elements

Hüseyin Emre Ilgın, Markku Karjalainen and Olli-Paavo Koponen

Abstract

Engineered wood products (EWPs) have been progressively more being utilized in the construction industry as structural materials since the 1990s. In the content of EWPs, adhesives play an important role. However, because of their petroleum-based nature, adhesives contribute to toxic gas emissions such as formaldehyde and Volatile Organic Compounds, which are detrimental to the environment. Moreover, the frequent use of adhesives can cause other critical issues in terms of sustainability, recyclability, reusability, and further machining. In addition to this, metal connectors employed in EWPs harm their end-of-life disposal, reusability, and additional processing. This chapter is concentrating on dovetail massive wood elements (DMWE) as adhesive- and metal connector-free sustainable alternatives to commonly used EWPs e.g., CLT, LVL, MHM, Glulam. The dovetail technique has been a method of joinery mostly used in wood carpentry, including furniture, cabinets, log buildings, and traditional timber-framed buildings throughout its rich history. It is believed that this chapter will contribute to the uptake of DMWE for more diverse and innovative structural applications, thus the reduction in carbon footprint by increasing the awareness and uses of DMWE in construction.

Keywords: Dovetail joint technique, dovetail massive wood elements, engineered wood products, building construction, sustainability

1. Introduction

Wood is an indisputably renewable, ecological, and environmentally friendly material that has been widely used throughout history [1, 2]. One cubic meter of growing wood can bind about one ton of CO_2 from the atmosphere, the mass of wood is about 500 kg/m^3, and about half of this mass is carbon = 250 kg/m^3 [3, 4]. Forests are carbon sink and wood products are carbon storage. According to FAOSTAT, 488 million m^3 sawn wood were produced globally [5].

Moreover, thanks to its numerous positive impacts on the environment and potential cost-effectiveness compared to traditional materials such as reinforced concrete and steel, accompanied by its technological advances; wood, in the form of engineered wood products (EWPs), has come back to break into modern building utilization e.g., multi-story construction after more than a century [6–8].

In this industry, as a growing market in Europe, EWPs e.g., cross-laminated timber (CLT – a prefabricated multi-layer EWP, manufactured from at least three layers of boards by gluing their surfaces together with an adhesive under pressure), glue-laminated timber (Glulam – made by gluing together several graded timber laminations with their grain parallel to the longitudinal axis of the section), laminated veneer lumber (LVL – made by bonding together thin vertical softwood veneers with their grain parallel to the longitudinal axis of the section, under heat and pressure), Massiv-Holz-Mauer® (MHM – a timber wall construction material consisting of dried soft wood joined with fluted aluminum nails that requires neither glue nor chemical treatment) have had an important position with the production capacity of more than 5 million cubic meter/year [9].

Particularly due to the easy coupling technique, airtightness, high rigidity, dimensional stability, and homogenous mechanical properties, EWPs e.g., CLT is competitive especially in multi-story wooden buildings [10, 11]. Similarly, Glulam external structural frame as a proven system for the buildings with over 10-story [12, 13] was used in the tallest wooden towers as in the cases of the 85 m and 18-story Mjøstårnet in Norway [14, 15], and 84 m and 24-story HoHo in Austria [16].

EWPs have been usually produced from the adhesive bonding of wood chips, flakes, veneer, or sawn wood sections, and/or the mechanical fastening of wooden sections to form larger sections, beams, panels, shear walls, or other structural members [17]. In these products, adhesives play an essential role particularly by helping save wood, making the structure light and strong, and restraining the contraction and expansion due to the inherent moisture. However, although there are advantages above associated with EWPs, the use of adhesives causes some concerns about their sustainability, recyclability, further machining, and broader environmental impact [9, 18].

More in particular, because of toxic gas emissions (e.g., formaldehyde and Volatile Organic Compounds) during their lifespan and while burning, resulting from their petroleum-based contents, the dominant use of adhesives has adverse effects on the environment e.g., climate change, air pollution, and human health [19–21]. Also, hardeners (e.g., amine and formaldehyde) in adhesives are irritating and skin sensitizing and are thus continually contact may cause allergic reactions [22]. Moreover, there are still critical questions about environmentally friendly biobased adhesives despite ongoing improvements in this research area [20, 23].

Besides several regulatory standards [24–26] addressing the points mentioned above, European Commission has a specific objective of improving air quality, which can also be achieved by reducing the use of harmful adhesives [27]. In addition to these detrimental substances, metal connectors used in EWPs have a negative impact on their end-of-life disposal, reusability, and recyclability [28]. It is worth noting here that the effect of moisture content on the mechanical and structural performance of DMWE should be particularly taken into account during design and construction phases because failure to control the moisture content of structural timber causes serious structural problems, such as excessive deflection of beams.

Therefore, this chapter is focusing on the dovetail joint technique and dovetail massive wood elements. Based on one of the oldest joining methods (see examples in **Figure 1**), these elements can offer a sustainable solution that is solid and completely pure wood enabling as healthy indoor air as possible (adhesive-&metal-connectors-free). It is believed that using the potential of DMWE will contribute to increase the competitiveness of large-scale industrial wooden construction and to create higher value-added circular economy opportunities as part of the bio-economy and sustainable development all over the world.

(a) (b) (c)

Figure 1.
Dovetail joints: (a) a finished dovetail joint, (b) dovetailed woodworking joints on a Romanian church, (c) stone pillar at the Vazhappally Maha Siva Temple (source: Wikipedia).

2. Historical background of dovetail joint technique

In history, the dovetail technique is a joinery method most used in woodwork (i.e. carpentry), including furniture, cabinets, log buildings, and traditional wood-framed structures. The history of the dovetail joint technique goes back to before Christ. Some of the earliest well-known examples of this technique were in ancient Egyptian furniture buried with mummies dating from the First Dynasty, stone pillars at the Temples in India (**Figure 2**) as well as Japanese and Korean traditional buildings [29, 30]. Besides these, this technique was utilized in Chinese ancient architecture [31, 32], where the dovetail joint was introduced - national building codes and construction methods in Song Dynasty in 1103 - as one of the primary joint methods employed in the oldest timber buildings in China [33]. Additionally, during the earliest times to the Middle Ages, in Egypt, the construction of cabinets was based on the mortise and tenon, dovetail, and mitred joints [34]. In Europe, the dovetail joint is also called a swallowtail joint, a culvertail joint, or a fantail joint [35].

The first residential constructions with wood-framed structures from the 13th century consisted of mortise and tenon joints, strengthened with wedges, notched joints with tenons, and dovetail joints [36]. Notable examples of connecting the roof rafter and beams involved making use of the dovetail joint (**Figure 3**) were churches in the 14th century [36]. The roof structure of the Church of St. Jacob in Torun (16th century) was one of the oldest preserved examples, which includes notched joints with dovetail tenons [37]. Moreover, as Polish churches, the Church in Cewków (**Figure 4**) and the Church in Chotylub (**Figure 5**) were among remarkable examples of wood-framed buildings with dovetail wall-corner joists from the 19th century.

Based on the skilled woodworkers' familiarity with design and manufacture, carpentry-type wood-to-wood joints were widely used in building construction till

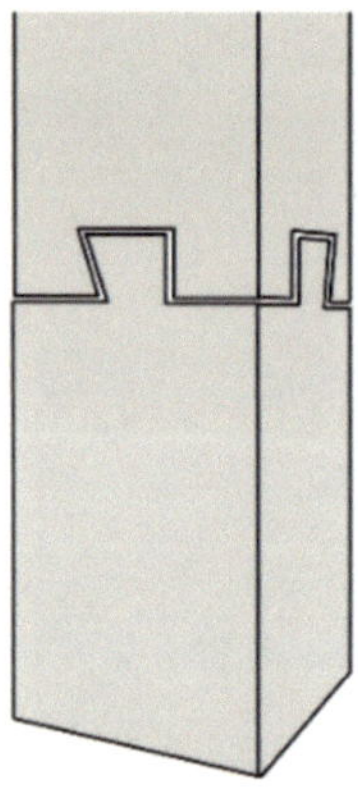

Figure 2.
A stone pillar at a temple in India (drawn by Emre ILGIN).

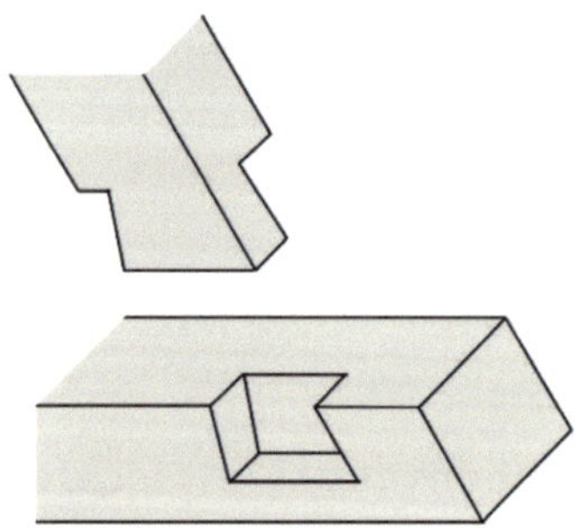

Figure 3.
Dovetail joint (drawn by Emre ILGIN).

Figure 4.
The Church in Cewków, Poland (left) (source: Wikipedia) with dovetail corner detail (right) (drawn by Emre ILGIN).

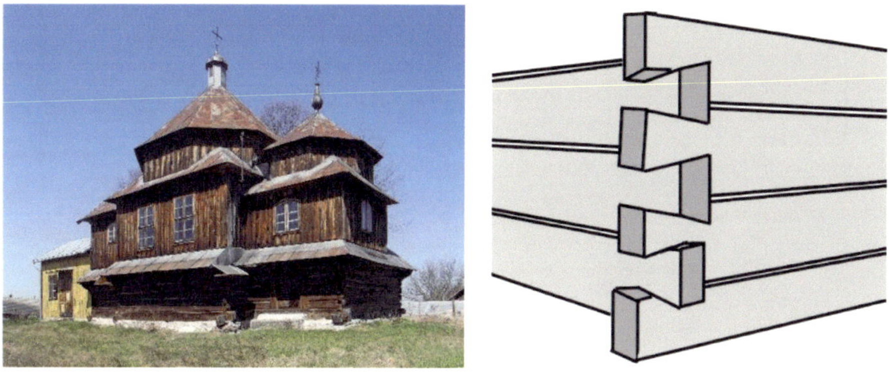

Figure 5.
The Church in Chotylub, Poland (left) (source: Wikipedia) with dovetail corner detail (right) (drawn by Emre ILGIN).

the mid-20th century [38]. Different dovetail designs in Europe and Asia often govern practical considerations [39]. However, high labor costs and inadequacies due to excessively traditional designs rendered these joints uncompetitive. Advancements in CNC wood processing machines re-established the cost-effectiveness for carpentry-type wood-to-wood joints.

3. The current state-of-the-art of dovetail massive wood elements

In the literature, thus far, there have been numerous studies regarding the technological aspects of timber in construction with different building solutions based on the utilization of engineered timber products such as CLT [40–43]. However, there is a limited number of researches (e.g., [44]) on dovetail massive wood elements (DMWE). To date, previous studies about DMWE is based on a few papers mostly about structural analysis and model testing of several types of joint details rather than even evaluating overall technical performance (e.g., structural, fire, sound) of a structural component such as a column, a beam, a shear wall or an entire structure.

Among these most prominent studies conducted in the last decade, Jeong et al. scrutinized the effects of geometric variables on the mechanical behavior of dovetail connection (**Figure 6**) through finite element method analysis together with experimental tests [45]. There different were parameters such as various tenon angles and tenon heights with three representative failure modes. The results showed that the geometry that maximizes the load-bearing capacity is the 57-degree tenon angle and the average allowable load for the dovetail joint is calculated as 21.4kN.

Also, failure modes of dovetail connection were dominated by tension perpendicular to the shear stress. Furthermore, planned failure criteria correlated with the critical stress played an important role in the projection of load-bearing capacity from dovetail connection.

Pang et al. studied the effects of size ratios on dovetail joints in Korean traditional timber building by examining moment resistance of various sizes of dovetail joints following experimental procedures together with dimensional analysis (**Figure 7**) [39]. It was observed that the average maximum and yield moment resistance was increased as the scale ratio was increased. As a result, moment resistance confirmed the similitude theory.

Tannert et al. presented various reinforcement methods (e.g., with self-tapping screws, with adhesive layer) to enhance the structural performance of rounded

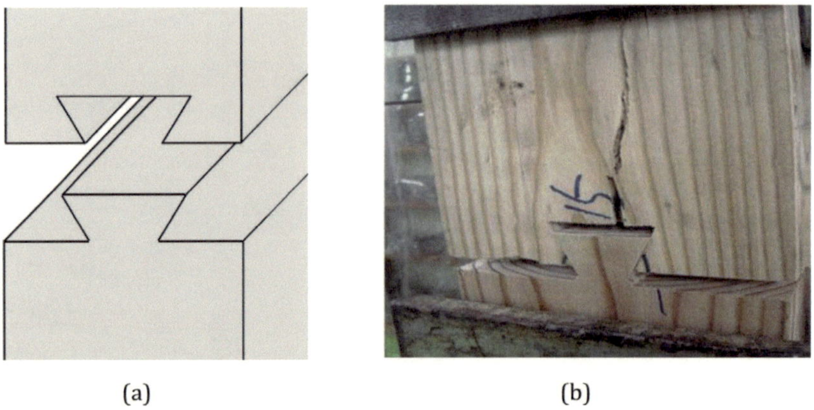

Figure 6.
(a) The dovetail joist for the test specimen (drawn by Emre ILGIN), (b) tension perpendicular to the grain failure at mortise in dovetail connection [45].

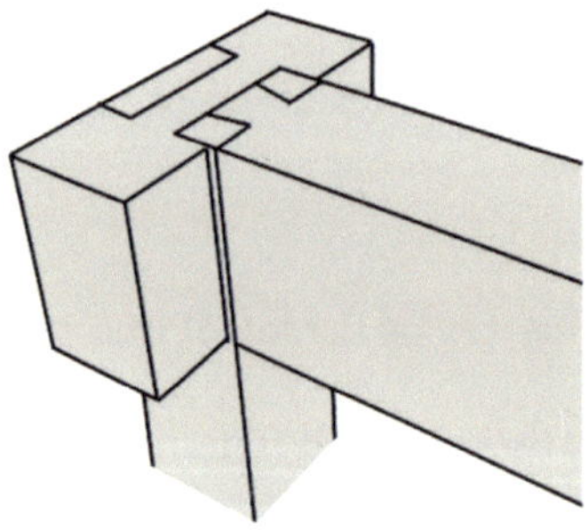

Figure 7.
The dovetail joist for the test specimen (drawn by Emre ILGIN).

dovetail joints (**Figure 8**) under static short-term shear loading [38]. Using the test series, comparisons between reinforced and non-reinforced joints were made to assess the potentials and limitations of different reinforcement methods. Based on the test results, adhesive-reinforced-rounded dovetail joints were proposed to improve structural performance under predefined loading conditions.

In the paper entitled 'Interlocking Folded Plate - Integral Mechanical Attachment for Structural Wood Panels', Robeller and Weinand built folded thin shell prototype consisting of timber panels by utilizing automatic fabrication of cabinetmaking joints, i.e. dovetail joints without adhesive (**Figure 9**) [46]. This interlocking arch prototype was constructed from 21 mm LVL panels and 12 mm

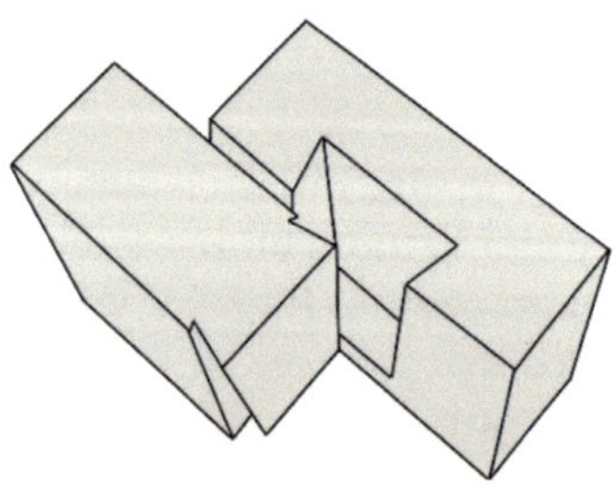

Figure 8.
The rounded dovetail joist for the test specimen (drawn by Emre ILGIN).

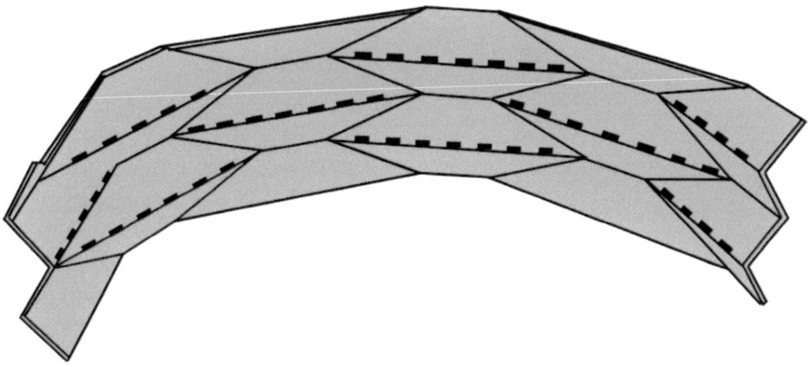

Figure 9.
Folded-plate arch prototype (drawn by Emre ILGIN).

plywood with a self-weight of 192 kg and a span of 3 meters to provide input on the load-carrying capacity of integrated joints. It was recommended that further research is needed for large-scale building applications.

Pozza et al. simulated and tested structural behaviors of three massive wooden shear wall configurations including the cross-laminated-glued wall, cross-laminated-stapled wall, and layered wall with dovetail inserts under seismic loads. According to the results, all configurations had good dispersion capacity and could be employed well for seismically vulnerable zones [47]. Similarly, Pozza et al. examined four massive wooden shear walls through experimental tests e.g., subjecting to compressive stress and numerical simulations. Analyzed shear wall configurations were CLT panels with glued interfaces together with massive timber panels adopting steel staples (stapled wall) or timber dovetail inserts to unite the layers (layered wall) [48]. Results indicated that all four variations offer a feasible construction technique for earthquake-prone zones.

Besides the abovementioned studies, other research showed that the critical aspects of the structure of material and failure behaviors without considering the effects of material properties and geometric configurations [49–51].

4. Types of dovetails

Throughout history, there are four most prevalent types of joints to be employed to fit walls together in building construction (**Figure 10**) [52, 53]:

i. double notch,

ii. half notch,

iii. "the lock", also called a "German" or "Saxon" joint (can be seen in different parts of the world e.g., the east of the Carpathian Mountains, northern Romania, Finland – *known as 'Hammasnurkka' or a 'toothed' corner joint*),

iv. full dovetail (fishtail) notch.

Joints were usually doweled, but round logs were often joined by undoweled 'saddle notched' joints. Although different types of joints were able to use in different parts of a wall, probably the most archaic type, the double notch was used

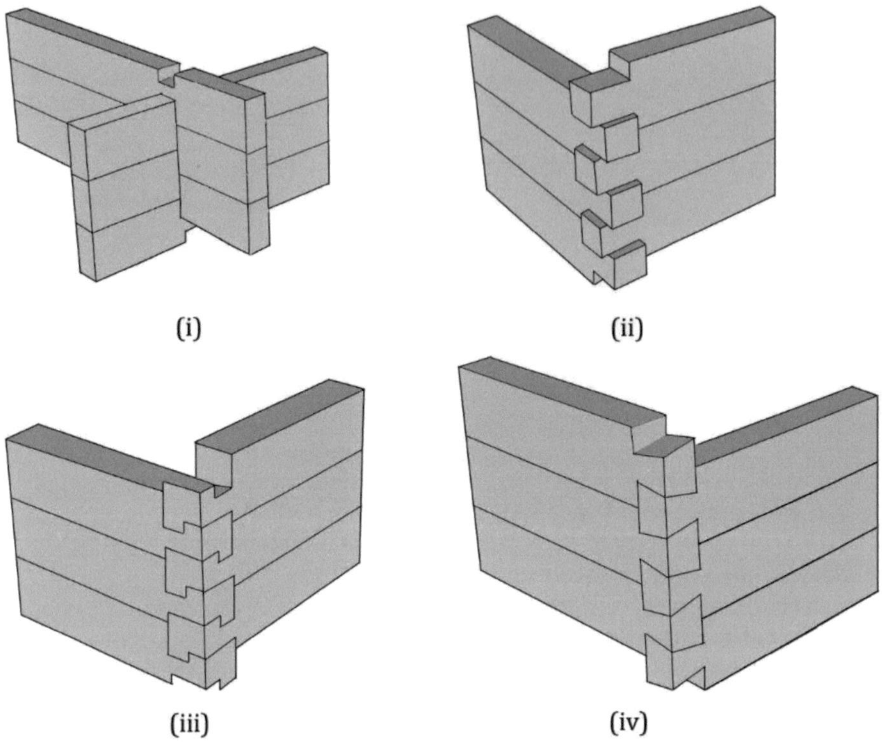

Figure 10.
Four types of carpentry joints: (i) double notch, (ii) half notch, (iii) the lock joint, (iv) full dovetail notch (drawn by Emre ILGIN).

directly under the deep eaves; the half notch was built over the eaves where the walls were to be planned or laid [53]. The full dovetail notch possibly began to be widely used in the 18th century.

Double notch is better than half notched joint as the beam is not weakened with respect to its load-bearing support, where skew angle and the notch depth are important parameters [54–56]. On the other hand, "the lock" is generally used for beams of large dimensions. As a rule, the four types of joints mentioned above can transfer both compressive and shear stresses, and dovetail connections can also transfer relatively small tensile stresses [57].

It is also worth mentioning here that in addition to the wood construction, throughout history, there are many different types of dovetail joint used in wooden furniture design, which can also be a source of inspiration for the construction industry of today, as follows (**Figure 11**) [34]:

(a) through dovetail, (b) lapped dovetail, (c) double lapped dovetail, (d) secret mitred dovetail. Joints used in carcases: (e) cross rails dovetailed into solid side, (f) cross rails dovetailed into framed side, (g) top dovetailed to side, (h) top dowelled to side, (i) top rebated, (j) housing, (k) tapered dovetail housing.

5. Conclusion

As one of our best allies in combating the climate crisis, timber is in the foreground due to its environmental-friendly features such as low carbon emissions in processing and carbon sequestration. In this sense, engineered wood products are increasingly used in the construction industry. However, adhesives used in their

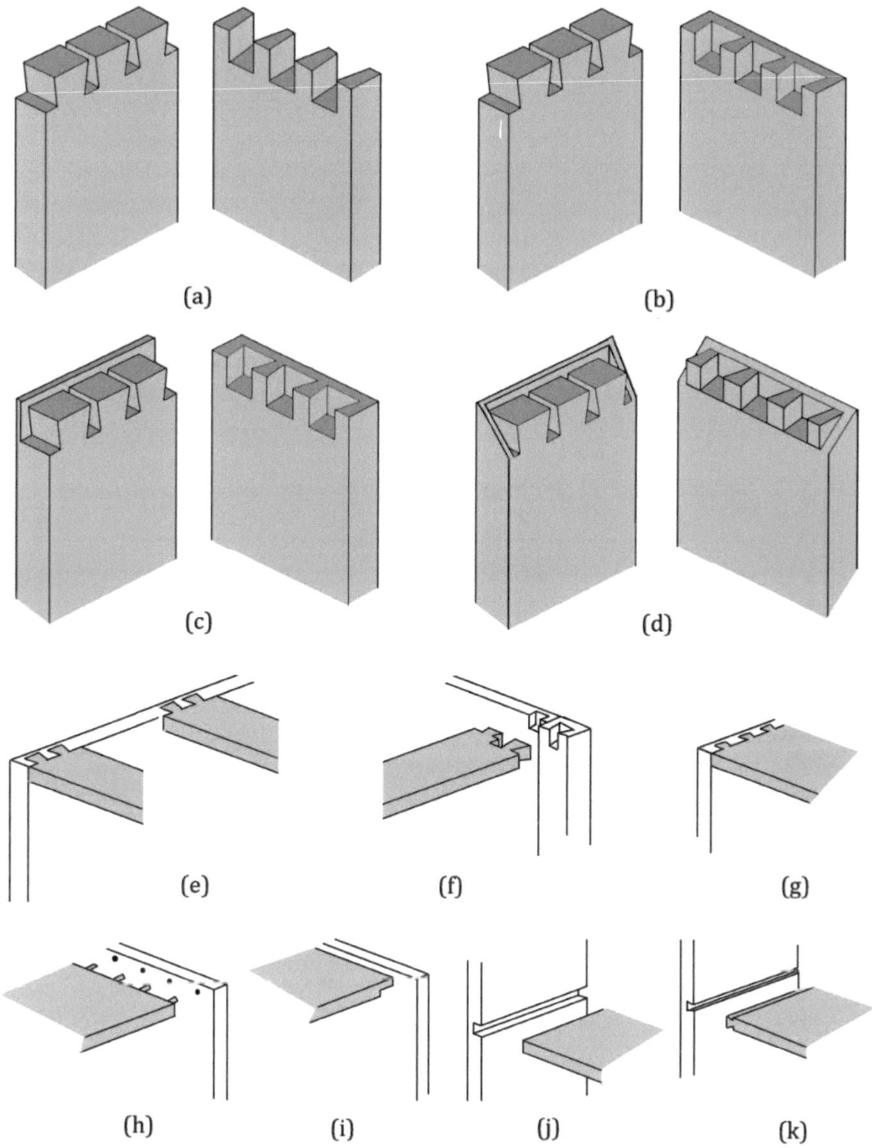

Figure 11.
Examples of dovetail joints in wooden furniture design: (a) through dovetail, (b) lapped dovetail, (c) double lapped dovetail, (d) secret mitred dovetail. Joints used in carcases: (e) cross rails dovetailed into solid side, (f) cross rails dovetailed into framed side, (g) top dovetailed to side, (h) top dowelled to side, (i) top rebated, (j) housing, (k) tapered dovetail housing (drawn by Emre ILGIN).

content contribute to toxic gas emissions, which are harmful to the environment and human health. Additionally, metal connectors utilized in EWPs threaten their end-of-life disposal and reusability. At this point, dovetail massive wood elements can be a sustainable alternative to commonly used EWPs due to their adhesive- and metal connector-free nature.

The history of the dovetail joint technique predates Christ. This technique has been widely used in many fields such as furniture design, cabinets, various traditional buildings e.g., churches in different parts of the world. To date, the state-of-the-art scrutinized DMWE only either at the member-based level or at most, small-scale-prototype level - not more than a connection detail - from a limited structural point of view and mostly in a theoretical framework. In other words,

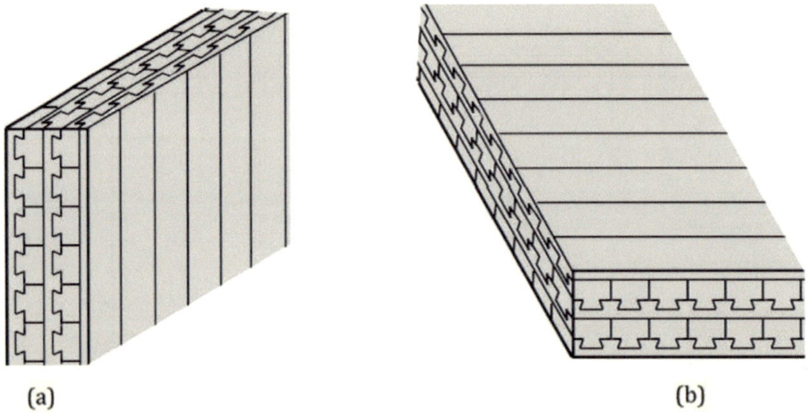

Figure 12.
Dovetail massive wood board elements: (a) vertical structural element, (b) horizontal structural element (drawn by Emre ILGIN).

literature about 'DMWE' is based on inadequate structural analysis and model testing of several types of jointing details rather than even evaluating the performance of a structural component e.g., a shear wall or a whole structure.

Although at present, the intake of DMWE for commercial and structural applications is limited, thanks to new research e.g., DoMWoB project (Dovetailed Massive Wood Board Elements For Multi-Story Buildings) (see acknowledgments) (**Figure 12**), the potential of groundbreaking 'innovative dovetail concept' can be further exploited in building construction e.g., multi-story or even tall buildings.

Acknowledgements

This project has received funding from the European Union's Horizon 2020 research and innovation programme under the Marie Skłodowska-Curie grant agreement No [101024593].

Author details

Hüseyin Emre Ilgın*, Markku Karjalainen and Olli-Paavo Koponen
Tampere University, Tampere, Finland

*Address all correspondence to: emre.ilgin@tuni.fi

References

[1] Bodig J, Jayne B. Mechanics of Wood and Wood Composites. Van Nostrand Reynold company, New York. 1982.

[2] Dinwoodie JM. Timber: its Nature and Behaviour, second ed. edn. E & FN Spon, London, 2000;272.

[3] Tolppanen J, Karjalainen M, Lahtela T, Viljakainen M. Suomalainen puukerrostalo: rakenteet, suunnittelu ja rakentaminen. Helsinki: Opetushallitus: Puuinfo. 2013.

[4] Aaltonen JA. Viewpoints of Residential Multi-Story Construction Industry's Productivity, Environmental Regulations and Future in Finland, MSc Thesis, Forest Economics and Marketing Faculty of Agriculture and Forestry, University of Helsinki, May. 2019.

[5] FAOSTAT-Forestry database, 2019, The Food and Agriculture Organization (FAO) Available from http://www.fao.org/forestry/statistics/80938/en/ [Accessed: 2021-03-22]

[6] Falk R. Wood as a Sustainable Building Material in Wood Handbook, Chapter 01, General Technical Report FPL-GTR-190. Madison, WI: U.S. Department of Agriculture, Forest Service, Forest Products Laboratory. 2010.

[7] Kremer PD, Symmons, MA. Mass Timber Construction as an Alternative to Concrete and Steel in the Australia Building Industry: A PESTEL Evaluation of the Potential, International Wood Products Journal. 2015;6:3.

[8] Toppinen A, Autio M, Sauru M, Berghäll, S. Sustainability-Driven New Business Models in Wood Construction towards 2030, in Filho, w. L. Pociovălișteanu, D. M., Borges de Brito, p. r. and Borges de Lima, I. (eds), Towards a Sustainable Bioeconomy: Principles, Challenges and Perspectives, Springer, Cham. 2018;499-516.

[9] Guan Z, Sotayo A, Oudjene M, El Houjeyri I, Harte A, Mehra S, Haller P, Namari S, Makradi A, Belouettar S, Deneufbourg F. Development of Adhesive Free Engineered Wood Products - Towards Adhesive Free Timber Buildings, WCTE 2018, August 20-23, Seoul, Korea. 2018.

[10] Harte A.M. Mass Timber-The Emergence of a Modern Construction Material. Journal of Structural Integration Maintenance, 2017;2:3:121-132.

[11] Karjalainen M. A Study of the Finnish Multi-story Timber Frame Buildings 1995-2018 S. ARCH 2019; 6th International Conference on Architecture and Engineering, Havana, Cuba; 5-7, March 2019.

[12] Abrahamsen RB, Malo K. Structural Design and Assembly of Treet - A 14-storey Timber Residential Building in Norway. In: Proceedings of the World Conference on Timber Engineering. 2014.

[13] Ramage MH, Burridge H, Busse-Wicher M, Fereday G. Reynolds T, Shah DU, Wu G, Yu L, Fleming P, Densley-Tingley D, Allwood J, Dupree P, Linden PF, Scherman O. The wood from the trees: The Use of Timber in Construction, Renewable and Sustainable Energy Reviews 68. 2017.

[14] Abrahamsen RB. Mjøstårnet - Construction of an 81 m tall timber building, 23. Internationales Holzbau-Forum IHF. 2017.

[15] CTBUH. Council on Tall Buildings and Urban Habitat. Illinois Institute of Technology, S.R. Crown Hall, 3360 South State Street, Chicago, Illinois, USA. Available from http://www.ctbuh.org [Accessed: 2021-03-22]

[16] HoHo. Available from http://www.hoho-wien.at/ [Accessed: 2021-03-22]

[17] Milner HR. Sustainability of engineered wood products in construction, Sustainability of Construction Materials, Woodhead Publishing Series in Civil and Structural Engineering. 2009;184-212.

[18] Chang W-S, Nearchou N. Hot-pressed Dowels in Bonded-in rod Timber Connections. Wood Fiber Sci. 2015;47:2:199-208.

[19] Stark NM, Cai Z, Carll C. Wood-based composite materials: panel products, glued-laminated timber, structural composite lumber, and wood-nonwood composite materials. In: Ross, R. (Ed.), Wood Handbook - Wood as an Engineering Material. U.S. Department of Agriculture Forest Service, Forest Products Laboratory, Madison, WI, 2010;11.1-11.28.

[20] Hemmila V, Adamopoulos S, Karlsson O, Kumar A. Development of Sustainable Bio-adhesives for Engineered Wood Panels-A Review. RSC Adv. 2017;7:61:38604-38630.

[21] Adhikari S, Ozarska B. Minimizing Environmental Impacts of Timber Products through the Production Process "From Sawmill to Final Products. Environ. Syst. Res. 2018;7:1:6.

[22] Frihart CR, Hunt CG. Adhesives with wood materials: bond formation and performance. In: Ross, R. (Ed.), Wood Handbook - Wood as an Engineering Material. U.S. Department of Agriculture Forest Service, Forest Products Laboratory, Madison, WI. 2010.

[23] Norström E, Fogelström L, Nordqvist P, Khabbaz F, Malmström E, Xylan A. Green binder for wood adhesives. Eur. Polym. J. 2015;67:483-493.

[24] California Air Resources Board, Airborne toxic control measure to reduce formaldehyde emissions from composite wood products, 2009. Available from https://www.arb.ca.gov/regact/2007/compwood07/fro-atcmfin.pdf [Accessed: 2021-03-22]

[25] World Health Organization. WHO Guidelines for Indoor Air Quality: Selected Pollutants. World Health Organization Regional Office for Europe, Copenhagen, Denmark. 2010.

[26] BS EN 13986. Wood-based Panels for Use in Construction Characteristics, Evaluation of Conformity and Marking. British Standards Institution, London. 2015.

[27] European Commission. A Roadmap for Moving to a Competitive Low Carbon Economy in 2050. 2011

[28] Sotayo A, Bradley D, Bather M, Sareh P, Oudjene M, El-Houjeyri I, Harte A, Mehra S, O'Ceallaigh C, Haller P, Namari S, Makradi A, Belouettar S, Bouhala L, Deneufbourg F, Guan Z. Review of state of the art of dowel laminated timber members and densified wood materials as sustainable engineered wood products for construction and building applications, Developments in the Built Environment 1. 2020;1-11.

[29] Sumiyoshi T, Matsui G. Wood Joists in Classical Japanese Architecture, Kajima Institute Publishing Co. Ltd., Japan. 1990.

[30] Pang S-J, Oh J-K, Park C-Y, Lee J-J. Influence of crossing-beam shoulder and wood species on moment-carrying capacity in a Korean traditional dovetail joint. Journal of Wood Science. 2011;57:195-202.

[31] Zhang J, Yixiang Xu Y, Mei F, Li C. Experimental study on the fire performance of straight-line dovetail joints, Journal of Wood Science, 2018;64:193-208.

[32] Zhang J, Wang Y, Linfeng L, Xu Q. Thermo-mechanical behaviour of dovetail timber joints under fire exposure, Fire Safety Journal. 2019;107:75-88.

[33] Yingzao Fashi QG. Twelfth-century Chinese building manual, Archit. Hist. 1998; 41:1-13.

[34] Rivers S, Umney N. Conservation of Furniture, Oxford, England: Routledge. 2003.

[35] Routledge French Technical Dictionary. 1. London: Routledge. 1994. (ISBN 9780415112253)

[36] Jasieńko J, Nowak T, Karolak A. Historical carpentry joints, Journal of Heritage Conservation, 2014;40.

[37] Kopkowicz F. Ciesielstwo polskie, Arkady, Warszawa. 1958.

[38] Tannert T, Keller N, Frei R, Vallee T. Improved Performance of Rounded Dovetail Joists, WTCE, World Conference on Timber Engineering, July 15-19, Auckland, New Zealand. 2012.

[39] Pang S-J, Oh J-K, Park C-Y, Lee J J. Effects of Size Ratios on Dovetail Joints in Korean Traditional Wooden Building, WTCE, World Conference on Timber Engineering, July 15-19, Auckland, New Zealand. 2012.

[40] Chiniforush A, Akbarnezhad A, Valipour H, Xiao J. Energy implications of using steel-timber composite (STC) elements in buildings. Energy & Buildings, 2018;176:203-215.

[41] Toivonen R, Lähtinen K. Sustainability - A Literature Review on Concealed Opportunities for Global Market Diffusion for the Cross-laminated Timber (CLT) in the Urbanizing Society. The manuscript for Bioproducts Business. 2019.

[42] Mohd Yusof N, Md Tahir P, Lee SH, Khan MA, Mohammad Suffian James R. Mechanical and physical properties of Cross-Laminated Timber made from Acacia mangium wood as function of adhesive types. J. Wood Sci. 2019.

[43] Li M, Zhang S, Gong Tian Z, Ren H. Gluing Techniques on Bond Performance and Mechanical Properties of Cross-Laminated Timber (CLT) Made from Larix kaempferi, Polymers. 2021;13:733.

[44] Drdácký M, Urushadze S. Retrofitting of Imperfect Halved Dovetail Carpentry Joints for Increased Seismic Resistance. Buildings. 2019;9:48; doi:10.3390/buildings9020048.

[45] Jeong GY, Park M-J, Hwang K, Park JS. Effects of Geometry on Mechanical Behavior of Dovetail Connection, WTCE, World Conference on Timber Engineering, July 15-19, Auckland, New Zealand. 2012.

[46] Robeller C, Weinand Y. Interlocking Folded Plate - Integral Mechanical Attachment for Structural Wood Panels, International Journal of Space Structures. June 2015. DOI: 10.1260/0266-3511.30.2.111

[47] Pozza L, Scotta R, Trutalli D, Pinna M, Polastri A, Bertoni P. Experimental and Numerical Analyses of New Massive Wooden Shear-Wall Systems. Buildings. 2014;4:355-374.

[48] Pozza L, Scotta R, Trutalli D, Polastri, A. Behaviour factor for innovative massive timber shear walls, Bull Earthquake Eng. 2015;13:3449-3469.

[49] Jeong GY, Hindman DP. Ultimate Tensile Strength of Loblolly Pine Strands Using Stochastic Finite Element Method. J. Mater. Sci. 2009;44:14:3824-3832.

[50] Jeong GY, Hindman DP, Zink-Sharp A. Orthotrpic Properties of Loblolly Pine Strands. J. Mater. Sci. 2010;45:21:5820-5830.

[51] Park CY, Lee JJ. Moment Carrying Capacity of Dovetailed Mortise and Tenon Joints with or without Beam Shoulders, J. Struct. Eng. ASCE. 2010;137:7:785-789.

[52] Nowak TP, Karolak A, Jasieńsko J. Historyczne Złącza Ciesielskie Historical Carpentry Joints. Wiadomości Konserwatorskieno - Journal of Heritage Conservation. 2014;40:58-82.

[53] Lubowiecka I, Zybała T, Bukalb G, Krajewskia M, Kujawa M, Kłosowskia P. On the Current State of Dovetail Wall-corner Joints in Wooden Greek Catholic Churches in Polish Subcarpathia with Structural and Sensitivity Analyses, International Journal of Architectural Heritage. 2019.

[54] Branco JM, Descamps T. Analysis and strengthening of carpentry joints. Construction and Building Materials. 2015;97:34-47.

[55] Branco JM, Piazza M, Cruz P.J.S. Experimental evaluation of different strengthening techniques of traditional timber connections. Engineering Structures. 2011;33(8):2259- 2270.

[56] Palma P, Cruz H, Mechanical behaviour of traditional timber carpentry joints in service conditions - results of monotonic tests. In From material to Structure - Mechanical behaviour and failures of the timber structures, XVI International Symposium, ICOMOS IWC, 2007.

[57] Jasieńko J, Nowak T, Karolak A. Historical carpentry joints, Journal of Heritage Conservation. 2014;40: 4058-4082

Chapter 18

Preservative Treatments on Wood and Their Effects on Metal Fasteners

Kelly Bossardi Dias and Ricardo Marques Barreiros

Abstract

Wood as a building material has characteristics that make it attractive environmentally compared to other materials. It is an economic, historical and sustainable material. Many species of wood are naturally resistant to the action of the organisms that degrade them. However, species with this natural resistance are unable to meet the demand for wood and wood-based products, which have been growing year by year. The scarcity of species resistant to biological degradation forced man to use other less durable species, mainly of rapid growth, from reforestation, such as some species of Eucalyptus and Pinus. These species have moderate or no resistance to attack by biological agents and require preservative treatments. And to increase the life span of these fast-growing woods, protecting them from fungi, insects and other xylophagous organisms, several preservative agents are used, these compounds being highly toxic to these biodeteriorating organisms. It is known that the effectiveness of traditional wood preservation systems is due to the biocidal effect of the products used, however, they pollute the environment. Thus, there is an increasing need to develop effective preservative chemicals, non-toxic to humans and the environment.

Keywords: Preservative treatment, xylophagous agents durability of wood, biodeterioration, treated wood in construction

1. Introduction

Wood as a building material has characteristics that make it attractive environmentally compared to other materials. Because it is a material that consumes little energy for its processing, helps to reduce the greenhouse effect and has good characteristics of thermal and electrical insulation. The natural durability of wood and its preservation are two factors that largely determine its use, especially in tropical countries. Although there is a series of works on this subject, it is difficult to compare the results achieved, due to the numerous and different conditions, under which the experiments are conducted (**Figure 1**).

Wood has a range of use in rural and urban environments. However, due to its structure and chemical constitution, it is attacked by several deteriorating organisms. Among organisms, fungi and termites are responsible for the greatest damage to wood [1]. However, no species of wood, not even those of recognized natural durability, is capable of resisting, indefinitely, the weather, variations in

Figure 1.
Treated wooden posts. Source: https://images.app.goo.gl/3eKubCdjmWMQdV2z6.

environmental conditions, the attack of microorganisms and the action of man himself. Physical, chemical and biological agents, acting together or separately on wood, accelerate its degradation process [2].

Reforestation woods, also known as fast-growing from planted forests, are not resistant and require preservative treatments. Currently, about 70% of the wood consumed worldwide comes from reforestation. In civil construction, specifically in the production of wooden housing, Pinus sp. and Eucalipto sp., among the species of fast growing wood from planted forest, in the form of sawn wood, agglomerated sheets, plywood and round pieces [1, 2]. Reforestation species have moderate or no resistance to attack by biological agents and require preservative treatments. And, to increase the life span of these fast-growing woods, protecting them from fungi, insects and other xylophagous organisms, several preservative agents are used, these compounds being highly toxic to these biodeteriorating organisms.

The chapter provides a review of traditional preservatives and the alternatives that have been researched for protection against biodeteriorating agents in wood. It also includes investigations into the corrosion of metal fasteners in wood treated with preservatives, which are limited and take into account only the most common commercial preservatives.

2. Preservation of wood

Wood, when used, is no longer alive, so it will be exposed to decomposition or deterioration like any living thing. This effect can be caused by several agents, among them: biological (bacteria, fungi, insects, mollusks and crustaceans); physical (fire, heat and humidity); mechanical (cracks, wear and permanent deformation) chemical (acidic, saline substances, etc.). The biological factors that are of great importance for their form of action are fungal microorganisms, which start their attack on trees that have not yet been cut. Other types of fungi prefer trees already felled during the following 10 processes: cutting, transportation, unfolding, storage and end use of the piece of wood. Bacteria, on the other hand, need moisture to carry out their attacks. For this to occur it is necessary that the pieces of wood when cut down are in contact with water, and may be submerged or moistened, if put to use in humid environments. The insects, in turn, are found in trees that have not yet been felled, while other insects opt for trees that have already been felled or are in an advanced degradation process [3]. Marine drills are biodegradation agents that cause depreciation in fixed or floating pieces of wood that are submerged in brackish or salt water. These agents are divided into two groups: mollusks and crustaceans [4].

Over time, history records some uses of wood and some preservation techniques, including several important passages. Perhaps the oldest is found in the Bible, in the passage that Noah was instructed to build an ark, large enough to house his family and animal couples to preserve. He also needed to store food for a period of at least 40 days. The bible contains instructions for caulking and preserving wood with bitumen. Bitumen was the oil that naturally flourished in the Middle East region. It was used as it appeared, in the form of tar. This method continued to be used by the Phoenicians on their sailing vessels. Since then, bitumen has established itself as a more traditional substance for the treatment of boat hulls, being perfected by the Greeks. Already, in the modern age, ships became the most important machines, being built with wood. Its preservation against degrading organisms required a lot of protection and conservation efforts. The most appropriate solution was reached at the end of the 18th century, when the hooves started to be covered with copper sheets, fixed on a hemp blanket and bitumen. Vegetable oils were used by the Romans to preserve and maintain the color of wood in civil works [5].

Wood preservation is defined as a set of products, methods and techniques designed to alter the durability of wood [3], which can be divided into:

- Indirect preservation - is the treatment of the environment in which the wood is being used;
- Biological preservation - involves the use of living organisms to preserve the attack of xylophagous organisms;
- Chemical preservation - it is the introduction of chemical products within the wood structure, aiming to make it toxic to the organisms that use it as a food source.

As the chemical preservation method is the most used method in construction, this will be detailed. Chemical preservatives were developed to be used in preservation methods in order to combat biological agents and thus prolong the life of the piece of wood and also, its resistance. Condoms are divided into oil-soluble products that use oil as a vehicle and water-soluble products that use water as a vehicle. The choice of the species, its use and the amount of wood will indicate the most appropriate treatment to be used, as well as the condom to be used [3]. Chemical preservative is the name given to certain substances or chemical formulations, of defined composition and characteristics, which applied to wood provide protection against biological degradation [4] and can be used preventively or curatively. Each condom is generally indicated to combat certain deteriorating agents and its dosage and treatment process to be used must be adequate [5].

According to [6], any substance capable of causing the poisoning of the cellular elements of the wood, making it resistant to the attack of fungi and insects is called a wood preservative. Condoms must have the following properties:

- Resistance to leaching and volatility: the product must have a long-lasting action on wood, it must be chemically stable and resist the risks of use which are leaching (rain, condensation water and soil water) and evaporation (heat action), in addition to not decomposing or changing when in contact with the constituents of the wood;
- Do not change the properties of the wood: the versatility of the use of wood is the result of its physical, chemical, mechanical, organoleptic and decorative characteristics, so the treated wood must not have its surface altered;

- Do not be corrosive: A corrosive product can cause esthetic damage and compromise the joints (straps, nails, screws, etc.);

- Do not increase the flammability of wood: one of the undesirable properties of wood is its ability to burn. Preservative products should not make it more flammable yet;

- Be affordable and available on the market: preserved wood must be competitive with other materials. It is not enough that the product is efficient, but that its use is viable, without compromising the final cost;

- Safe in relation to man and the environment: the toxicity of the condom must be restricted to xylophagous organisms, avoiding the intoxication of men and animals, as well as changes in the ecological balance. It should also not present odors when in contact with humans and animals.

It is unlikely that a chemical preservative will be able to combine all these related properties. As far as possible, the choice should fall on a product that has the greatest number of properties. The choice of this product will also depend on the situation in which the wood is used. For example, creosote has always been considered a very efficient condom, but it makes impractical the subsequent application of paints and varnishes on wood, as well as woods that will be close to man and animals, such as housing, furniture, food storage boxes, etc. Other products, such as boron derivatives, are very efficient preservatives and have low toxicity to humans and animals, but are highly leachable and do not stick to wood. Copper-derived products, on the other hand, have the same advantages, but are corrosive to metals, destroying parts in contact (nails, staples, hinges, etc.) [6].

Historically, the oldest synthetic chemical preservative consists of tar, which consists of a by-product of the carbonization of wood, peat, lignite, oil shale and coal. Creosote was patented by the Englishman John Bethell in 1838 for the treatment of wood exposed to the weather, such as sleepers and transmission poles [7]. Creosote is defined as a tar distillate, extracted from stone coal at high temperatures; and it can also be produced from oil [8].

CCA's history began in 1933, when a wood engineer, Sonti Kamesam, made a discovery that saved the lives of countless coal miners: the injection of arsenic and copper into wooden beams prevents rot. Arsenic, a classic poison kills insects that feed on wood, while copper kills fungi. Kamesan's discovery was to add chromium to this formula, thereby linking the two toxic metals in the cell walls of the wood. The result was to obtain stronger props in the humid underground tunnels, through which they extract the coal. Kamesam's invention not only extended the miners' life expectancy, but also reduced deforestation [9]. CCA is highly effective in protecting wood against a wide variety of wood-degrading organisms, as well as being inexpensive, water-soluble, and resistant to leaching. Since its discovery, many types of CCAs have been introduced, and today the most common formulation contains a mixture of arsenic pentoxide (As2O5), chromium trioxide (CrO3), and cupric oxide (CuO), which can differ in the amount of each component. Lately it has suffered serious restrictions, due to occupational risks resulting from chronic exposure to CCA, and is being banned in several countries, such as Germany, France, England and, recently, the United States [7].

CCB is a mixture of copper sulphate, boric acid and potassium dichromate, which was created to replace CCA, with the advantages of less environmental impact and risk to operators, and the possibility of treatment in open tanks; but highly leachable [7]. The preservatives based on copper, azoles and quaternary

Preservative	Benefits	Disadvantages
Creosote	• Low Cost; • Not Leachate – insoluble in water.	• Exudation problem; • Cannot be used in applications requiring finishing.
CCA	• Highly effective; • Cheap; • Soluble in water; • Resistant to leaching.	• High toxicity due to the presence of arsenic and chromium; • Environmental restrictions.
CCB	• Effective; • Cheap; • Soluble in water; • Less environmental impact compared to CCA.	• Greater leaching; • Toxicity due to the presence of chromium.
Azoles AND quaternary ammonium	• Low toxicity; • Resistant to leaching.	• High cost; • Greater susceptibility to moldy fungi.
Boron compounds	• Effective; • Cheap; • Soluble in water.	• Highly leachable.

Table 1.
Advantages and disadvantages of the main preservative products.

ammonium have variations in their formulations, do not contain chromium and arsenic and can be alternatives in situations where the CCA has restrictions. The azoles and ammonium quaternaries were mentioned as potential substitutes for CCA at the 39th International Forum on Wood Preservation, held in Istanbul, Turkey, in 2008. The international event is considered a reference in the wood preservation sector, presenting the main trends and the most important research results worldwide. These condoms, however, when compared to CCA, present some problems, such as high cost, less fixation power and consequent greater leaching, in addition to greater susceptibility to moldy fungi, requiring the application of additives to control this type of fungus.

Borates are a group of preservatives that were rediscovered in the 1980s, including their salts, such as sodium octaborate, sodium tetraborate, and sodium pentaborate, which are water-soluble. Boron compounds offer the most effective wood preservative systems available today, combining the broad-spectrum properties of effectiveness against bacteria, fungi and insects with low toxicity in mammals. The obstacle is that they remain soluble in water and are easily leached from treated wood [9]. Currently, the fight against fungi in wood has been driven by the use of fungicides such as creosote, CCA (chrome-plated copper arsenate) and CCB (chrome-plated copper borate). Although such compounds have efficient fungicidal actions, the tendency is to be replaced, due to their high toxicity. **Table 1** shows the advantages and disadvantages of the main preservative products used.

3. Sustainable alternatives as preservatives for wood

Since the early 1990s, the search for viable alternatives in the area of wood protection has intensified around the world. Historically, the wood preservation industry has used three major conservation systems for the pressure treatment of wood: pentachlorophenol, creosote and water-based arsenic. The main factors that drove changes in treatment technology and conservation systems worldwide were: environmental concerns, including air and water quality standards, the effect of treated wood on humans and other organisms, and the energy crisis, especially when it comes to oil-based preservative systems. Of these, environmental concerns are prevalent [10].

The classic concept of wood preservation is based on the principle of toxicity, that is, impregnation with biocides to prevent biological degradation. However, chemical control can induce the resistance of fungi, bacteria and insects to biocides, as well as potential health and environmental risks. Effective traditional preservatives are based on metals, such as copper, chromium, zinc, arsenic, boron and fluorine, and compounds such as creosote and amines [11]. For environmental reasons, both the preservation of traditional wood and the use of resistant wood species are subject to political and consumption restrictions. It is known that the effectiveness of traditional wood preservation systems is due to the biocidal effect of the products used, however, consequently, they pollute the environment. In addition to the risks involved in the use of such materials, there is a growing concern with the problems arising from the flow of wood at the end of its commercial life [9]. Thus, there is a growing need to develop effective antifungal chemicals, non-toxic to humans and the environment.

In addition to the risks involved in the use of wood treated with toxic products, there is a growing concern about the problems that arise in the elimination of wood after the end of its commercial life. Systems for improving the durability of wood must be sustainable in production and use. In addition to this, treated wood products must, at the end of their useful life, be suitable for the production of energy by combustion or composting or for use as a source of secondary fibers from related industries, without presenting any problems related to the residual chemicals from their treatment [9].

Derivatives of plants, such as oils, have been used for generations to improve the appearance and to prolong the useful life of wooden products, such as furniture, walking sticks, etc. However, the use of products derived from plants became less attractive for wood protection when synthetic and inorganic compounds were introduced, as they were shown to be more effective against wood decaying organisms [12]. The development of research on wood preservatives is in a critical phase, being necessary its direction for the analysis of products with less environmental impact and competitive cost [13].

The deterioration of wood is, therefore, caused by a combination of chemical, biological and physical processes, and water that plays an important role in all cases. Deterioration and discoloration caused by fungi, and to a lesser extent by bacteria, are an important source of losses [9]. For wood to be attacked by fungi and insects, four factors must be present: water, oxygen, temperature and nutrients. So, to avoid the attack of these organisms, some researchers have been aiming at removing one of these factors [14]. Searches for alternatives range from substances of natural origin and/or plant extracts, to systems that inhibit one of the factors that favor the development of these organisms. Despite the proven efficiency of some environmentally friendly alternatives for the treatment of wood, economic viability makes its use difficult.

In more significant quantities, many studies exploring natural extracts as preservatives for wood. Derived from a number of plants and various parts of the plant, such as bark, wood, leaves, seeds and fruits, they have been examined for their wood-protecting properties in many studies. Among plant extracts, there are essential oils from aromatic plants, extracts from poisonous plants [15], and oils extracted from seeds/grains. And yet, the extracts of the wood itself such as tannin, starch, dyes, oils, resins, waxes and fatty acids. Isolated, or in combination with solvents and other additives, some natural products can perform well in preserving wood [5].

Among the studies on extractives, cinnamon leaf extracts have proven to be highly effective against fungi and wood termites and can potentially be developed into excellent organic preservatives [16–19]. According to the authors, cinnamon oil proved to be highly effective when used in ethanol, but its activity decreased when mixed with water. The addition of surfactants to the cinnamon oil/water combination did not produce a completely stable solution.

Hashim et al. [20] obtained promising results in the evaluation of the resistance of the extracts of the leaf, fruit, wood, bark, seed and flower of *Cerbera odollam* to deterioration by fungi and termites. Further studies are also needed to compare properties with commercial synthetic preservatives; Jain et al. (2011) verified the effectiveness of methanolic extracts from leaves and barks of Cleistanthus collinus and *Prosopis juliflora* to inhibit the growth of white rot and brown rot fungi.

Among natural oils, flaxseed oil is considered one of the natural treatments with the best result, as it is dry, providing good impermeability and protection. In addition to isolated flaxseed oil [5], studies have shown that combinations with copper-chromium [21], and with boron [22] provide better protection results, as it reduces the leaching of the incorporated biocides.

In addition to flaxseed oil, other sources of study as a preservative for wood are oils extracted from the seeds of neem (*Azadirachta indica* A. Juss) and castor oil (*Ricinus communis*). All neem-based products are completely natural, being non-toxic to humans, pets and the environment. The fruits, seeds, oil, leaves, peels and roots of the neem have the most varied antiseptic and antimicrobial uses. Research shows that neem oil is effective against fungi, parasites, insects, some bacteria and viruses [8, 11, 23]. Castor oil, on the other hand, has been studied to improve the persistence of neem oil in wood, since, after twenty days in contact with the soil, neem oil deteriorates, making it difficult to use for the treatment of wood, in which the active ingredients of the substances used for this purpose, must persist for a long time in the treated parts [11, 24, 25].

Bark of many tree species is a rich source of antioxidant and antimicrobial agents, such as waxes, resins, tannins and other extracts [12]. Heartwood extracts from a wide range of plant and tree species show activity against fungi and insects, and many potentially can serve as wood protection agents, alone or in combination [26]. Vegetable or natural tannins are extracts that can be found in various parts of plants, such as wood (heartwood) and bark. Tannins, due to their antifungal properties, have been tested as alternatives to immunizers for wood. They already have applications in the tanning of hides, in the oil industry, as a dispersing agent to control the viscosity of clays in the drilling of wells, being also used in the treatment of water supply and waste, and in the manufacture of paints and adhesives [1]. The research covers the effectiveness of different types of tannins that have been proven to successfully inhibit wood decay fungi, such as wood immunizers, alone [2, 4, 27–31]; combined with other chemicals with established antifungal efficacy [26]. However, one of the biggest problems observed with tannins and their derivatives is that they are difficult to stick to the wood after treatment, although attempts have been made to retain them using additives [12, 28].

When considering the bark as a source of organic biocides, remember that the bioactivity of bark extracts from different sources varies, as has been shown in studies involving the evaluation of antifungal properties of the bark from various species. For the heartwood, in addition to the chemical composition of the extractable products, the durability of the heartwood is related to the quantity and distribution of extractable products within the wood fabrics. In highly durable species, extractives are not only present in cell lumens, but also impregnate cell walls and membranes and it is the combination of chemical substance and physical factors that determine the durability of wood species [32].

Extracts that exhibit antifungal activity, such as the extracts of mimosa bark (Acacia mollissima) and quebracho (*Schinopsis lorentzii*) [33], as well as the essential oils of *P. graveolens* and their fractions [34], and the mistletoe leaves and extracts of lichen [35], extracts from the heartwood of Teak (*Tectona grandis*) [36], extracts of propolis [37] and extracts from the peel of fruits and vegetables [38] have great potential with preservatives for wood. As can be seen, research with natural extracts has been extensive and has proven to be effective in resisting biodeterioration, but its leaching of wood into the environment remains a challenge to be overcome.

In the literature, some alternatives of by-products under study with interesting characteristics such as wood preservatives are:

- Chitosan - By-product of the crustacean processing industries such as shrimp, crab and lobster;

- Okara – Organic by-product produced from the manufacture of soybeans and tofu;

- Extracts of residues related to coffee - a by-product of the coffee roasting process;

- Crude Tall Oil (CTO) and its derivatives - By-product of Kraft pulping.

Research shows that chitosan has been shown to minimize the attack of fungi, however, few studies have been carried out on the application of chitosan to wood [21, 39–43]. According to the authors, chitosan is a low-cost, naturally occurring, nontoxic, edible and biodegradable polysaccharide, which has been found in a wide variety of natural sources (crustaceans, fungi, insects, annelids, mollusks, coelenterates, etc.). It is usually a by-product of the crustacean processing industries (shrimp, krill, crab and lobster). According to [43], the fungicidal effect of chitosan is well documented in the literature, the most accepted model being the one related to the polycationic nature of the polysaccharide that interacts with anionic sites of the cell walls of the fungi. Such interaction is mediated by electrostatic forces, causing changes in the permeability of cell membranes and osmotic instability.

Another by-product that has shown satisfactory environmentally favorable results is the hydrolyzed enzymatic okara (OK), which is an organic waste produced from the manufacture of soy milk and tofu. The formulation studied as a preservative for wood is Okara combined with other products such as copper chloride, sodium borate or ammonium hydroxide [44].

Another alternative presented with the capacity to reduce the growth of fungi that rot the wood was the silver skin of the coffee [45]. The silvery skin of the coffee is a by-product of the coffee roasting process and this has caused a significant suppression of the growth of all fungi. This was not just because of caffeine, but also because of other chemicals present in the residual extracts, which demonstrated that spent coffee can be used as a source of green chemicals in wood preservative formulations.

Tall Oil Crude (CTO), Tall Oil or Talol or resin oil is the generic name for products derived from residual, smelly, gummy and black liquor. It is found and extracted from the residual liquor from the Kraft cooking process for the production of paper and cellulose, known as "black liquor" and considered one of

Preservative under study	Benefits	Disadvantages
Natural extractives	• Resistance to proven biodeterioration.	• Easily leached; • Many types studied and few with more detailed information.
Chitosan	• Resistance to proven biodeterioration; • Biodegradable; • Inexpensive – by-product of the crustacean processing industries.	• Little investigated.
Okara	• Resistance to proven biodeterioration; • Inexpensive – Organic by-product produced from the manufacture of soybeans and tofu.	• Easily leached; • Little investigated.
Coffee–related waste extracts	• Resistance to proven biodeterioration; • Inexpensive – a by-product of the coffee roasting process.	• Little investigated.
+*Crude Tall Oil* (CTO)	• Resistance to proven biodeterioration; • Inexpensive – Kraft pulping by-product.	• Little investigated; • Exudation problem.

Table 2.
Some advantages and disadvantages of the main preservative products under study.

the cheapest, renewable natural oils on the world market, as it is an industrially generated product, not depending on weather and soil, but on the production of cellulose and kraft paper. The yield and composition of Tall Oil may vary, as they are influenced by the quantity of extracts, the quality and species of the wood, and the storage time before cooking. It is not composed of pure triglycerides, like other vegetable oils, but a mixture of fatty acids, resin and unsaponifiable acids (for example, sterols, waxes, hydrocarbons). The amount of these components varies with age, wood species, geographic location, and also with all operations before and during the pulping process.

Investigations made with Tall Oil indicate its potential as a protective agent for wood. Those surveyed who focused on the search for an alternative as a preservative for wood in Tall Oil indicated that the preventive effect is related to the hydrophobic properties, with the high concentration of oleic acid, with the presence of unsaponifiables, with the stability when the acidity index and the saponification index [9, 46–53]. These factors may be indicative of increased resistance to biodeterioration to a greater or lesser degree.

Table 2 shows the advantages and disadvantages of the main preservative products being researched. Much information about these alternatives that are being researched still needs to be investigated.

4. Corrosion of metal fasteners embedded in treated wood

Corrosion of metals in contact with wood has been studied for more than 80 years. However, the durability of metal components attached to wood treated with preservatives has become a concern. The rate and amount of corrosion depends on the metal, the conductivity of the wood, the duration of wet conditions and the type of preservative used. Studies on the corrosivity of wood preservatives in metal fastening components are limited and take into account only the most common commercial preservatives such as Chromed Copper Arsenate (CCA),

Ammoniacal Copper Arsenate (ACA), Alkaline Copper Quaternary (ACQ), Creosote and chromed copper borate (CCB) (**Figure 2**).

Wood and metal are compatible in most construction and furniture uses and few corrosion problems occur. However, if there is sufficient moisture at the wood-metal interface, corrosion with susceptible metals can be expected. Corrosion of metal in contact with wet wood is an electrochemical process. The rate and amount of corrosion depends on the metal, the conductivity of the wood and the duration of wet conditions. Factors that can increase the rate of metal corrosion in wood include wood species, the presence of external corrosive contaminants and wood preservative treatment.

The oldest study found was from 1934, which studied the corrosion of wire claws on wood treated with zinc chloride and exposed to moderately humid conditions in Madison. He declared that the corrosion of the wire claws in the wood treated with zinc chloride was significant, and some efforts to protect the wood, such as painting and adding sodium chlorate, were ineffective in reducing corrosion rates [54]. After a few decades, Baker [55] proposed a mechanism for corrosion of metals in contact with wood treated with a preservative. The wood preservatives that Baker studied contained cupric ions that are thermodynamically unstable in the presence of steel and galvanized steel. The mechanism involves transporting cupric ions through the wood to the metal surface where they are reduced. In addition, the mechanism was aqueous, since the corrosion rate depended heavily on the moisture content of the wood.

According to Graham [56] since 2004, wood preservatives seek to remove wood treated with chromed copper arsenate (CCA) from the market, citing concerns about arsenic and chromium contained in CCA. Wood preservatives have introduced a number of preservative treatment substitutes, including alkaline copper type C (ACQ-C), alkaline copper type D carbonate (ACQ-D carbonate), micronized copper quantum (MCQ), copper azole (CuAz), copper and zinc ammonium arsenate (ACZA), sodium octaborate sodium borate tetrachloride (SBX / DOT) and zinc borate, which are currently in commercial use. Published studies indicate that most preservative treatments are, in general, more corrosive than CCA. For example, the carbonate ACQ-D, CuAz and ACZA showed more than twice the corrosivity of CCA [57]. SBX/DOT and zinc borate treatments have been shown to be slightly less corrosive than CCA [56].

Of the alternative treatments for CCA two of the most used are copper azole (CuAz) and acrylated copper quaternary (ACQ). The increased use of alkaline copper quaternary (ACQ) and copper azole (CuAz) as wood preservatives, replacing chrome-plated copper arsenate (CCA), for residential construction has raised concerns about the corrosion performance of fasteners. ACQ and CuAz are believed

Figure 2.
Embedded corroded metal fasteners. Source: https://www.pikist.com/free-photo-socdl/pt.

to be more corrosive than CCA due to the higher percentage of copper in these preservatives and the absence of chromium and arsenic; both are known as corrosion inhibitors [57].

In general, boron-based wood preservatives have been proposed as environmentally friendly alternatives to copper-based preservatives, as they have relatively low toxicity, and do not contain potentially harmful heavy metals, and their handling is safer. Research has shown that boron compounds are less corrosive than copper compounds. Based on these considerations, wood preserved with boron does not present a higher risk of corrosion for steel fasteners than copper-based preservatives and there is a reasonable expectation that the long-term corrosion rate will be lower [58].

Based on the various studies addressed, it can be said that wood preservatives, ACQ and CuAz, presented as alternatives to CCA, are up to 2 times more corrosive than CCA due to the higher percentage of copper in these preservatives and absence of chromium and arsenic; both are known as corrosion inhibitors. CCB is less corrosive than CCA. These results show that research for the development of sustainable wood preservatives must cover the corrosivity of the fixing components that are used in wood, both for civil construction and furniture.

5. Concluding remarks

Based on the various studies cited, the search for alternatives to current condoms has been efficient, but not effective, that is, a viable alternative for existing products has not yet been found. Many sustainable alternatives have high efficiency against biodeteriorating fungi, however they are easily leachable, that is, they do not stick to wood. Those that have a good fixation on the wood, however, do not present the desired resistance. Thus, research is still needed to develop a sustainable product, with high resistance to decaying fungi and with excellent fixation on wood.

In parallel, corrosion of treated wood fasteners can limit applications, reduce consumer satisfaction and generate costly returns or increase fastening costs if stainless steel or other costly fastening devices are specified to avoid future problems. This could lower the timber from the market for some applications.

Author details

Kelly Bossardi Dias* and Ricardo Marques Barreiros
Universidade Estadual Paulista "Júlio de Mesquita Filho" (UNESP),
Itapeva, SP, Brazil

*Address all correspondence to: kbossardi@gmail.com

References

[1] Paes J, Diniz C, Marinho I, Lima C. Avaliação do potencial tanífero de seis espécies florestais de ocorrência no semi-árido brasileiro. Cerne, v. 12, n. 3, p. 232-238, 2006.

[2] Oliveira J, Tomasello M, Silva J. Resistência natural de sete espécies de eucalipto ao apodrecimento. Revista Árvore, v. 29, n. 6, p. 993-998, 2005. DOI: 10.1590/S0100-67622005000600019.

[3] Castro, V G de, Guimarães, P P. Deterioração e preservação da madeira — Mossoró: EdUFERSA, 2018.

[4] Costa, A F, Silva, G F, Escudeiro, M C. Estudo comparativo entre produtos químicos preservantes e licores pirolenhosos na inibição de fungos emboloradores. Brasil Florestal, n. 75, jan. 2003.

[5] Gonzaga, A L. Madeira: uso e conservação. Brasília, DF: Iphan/Monumenta. 246 p., 2006.

[6] Mendes, A S, Alves, M V S. A degradação da madeira e sua preservação. Brasília, IBDF/DPq-LPF, 1988.

[7] Stumpp, E, Rech, V, Sattler, M A, Barros, N M, Abitante, A L. Avaliação de sustentabilidade e eficácia de tratamentos preservantes naturais de madeiras de florestas plantadas no RS para o controle do cupim. Ambiente Construído, v. 6, n. 2, p. 21-31, 2006.

[8] Paes, J B, De SOuza, A D, De Lima, C R, Neto, P N de M. Eficiência dos óleos de nim e mamona contra cupins xilófagos em ensaio de alimentação forçada. Cerne, v.16, n.1, p. 105-113. 2010.

[9] Koski, A. Applicability of crude tall oil for wood protection. Acta Universitatis Ouluensis C Technica 293, Oulu, 2008.

[10] Barnes, H M. Wood: preservative treated. In: Buschow, K H J, Cahn, R W, Flemings, M C, Ilschner, B, Kramer, E J, Mahajan, S. (eds.). Encyclopedia of Materials: Science and Technology. Oxford: Elsevier Science, v.10, p. 9683-9688, 2008.

[11] Machado, G O, Calil Júnior, C, Polito, W, Pawlicka, A. Preservante natural de madeira para uso na construção civil – óleo de neem. Minerva, v. 3, n.1, p. 1-8. 2006.

[12] Singh, T, Singh, A P. A review on natural products as wood protectant. Wood Science Technology, n. 46, p. 851-870, 2012. DOI: 10.1007/s00226-011-0448-5

[13] Brand, M A, Anzaldo, J, Moreschi, J C. Novos produtos para o tratamento preservante da madeira: "Perspectiva da pesquisa e utilização". Floresta, v. 36, n. 1, p. 129 - 138. 2006.

[14] Aaserub, J, Larnoy, E, Glomm, W R. Alternative systems for wood preservation, based on treatment with silanes. In: BERGSTEDT, A. Proceedings of the 5th meeting of the Nordic-Baltic Network in Wood Material Science and Engineering. Copenhagen: Denmark, n. 43, p. 21-26, 2009.

[15] Goktas, O, Mammadov, R, Duru, M E, Ozen, E. A research on the usage of extracts from a poisonous plant (*Ornithogalum alpigenum Spapf*) as a wood preservative. Abstracts/Journal of Biotechnology, n. 136S, p. S672, 2008.

[16] Wang, S Y, Chen, P F, Chang, S T. Antifungal activities of essential oils and their constituents from indigenous cinnamon (Cinnamomum osmophloeum) leaves against wood

decay fungi. Bioresour Technology, n. 96, p. 813 – 818, 2005. DOI: 10.1016/j.biortech.2004.07.010.

[17] Cheng, S S, Liu, J Y, Hsui, Y R, Chang, S T. Chemical polymerization and antifungal activity of essential oils from leaves of different provenances of indigenous cinnamon (Cinnamomum osmophoeum). Bioresour Technology, n. 97, p. 306-312, 2006.

[18] Li, S, Freitas, C, Morrelli, J J. Preventing fungal attack of freshly sawn lumber using cinnamon extracts. Forest Products Journal, n. 58, p.71 – 81, 2008.

[19] Lin, C Y, Wu, C L, Chang, S T. Evaluating the potency of cinnamaldehyde as a natural wood preservative. The internacional research group on wood preservation. IRG/WP 07- 30444, 2007.

[20] Hashim, R, Boon, J. G, Sulaiman, O, Kawamura, F, Lee, C Y. Evaluation of the decay resistance properties of *Cerbera odollam* extracts and their influence on properties of particleboard. International Biodeterioration & Biodegradation. v. 63, p. 1013-1017, 2009.

[21] Treu, A, Larnoy, E, Militz, H. Leaching of new environmental friendly wood protection agents. In: BERGSTEDT, A. Proceedings of the 5th meeting of the Nordic- Baltic Network in Wood Material Science and Engineering. Copenhagen: Denmark, n. 43, p. 33-40. 2009.

[22] Lyon, F, Thevenon, M F, Hwang, W J, Imamura, Y, Gril, J, Pizzi, A. Effect of an oil heat treatment on the leachability and biological resistance of boric acid impregnated wood. Annals of Forest Science, n. 64, p. 676-678, 2007.

[23] Rodrigues, M, Paiva, R, Nogueira, R C, Martinotto, C, Silva Jr, J M. Morfogênese *in vitro* de nim a partir de explantes cotiledonares. Revista Árvore, v. 33, n. 1, p. 21-26, 2009.

[24] Araújo, L V, Rodrigues, L C E, Paes, J B. Características físico-químicas e energéticas da madeira de nim indiano. Scientia Forestalis, n. 57, p. 153-159. 2000.

[25] Paes, J B, Diniz, C E F, Marinho, I V, Lima, C R. Avaliação do potencial tanífero de seis espécies florestais de ocorrência no semi-árido brasileiro. Cerne, v.13, n.2, p. 160- 169. 2007.

[26] Sen, S, Tascioglu, C, Tirak, K. Fixation, leachability, and decay resistance of wood treated with some commercial extracts and wood preservative salts. International Biodeterioration & Biodegradation, n. 63, p. 135-141, 2009. DOI: 10.1016/j.ibiod.2008.07.007.

[27] Barbosa, A P, Nascimento, C S, Morais, J W. Estudos de propriedades antitermíticas de extratos brutos de madeira e casca de espécies florestais da Amazônia central, Brasil. Acta Amazonica, v.37, n.2, p. 213-218, 2007. DOI: 10.1590/S0044-59672007000200006.

[28] Barbero-López, A, Akkanen, J, Lappalainen, R, Peräniemi, S, Haapala, A. Bio- based wood preservatives: Their efficiency, leaching and ecotoxicity compared to a comercial wood preservative. Science of The Total Environment, v. 753, 2021. DOI: 10.1016/j.scitotenv.2020.142013.

[29] Anttila, A K, Pirttilã, A M, Häggman, H, Harju, A, Venäläinem, M, Haapala, A, Holmbom, B, Julkunen-Tiitto, R. Condensed conifer tannins as antifungal agents in liquid culture. Holzforschung, v.67, p. 825-832. 2013. DOI: 10.1515/hf-2012-0154.

[30] Da Silveira, A G, Santini, E J, Kulczynski, S M, Trevisan, R,

Wastowski, A D, Gatto, D A. Tannic extract potential as natural wood preservative of Acacia mearnsii. Anais da Academia Brasileira de Ciências, v. 89, p. 3031-3038. 2017. DOI: 10.1590/0001-3765201720170485.

[31] Tomak, E D, Gonultas, O. The wood preservative potentials of valonia, chestnut, tara and sulphited oak tannins. J. Wood Chem Technol, v. 38, p.183-197. 2018. DOI: 10.1080/02773813.2017.1418379.

[32] Yang, D Q. Potential utilization of plat and fungal extracts for wood protection. Forest Products Journal, n. 59, p. 97 – 103, 2009.

[33] Tascioglu, C, Yalcin, M, Sem, S, Akcay, C. Antifungal properties of some plant extracts used as wood preservatives. International Biodeterioration & Biodegradation, v. 85, p. 23-28, 2013. DOI: 10.1016/j.ibiod.2013.06.004.

[34] Moutaouafiq, S, Farah, A, Ezzoubi, Y, Ghanmi, M, Satrani, B, Bousta, D. Antifungal activity of Pelargoniumgraveolens essential oil and its fractions against wood decay fungi. Journal of Essential Oil Bearing Plants, v.22, p. 1104-1114. 2019. DOI: 10.1080/0972060X.2019.1646164.

[35] Yildiz, U C, Kilic, C, Gurgen, A, Yildiz, S. Possibility of using lichen and mistletoe extracts as potential natural wood preservative. Maderas Ciencia y tecnologia, v. 22(2), p. 179 - 188, 2020. DOI: 10.4067/S0718-221X2020005000204.

[36] Brocco, V F, Paes, J B, Da Costa, L G, Brazolin, S, Arantes, M D C. Potential of teak heart wood extracts as a natural wood preservative. Journal of Cleaner Production, v. 142, p. 2093-2099, 2017. DOI: 10.1016/j.jclepro.2016.11.074.

[37] Wozniak, M, Kwasniewska-Sip, P, Waskiewicz, A, Cofta, G, Ratajczak, I. The possibility of propolis extract application in wood protection. Forests v. 11, n. 465. 2020. DOI: 10.3390/f11040465.

[38] Barbero-Lopez, A. Antifungal activity of several vegetable origin household waste extracts against wood-decaying fungi in vitro. Waste Biomass Valorization. 2020. DOI: 10.1007/s12649-020-01069-3.

[39] Maoz, M; Morrel, J J. Ability of chitosan to limit wood decay under laboratory conditions. The international research group on wood protection. IRG/WP 04-30339, 2004.

[40] Eikenes, M, Alfredsen, G, Christensen, B E, Militz, H, Solheim, H. Comparasion of chitosans with different molecular weights as possible wood preservatives. Journal Wood Science, n. 51, p. 387-394, 2005.

[41] Torr, K M, Singh, A P, Franich, R A. Improving stiffness of lignocellulosics through cell wall modification with chitosan melamine co-polymers. New Zealand Journal of Forestry Science, n.36, p. 87-98, 2006.

[42] Singh, T, Vesentini, D, Singh, A P, Daniel, G. Effect of chitosan on physiological, morphological, and ultrastructural characteristics of wood-degrading fungi. International Biodeterioration & Biodegradation, n. 62, p. 116-124, 2008.

[43] Sattolo, N M S, Britto, D de, Assis, O B G. Quitosana como fungicida em madeiras *Pinus* sp. empregadas na confecção de caixas "K". Brazilian Journal of Food Technology, v. 13, n. 2, p. 128-132. 2010.

[44] Ahn, S H, Oh, S C, Choi, I, Han, G, Jeong, H, Kim, K, Yoon, Y, Yang, I. Environmentally friendly wood preservatives formulated with enzymatic-hydrolyzed okara, copper and/or boron salts. Journal of

Hazardous Materials, v. 178, p. 604-611, 2010.

[45] Barbero-Lopez, A, Monzo-Beltran, J, Virjamo, V, Akknen, J, Haapala, A. Revalorization of coffee silverskin as a potential feed stock for antifungal chemicals in wood preservation. International Biodeterioration & Biodegradation, v. 152, 2020. DOI: 10.1016/j.ibiod.2020.105011.

[46] Paajanen, L, Ritschkoff, A C. Effect of crude tall oil, linseed oil and rapeseed oil on the growth of decay fungi. The International Research Group on Wood Preservation, IRG/WP 02-30299, 2002.

[47] Alfredsen, G, Flaete, P O, Temiz, A, Eikenes, M, Miltiz, H. Screening of the efficacy of tall oils against Wood decaying fungi. The internacional research group on wood preservation. IRG/WP 04-30354, 2004.

[48] Anita, S H, Fatriasari, W, Zulfiana, D. Utilization of biopulping black liquor as preservative to fungal attack on jabon wood (*Anthocephalus cadamba* Miq.). Teknologi Indonesia, n.37, v.3, p. 147-153, 2014.

[49] Hyvonen, A, Piltonen, P, Niinimaki, J. Tall oil/water – emulsions as water repellents for scots pine sapwood. Holz als Roh-und Werkstoff, n. 64, p. 68-73, 2006.

[50] Vahaoja, P, Hyvonen, A, Piltonen, P, Niinimaki, J, Kuokknen, T. Biodegradability studies of certain wood preservatives in groundwater as determined by the respirometric bod oxitop method. Water, Air and Soil Pollution, n. 165, p. 313-324, 2005.

[51] Temiz, A, Alfredsen, G, Eikenes, M, Terziev, N. Decay resistance of wood treated with boric acid and tall oil derivates. Bioresource Technology, n.99, p. 2102-2106. 2008.

[52] Durmaz, S, Erisir, E, Yildiz, U C, Kurtulus, O C. Using kraft black liquor as a wood preservative. Procedia - Social and Behavioral Sciences. n. 195, p. 2177 – 2180. 2015.

[53] Sivrikaya, H, Can, A. Effect of weathering on wood treated with tall oil combined with some additives. Maderas. Ciencia y tecnología n. 18, v.4, p. 723-732. 2016

[54] Baechle, R H. Corrosion of metal fastenings in zinc-chloride-treated wood. Industrial Engineering Chemistry, v. 26, n.12, p.1336-1338, 1934.

[55] Baker, A J. Corrosion of metal in wood products, in: Durability of Building Materials and Components. ASTM STP 691, American Society for Testing and Materials, p. 981-993, 1980.

[56] Graham, M S. Fastener corrosion and treated wood. 2009. http://docserver.nrca.net/technical/9340.pdf.

[57] Zelinka, S L, Rammer, D R. Review of test methods used to determine the corrosion rate of metals in contact with treated wood, Gen. Tech. Rep. FPL-GTR-156. 13 Madison, WI: U.S. Department of Agriculture, Forest Service, Forest Products Laboratory. 23 p. 2005.

[58] Morris, J. Effect of borate wood preservatives on corrosion of steel fasteners. Opus. Disponível em: http://www.stepupgroup.co.nz/wp-content/uploads/2011/09/52041.pdf.

Chapter 19

Examination of Lightning-Induced Damage in Timber

Jingxiao Li and Jing Li

Abstract

The ancient Chinese architectures were constructed using timber as the main building material. Considering that the lightning strike is the primary natural cause of damage to ancient building, the lightning strike damage mechanism of ancient building timber and the related influencing factors are investigated using the representative timber materials from the ancient building. The burning of timber was mainly caused by the heat of lightning arc. The splitting and damage pit of timber were mainly caused by the mechanical force generated by the temperature rise of the injected by lightning current and air shock wave effects of the lightning. These ways all played in different roles under different conditions. The higher the water content of timber was, the easier it was to crack, and the greater the damage depth and the larger the damage area were. It was easy to burn for the dry timber or the loose timber with low density, but it was difficult for the thick timber. When the wood was too thin, the lightning air shock wave could cause damage. This research may provide reference for protection of ancient timber architecture from possible damage caused by lightning.

Keywords: timber, ancient Chinese architecture, lightning strike, water content, fire

1. Introduction

1.1 The timberwork in ancient Chinese architectures

Ancient Chinese architectures have a unique contribution to the world architectural heritage and is part of the three largest global building systems, along with European and Islamic architectures [1]. Ancient Chinese architectures with thousands of years of culture and art have important historical, cultural and artistic values. At the same time, ancient Chinese architectures are an important national human tourism resources and precious cultural heritage, which are irrecoverable. A large number of precious ancient building relics are preserved in China. There were a total of 5,058 national key cultural relics protection units in 2019, including 2160 ancient building relics, accounting for 42.7%, nearly half of the total. It is of great significance to protect these ancient architecture and strengthen the safety for the continuation of ancient Chinese civilization. So the safe protection of these ancient buildings is very important. Preventing the destruction of ancient building by natural disasters, especially protecting ancient building from lightning damage, is an important task of ancient building protection [2].

Timber is widely used as the main building material for most of the ancient Chinese architecture, with the roof trusses of timberwork, such as beams, columns, buckets, arches, purlins, rafters, windows, and other components. As shown in **Figure 1(a)**, the

Sakyamuni Pagoda of Fogong Temple in Shanxi Province with the height of 67.31 m and the bottom diameter of 30.27 m is the highest timber pagoda in the world. The Sakyamuni Pagoda, the Leaning Tower of Pisa and the Eiffel Tower in Paris are known as the "Three Wonders Towers of the World". Generally speaking, ancient architecture shows a per square meter of construction area with 1 m^3 of timber, which is somewhat greater than the timber standards used in modern architecture (not greater than 0.03 m^3). In the Palace Museum in Beijing, one of the largest and best preserved ancient timber architectures in the world, all the palaces are made of timber with about 2 m^3 of timber for a per square meter of construction area, which is shown in **Figure 1(b)**.

The main reason for the heavy use of timber is that the timberwork has the advantages of material and structure. In China, it is easier to obtain many kinds of timber from local areas, and people are good at the process of timber. Compared with stone materials, timber is light with high toughness and good flexibility, which could be widely used in palaces, temples, towers and so on. And even the bracket joint and tenon-and-mortise connection structure, as shown in **Figure 1(c)**, are used in the joint of the timberwork, which increases the toughness of the timber structure. **Figure 1(d)** shows that, in the aspect of structural mechanics, the distinction and cooperation between the load-bearing part and the non-load-bearing part of the timberwork increase the stability and flexibility for the timber frame structure. So the timber structure is widely adaptable which can exist in different forms of buildings in the south and north of China where the climate conditions are significantly different.

However, under the condition of ancient construction, it is difficult to deal with the timber by anti-corrosion and fire prevention technology, which makes it difficult to preserve the ancient architecture. Timber in the ancient architecture is prone to fire, which is the fatal defect of timberwork. In particular, the burning caused by lightning is very dangerous because it is difficult to control and prevent. The material of timber is an organic matter composed of cellulose and lignin, and its main chemical components are carbon, hydrogen and oxygen. Most of ancient

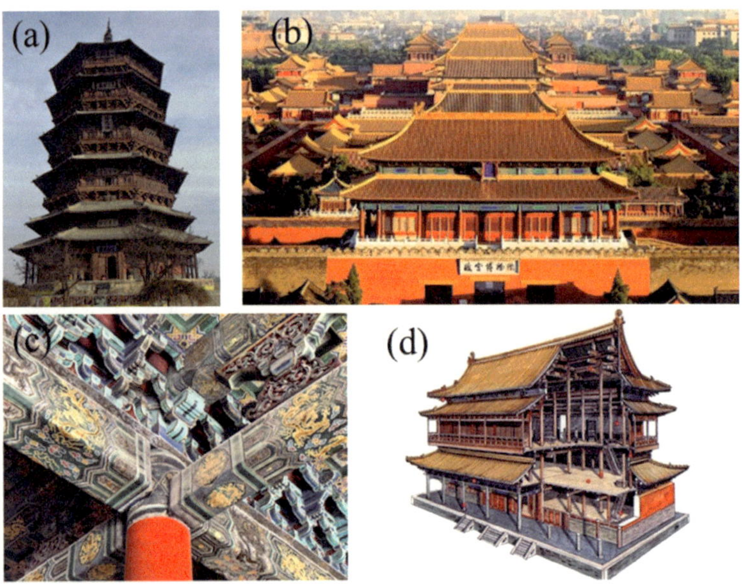

Figure 1.
(a) The Sakyamuni Pagoda of Fogong Temple in Shanxi Province built in 1056 is the highest timber pagoda in the world. (b) The Palace Museum in Beijing with the history of 600 years is one of the largest and best preserved ancient timber architectures in the world. (c) The bracket joint and tenon-and-mortise connection structure in the Palace Museum in Beijing. (d) The diagram for the timber framework with load-bearing part and the non-load-bearing part ((d) from the Internet).

architectures used high quality timber with high oil content, such as pine, cypress, camphor, nanmu, etc. The timber of ancient building have experienced long histories of possibly hundreds of years, so the moisture content of the timber is far lower than that of the normal dry wood (normal value of 12% - 18%), but the oil content is still very high. Furthermore, the timber in ancient architecture becomes rotten and loose in texture, so it is easier to burn after being struck by lightning.

In addition, the layout of ancient architecture with high building density emphasized the balance and symmetry of architectural design. And the form of "courtyard" and "courtyard house" are always adopted, which results in the lack of fire separation and enough safety space. If one of them is struck by lightning, it is easy to spread horizontally, which results in the burning for all of them. Moreover, this kind of structure characteristics also makes it difficult to put out the fire for the ancient architecture.

Due to the material, structures and other characteristics of ancient architecture, events such as lightning-induced fires or lightning strike damage occasionally occur [3, 4]. **Figure 2(a)** shows that, on August 27, 1987, the Jingyang Palace of the Palace Museum in Beijing was caught fire after struck by lightning and the hooks on the timber fell off, causing the plaque to fall to the ground. In the summer of 1992, the Xianling Building of Ming Tombs in Beijing was completely burned by lightning. **Figure 2(b)–(d)** show that, on May 11, 2004, the Great Buddha Temple in Shanxi Province was directly struck by lightning, resulting in a fire, and some buildings and important cultural relics were destroyed. On August 2, 2005, the Dragon Pavilion of Baiquan scenic spot in Henan Province was completely destroyed because of fire after being struck by lightning. On August 11, 2008, the Buddhist Sutra of Chongning Temple in Jiangsu Province was destroyed partially on fire caused by lightning strike.

Most ancient buildings are nationally important cultural tourism resources, as well as precious cultural heritage sites with irrecoverable properties. The timber of ancient buildings is easy to burn after lightning [5], which leads to the rapid and complete disappearance of many specimen components with historical information. At the same time, after lightning damage, the mechanical support performance of timber decreases, which brings hazards to the safety of the overall structure of ancient buildings. Thus it is of great significance to investigate the causes and ways of lightning damage to the ancient buildings timber.

Figure 2.
(a) The Jingyang Palace of the Palace Museum in Beijing after the lightning strike in 1987. (b), (c), (d) The Great Buddha Temple in Shanxi Province after the lightning strike.

1.2 The research process of the lightning strike

At present there are few research results on the damage caused by lightning to the ancient buildings timber. In addition, previous studies on lightning damage and destruction mainly focused on artificial composite materials, metal objects or lightning forest fire. The direct damage effects of lightning currents on graphite and epoxy composite laminates were studied experimentally, and the change laws of the fiber damaged areas, and the maximum damage thicknesses over the electric charge quantity, current peak value, and action integral were analyzed by Hirano et al. [6]. The combustion damage of carbon fiber composites caused by lightning stroke and the residual strength after lightning strike are theoretically analyzed and numerically simulated by Wang et al. [7]. Furthermore, they also conducted lightning impulse tests on the composite materials [8]. The experimental results showed that the high potential, temperature and thermal stress generated at the moment of lightning strike mainly expand symmetrically along the maximum direction of conductivity at the top of the composite.

In the research of forest fire caused by lightning, a small lightning fire simulation experimental platform with local forest combustibles as experimental materials was built firstly by Latham et al. to obtain the ignition probability logistical equation of different combustibles [9]. A lightning test bench to calculate the energy for lightning-induced inflammable ignition was also developed by Darveniza et al. [10]. The discharge process of ground lightning using an indoor artificial arc was simulated by Zhu et al., in order to analyze the continuous discharge time and inflammable water content, along with the structures influencing the ignition during lightning-induced forest fires [11]. The conditions of lightning-induced forest fires were examined in accordance of the formation and development of lightning and its energy and power [12].

As for lightning and metal objects, Metwally used a long-duration current component after a lightning strike and a simulated lightning voltage with a waveform of 1.2/50 μs to analyze lightning-strike metal damage [13]. A simulated lightning current with a waveform of 10/350 μs was adopted to investigate the heating properties of lightning-strike round metal conductors by Paisios [14]. Considering that the damaged area largely depended on the lightning current amplitude and the damage depth mainly depended on the transferred charge, Liu et al. also conducted an experimental study on lightning damage to metal storage tanks using impact current with the waveform of 30/80 μs [15]. In addition, Mi et al. performed lightning current-induced round steel damage experiments, and the results showed that the high temperature effects of lightning arc discharges would cause damage to metal structures [16]. These studies have positive reference significance for the study of the lightning damage mechanism of ancient building timber.

2. Experimental method and equipment

The action mechanism of lightning damage to the ancient building timber is known to be complicated. In this study, according to the recommendation by the standard of IEC 62305-2010, an indoor lightning simulation device using a 10/350 μs first-lightning current waveform was adopted, along with the representative timber materials of ancient buildings. The lightning damage mode and influencing factors for the ancient building timber were examined. Based on the timber properties and lightning current parameters of the ancient buildings, the relationship between the lightning damage effects on the ancient building timber and the water content and thicknesses of the ancient buildings timber was analyzed, as well as the influence of lightning current on the lightning stroke. The thermal effect of lightning on ancient building components and the energy dissipation mode at lightning strike point were studied to reveal the lightning-induced timber damage mechanism and characteristics, which could provide guidance and reference for protection of ancient buildings from future damage caused by lightning.

Figure 3 shows the schematic diagram for the experiment of the lightning strike timber in this study. A waveform generator was firmly connected with Screw 1 and Screw 2, which were respectively inserted into the end of an ancient building timber board. The distance was about 100 mm for the two screws connected by 1.25 mm- diameter copper wire, which was placed into the 1.5 mm-deep groove and then fixed by insulating tape. One end of the copper wire was firmly connected with Screw 1, and the other end was kept at a certain distance from Screw 2, so as to generate arc discharge in the discharge process according to the action process of lightning effect. A larger space could be set with the increase of the impulse voltage. In order to ensure the safety of the experimental process, Screw 2 was well grounded and a grounding rod was used to release the residual current of the experimental device for each experiment.

Figure 4 shows the waveform generator (GTPS30-20kV, GrandTop Eletric, China) which is used to produce 10/350 μs of direct lightning current waveform with a maximum of 25 kA. The experimental current range was about 10 kA to 20 kA. Because the lightning hitting the timber was not generally the lightning induced wave, a 10/350 μs direct lightning current waveform was selected. Another reason is that there is no changes for the timber surfaces in the case of the selected 8/20 μs induced current waveform. The electrical generator mainly includes a charging device, a capacitor unit, a controllable trigger discharge device and a control system. The current voltage and charge–discharge state are displayed on the screen in real time. The digital fluorescent oscilloscope of a lightning current waveform (DPO3054, Tektronix, USA) can not only precisely display the lightning current peak, wave front time, transfer electric charge and other parameters, but also directly store the lightning current waveforms. In addition, a video camera was used to record the entire process of the lightning strike. For the experiment carried out in Beijing Lightning Protection Device Test Center, the temperature was 26 °C - 28°C and the relative humidity 42% - 45% in the laboratory with altitude of 32.5 m.

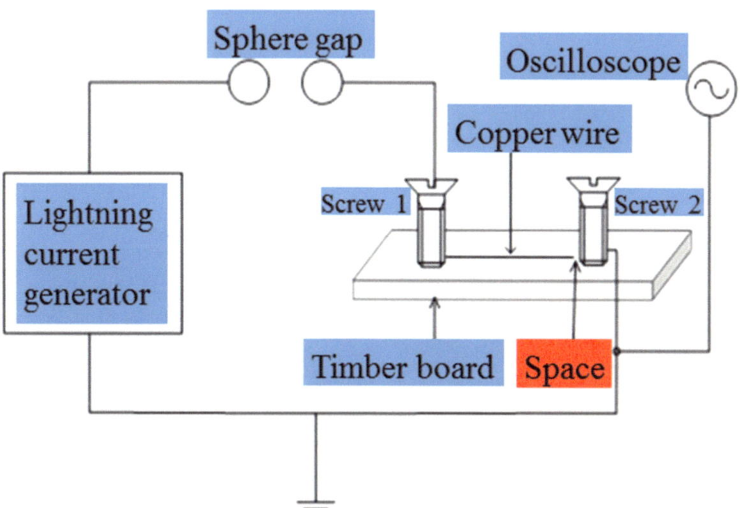

Figure 3.
The schematic diagram for the experiment of the lightning strike timber.

Figure 4.
The multi-waveform lightning generator with the voltage of 20 kV.

Just as shown in **Figure 5**, a voltage divider (PSURGE6.1, HAEFELY, Switzerland) with a 1.2/50 μs voltage wave was connected with timber, which was used to obtain firstly the specific value of the breakdown voltage of the timber as well as the spacing between the copper wire and Screw 2. In this experiment, ancient building rafters and sheathing components (white pine and yellow-flowered pine) were selected as the specimen and the moisture content of timber ranged from 30% to 55% changed by spraying. The timber surfaces were relatively rough with burrs because of the natural aging caused by long-term exposure to sun, wind, and rain. In the test of different water content and different current intensities, the thicknesses of the timber boards with the same material was kept at 20 mm. When assessing the damage to the board with the same material of yellow-flowered pine, the boards with thicknesses of 10 mm, 20 mm, 30 mm, and 40 mm were selected, just as shown in **Figure 6**, which were removed from the ancient building for maintenance. Each board had only one discharge position which was the same for each experiment.

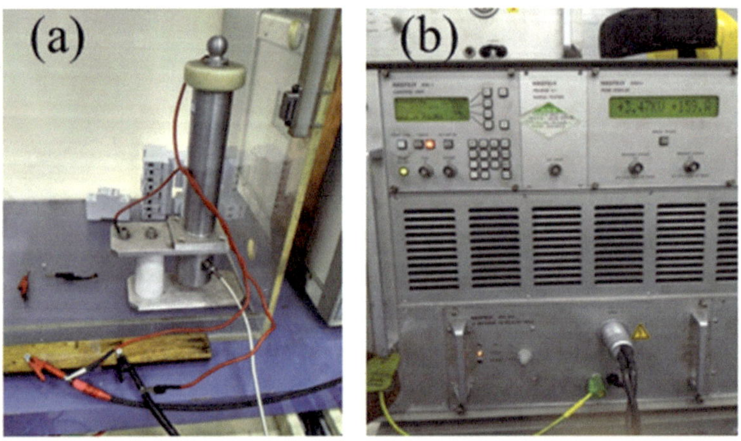

Figure 5.
(a) The diagram for the connection of the voltage divider with timber. (b) The operation interface of the voltage divider.

Figure 6.
The timber board of ancient building with different thicknesses.

3. Experimental results and analyses

Lightning may directly strike the timber artifacts of ancient buildings (architraves, beams, windows, brackets, and so on) if there is no lightning protection systems for the ancient buildings. In the cases there is lightning protection systems but the protection is not perfect, lightning may strike the parts beyond the scope of the protection [3]. Especially when the eaves or sheathing have become wet due to thunderstorm conditions, lightning may directly strike the eave rafters or sheathing of ancient buildings, which is shown in **Figure 7**. In this study, the damage of lightning strike to ancient building timber were studied.

3.1 The experimental phenomena for the process

Figure 8 shows the process for the timber after the lightning strike within several tens of milliseconds. For the experiment, capacitance charging was first conducted on lightning impulse equipment, and then an automatic triggering of the switch produced, which simulated a thunder-release lightning current. At that moment, as shown in **Figure 8(a)**, a strong white light instantly appeared and then disappeared, along with the sound of a loud explosion ("pow!"). And even the door of test bench where experimental specimens were placed was burst through. The white light instantly appeared and then disappeared quickly, which was the simulated lightning. Followed by the white light, there were a large amount of dazzling scattered sparks near the surface of the timber, which is shown in **Figure 8(b)**. After that, the combination of **Figure 8(c)** and **Figure 8(d)** shows that the brightness of sparks slowly weakened and its number slowly decreased. Based on the preliminary analysis, after lightning struck the timber, the timber skin would first be shattered into many tiny timber chips, which was followed by the lightning arc igniting the timber chips or the timber surface burrs. So, as shown in **Figure 8(b)**, a lot of sparks were induced. From the surface of the copper wire between the two screws, it could be seen that the surface of the copper wire was blackened, and there were black substances attached to it, which were the components decomposed by high temperature of the timber and the adhesion of carbonized substances. At the same time, the end of the copper wire near the screw 2 is etched with a sharp tip. Meanwhile, if the space (seeing the red mark in **Figure 3**) was zero, which meant that the copper wire was connected with to Screw 2, the copper wire would melt

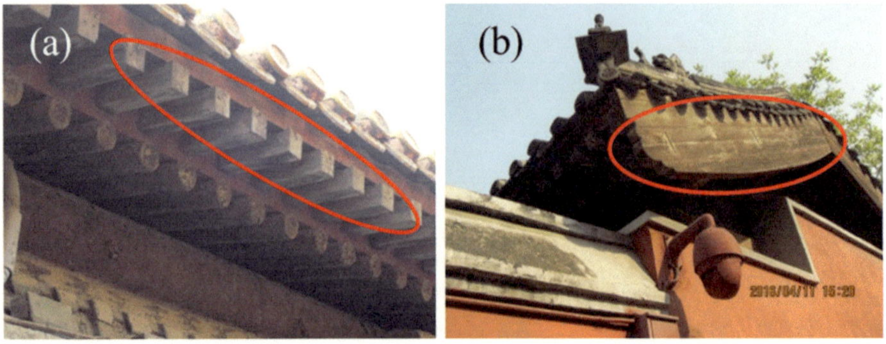

Figure 7.
The components prone to lightning strike for the ancient building timber. (a) the rafters and (b) the sheathing of Yangxin Hall in the Palace Museum in Beijing (the red ellipse locations highlighted in the figure).

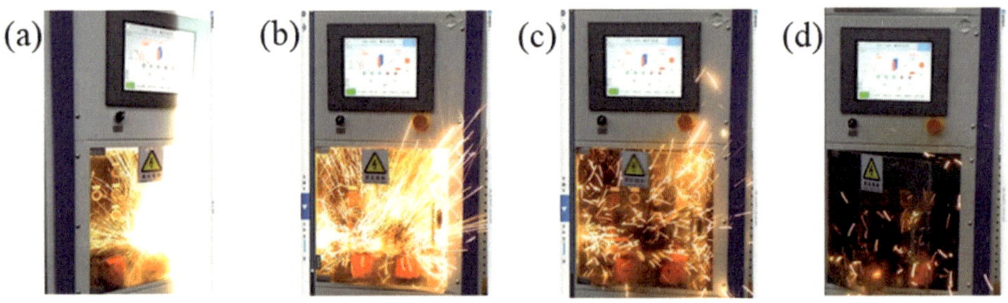

Figure 8.
The process for the timber being struck by lightning. (a) A strong white light instantly appeared and then disappeared quickly. (b) Then a large amount of dazzling scattered sparks near the surface of the timber. After that, the sparks slowly weakened and its number slowly decreased in (c) and (d). Each picture was separated by 10-20 milliseconds.

and crack into multi-segments. There were some burning spots on the surfaces of the timber after struck lightning. And even there were some cracks which affected the support strength of ancient building timber, or it was easy to get rotten or attract insects.

3.2 Effect of the water content on the timber after lightning strike

As the water content in the transverse direction is generally smaller than that in the longitudinal direction, the longitudinal direction is selected and the insertion depth of the water content tester (GM610, BENETECH, China) was kept at 2 mm in this experiment. During the measurement process, it was found that the water content of the same timber at different locations was not completely consistent, which was consistent with the actual situation of the ancient building timber. In thunderstorm weather, rain will pour on the rafters of the ancient building, and the timber water content may be uneven and variable. In order to solve this problem, the water content of the same timber at the lightning impulse position was measured three times and then the calculated average value was adopted. The water content of the ancient building timber was successively increased (about 5% each time) with the same thickness and material of timber. The lightning impulse results for the different water content are shown in **Table 1** and **Figure 9**. After the lightning strike, the position of the timber that was directly hit turned black and the water content obviously decreased with the evaporation of water. In addition, the tape used for

No.	Water content of the timber prior to impulse	Water content of the timber posterior to impulse	Timber state
1	32.3%	26.2%	Blackened burn marks; a square damage pit posterior to the impulse with a depth of approximately 3 mm, length of 8 mm, and width of 3 mm
2	37.8%	34.4%	Blackened burn marks; a quasi-square damage pit posterior to the impulse with a depth of 5 mm, length of approximately 12 mm, and width of 6 mm
3	42.2%	36.7%	Blackened burn marks; an elliptical damage pit posterior to the impulse with a depth of approximately 6 mm, length of 28 mm, and width of 8 mm
4	47.6%	38.1%	Timber surface is hit to an oblique cleavage (a length of approximately 55 mm); a strip damage pit occurred with a depth of 4 mm, maximum width of 10 mm, minimum width of 2 mm; and with obvious water evaporation in surrounding area
5	52.7%	38.8%	A long crack occurred along the timber fiber direction, with a depth of 20 mm (for example, an impulse crack), length of approximately 120 mm, and width of 3 mm

Table 1.
Timber state changes in the cases of different water content posterior to impulse (the thickness of 20 mm, the current of 20 kA).

Figure 9.
Impulse damage map of the timber with the water content of (a) 42.2%, (b) 47.6%, (c) 52.7%. (The red ovals represent the damage locations, and the red arrow faces toward Screw 2).

fixing the copper wire turned black due to high temperature. The instantaneous high current hits resulted in the elliptical damage pits or crack at its outflow position (i.e. at screw 2), of which the damage pits were commonly elliptical or square. With the increase of the water content, the hit damage pit or crack became deeper and the depth for the lightning current entering into the timber increased.

The timber board morphology after the lightning strike was used to observe the lightning damage. As shown in **Figure 9**, for the dry timber with low water content, it was easily burnt when encountering the lightning arc. And it was more likely to show damage pits or splitting for the timber with high water content. With the increase of the water content, the hit damage pit or crack became deeper and the depth for the lightning current entering into the timber increased. The reason might be that the energy conversion process at the contact between the lightning channel and the timber could be similar to the phenomenon of spark-gap arc heating, and the temperature of the lightning arc was very high, resulting in the burn marks on timber. However, due to the short action time, the energy was limited and the instantaneous high temperature might not ignite the timber. In the cases of low water content, the heating of the lightning struck timber was mainly decided by the arc heat based on the burning effect of lightning arc spark. Because of the poor electrical conductivity, the lightning current into timber with low water content was found to be small, and the heat was also very small. Although it is easy to ignite when the moisture content of combustible materials is low, it is difficult to ignite only by the arc temperature at the moment of lightning. However, when the water content was high, the surface resistivity of timber decreased, and the lightning current entering the lighting point of timber also increased, producing more heat energy. These situations are basically referred to as resistance heating. Lightning currents travelling through timber induced the generation of heat which produced interior gases. The impact force formed by the instantaneous expansion of gas would knock the timber out of the damage pit or crack. The generated gas was not only from the instantaneous evaporation of timber moisture at high temperatures, but also from the decomposition of timber material at high temperatures. And then the gas expanded rapidly at high temperatures, producing a great mechanical force. Under this section, the timber fiber peeled, split and decomposed instantaneously, and finally form pits or bursts. In the cases of high water content, the heat generated by the lightning generally caused more water to evaporate. Therefore, in the cases of high water content, splitting or damage pits were more likely to occur.

Figure 10 shows that the impulse damage pit depth and area for the timber with different water content. It could be found that the depth and area of the impulse damage pit became large when the timber water content increased. And the change of the damage pit area was more significant with the increase of water content. There was a sudden increase of the damage depth at the water content of 52.7% (just as shown in No. 5 in **Table 1**), and the complete longitudinal split of the timber was observed. The conductivity of timber increased rapidly with the increases of water content. Even a small amount of moisture would lead to a significant increase in conductivity. The conductivity of timber with 30% water content was about 10 million times that of completely dried timber [17]. Lightning experiments for the carbon fiber composite materials show that the conductivity has a maximum influence on composite material damage [18], and there are more similarities for the timber and carbon fiber composite materials. The most important factor affecting the conductivity of timber may be the water content. It is believed that the water content is the biggest internal cause of timber damage by lightning. What's more, the conductivity in the timber fiber direction (longitudinal) is greater than that in the transverse direction or thickness direction, due to the pore structure of timber

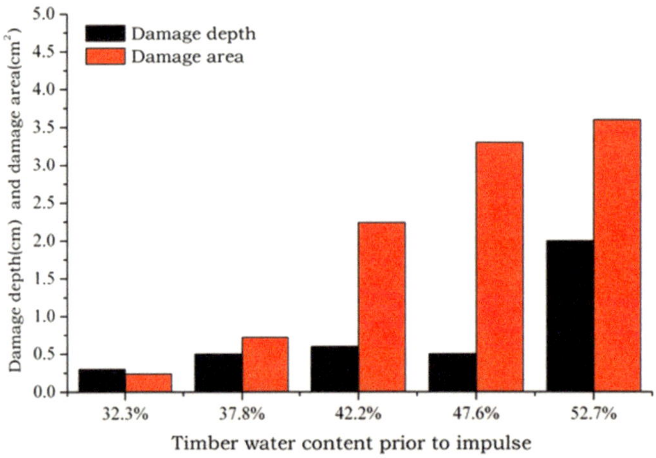

Figure 10.
The impulse damage pit depth and area of timber with different water content.

with more water. For example, the conductivity in the longitudinal direction is about twice that in the transverse direction. Lightning current propagates rapidly along the longitudinal direction and is blocked along the transverse and thickness directions. Therefore, in the case of lightning, the timber is more likely to crack along the longitudinal direction.

Timber with high water content is equivalent to the semi-insulator material with a large electrical resistivity and small conductivity. For example, the resistivity of fully-wet timber at room temperature is about 10^2 Ω•m to 10^3 Ω•m, which belongs to semi-insulator material. The lightning current into the timber causes heating, resulting in the splitting or damage pits of timber, which is different from the metal body. The metal tank lightning experiment shows that the arc temperature is the main factor of lightning-induced metal body damage [15]. The resistance of the metal body is small, and the temperature rise caused by lightning current is low. However, because of the poor electrical and thermal conductivities of timber, the current density near the lightning strike points (lightning attachment points) will be the maximum, along with the fastest heating and the highest temperatures. In these regions, heat accumulates with internal pyrolysis, resulting in gas pressure, which leads to splitting.

Figure 11 shows the damage state of fresh willow branch after lightning strike in order to further study the relationship between lightning and timber water content. Because willow material with high flexibility was also used in the ancient building timber, fresh willow branches were selected for the experiment of high water content timber. Prior to the experiment, the surface resistivity of the willow branch with the measured water content of 62.4% (higher than those shown in **Table 1**) was about 150 Ω•m, because the free ion concentration (the quantity of charge carrier) was high for timber with high water content. Just as shown in **Figure 11**, the fresh willow branches presented a bark peeling effect when exposed to the lightning strike conditions. The peeling region was approximately a rectangular shape, with a maximum length of approximately 25 mm and a maximum width of approximately 18 mm, and there are black burnt marks in this area. Under the condition of lightning current, willow twig bark produced high-temperature evaporation, and the released gas had great mechanical force, resulting in willow bark peeling. It was found that fresh trees (living trees) with high water content were more likely to experience splitting when struck by lightning. However, as for the dead trees with low water content, they were more prone to fire under lightning strike.

According to the above analysis, the damage of timber caused by lightning was mainly determined by the heat generated by lightning channel arc and the heat generated by lightning current injected into timber with a certain moisture content. The potential difference U between the lightning channel top and the ground can be approximately 10^7 V to 10^9 V [19]. If U is set as 10^8 V, the transmission charge is set as

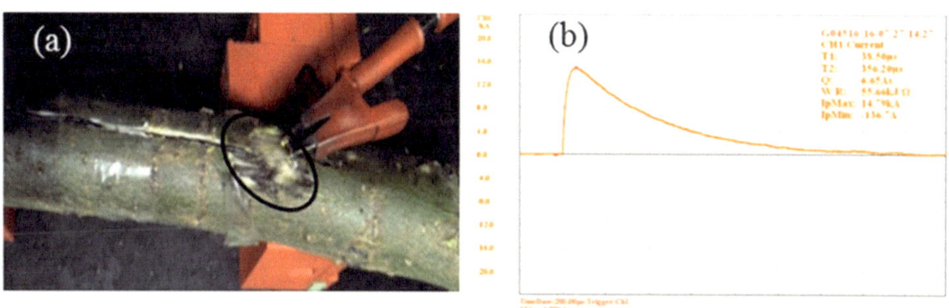

Figure 11.
(a) The damage state of fresh willow branch after lightning strike (the black oval represents the peeling region), (b) the corresponding lightning current waveform.

the typical value Q of 20 C, and then the lightning energy W of 2.0×10^9 J is obtained according to the formula $W = Q{\cdot}U$. If the lightning discharges to the ground at an altitude of 5 km, the energy per unit length of the discharge channel is about 0.4 KJ/mm, which is the energy generated by the lightning arc. If the resistivity of timber R is 0.1 Ω•mm at the lightning strike point and the lightning current i is set as 20 kA, the energy per unit length of timber with lightning current is approximated as 14 kJ/mm according to the formula $W = R\int i^2 dt$. Therefore, it can be seen from the approximated estimation that the energy per unit length of timber with lightning current is somewhat greater than the energy produced by the lightning arc.

3.3 Effect of the thickness on the timber after lightning strike

Table 2 shows the experimental results for the timbers with different thicknesses under equal current strength, and the water content of timber was fixed at 42.5%. It was found that there was bursting for thin timber, which was related to the lightning air shock wave and the rising temperatures from joule heat generated by the lightning current through the timber with certain water content. During the lightning discharge process, the temperature for the lightning channels increased abruptly and the air volume rapid swelled, which diffused around at the speed of ultrasonic waves to form shock waves. The generated pressure could be as high as dozens of atmospheric pressures, which led to the timber damage. The glass door of the test bench where the experimental samples were placed was observed to vibrate because of the lightning shock wave effects. And even, the vibrations or collapses of the door, window and wall for building in the cases of large lightning current strength very close to impulse points [20]. The thinner the timber board was, the smaller the withstanding impact strength would be, and the more easily cracking would occur under the action of the lightning surges. Meanwhile, it was prone to blackening and burning when the timber was thick. The thicker the timber board was, the larger the withstanding impact strength would be. And the large thicknesses resulted in the boards being more difficult to burst under impact pressures. Therefore, under the same heat conditions, the thicker boards were found to be more difficult to light on fire, and the fire burn rate was also slower.

Figure 12 shows that it's more easily for the broadsides of the timber to be caught on fire after the lightning strikebecause there was more contact with oxygen. Similarly, just as shown in **Figure 7(a)**, the timber with fissures, cracks, or decaying were prone to fire. Over long periods of time, the exposed portions of eave rafters become decayed with cracks, which leads to larger oxygen contact, making them more prone to burning during lightning strikes.

No.	Timber thickness	Current peak	Timber state
1	10 mm	20 kA	Blasting occurs at the lightning strike point of the timber, which leads to the timber cracking into two parts
2	20 mm	20 kA	Approximated square damage pit occurs on the timber surface, with a pit depth of approximately 3 mm
3	30 mm	20 kA	Square damage pit occurs on the timber surface with a pit depth of approximately 4 mm, and some parts display blackened burnt marks
4	40 mm	20 kA	Timber is burnt black and shows carbide markings

Table 2.
State changes of timber with different thicknesses after lightning strikes (water content of 42.5%).

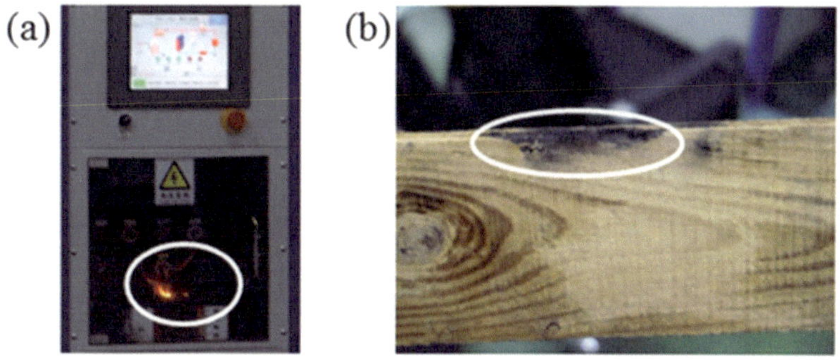

Figure 12.
(a) Experimental onsite for the fire of timber with the water content of about 52.3% and the thickness of 20 mm after lightning with lightning current of 20 kA (the white oval represents the location of timber). (b) The timber after fire (the white oval represents the location of fire).

No.	Wave-head time (μs)	Half-peak time (μs)	Charge (As)	Ip_{max} (kA)	Ip_{min} (A)	Unit energy (kJ/Ω)
1	40.5	361.2	4.71	10.25	125.0	26.93
2	38.0	342.4	5.08	12.45	181.2	36.87
3	38.8	364.8	6.70	14.85	133.8	56.74
4	38.8	348.4	7.44	16.32	128.7	67.52
5	38.5	349.2	8.00	17.91	131.9	78.49
6	38.9	349.5	9.10	20.15	140.6	91.32

Table 3.
Energy parameters of the different lightning currents.

3.4 Effect of the lightning current on the timber after lightning strike

In order to study the influence of lightning current on the lightning stroke for timber, the lightning current value was successively increased (about 2.2 kA each time) and the lightning current parameters in this experiment were shown in **Table 3**. With the increase of the peak current which was an important parameter of lightning current, the sound of explosion increased when lightning strikes, and the damage area and the damage depth for the timber also increased. This is consistent with the fact that the higher the water content of timber, the more serious the lightning damage. When the peak current increased, the impact energy of lightning current increased, so the damage for the timber was more serious. **Table 3** also shows that, when the peak value of lightning current increased in turn, the transferred charge and unit energy increased accordingly. From the perspective of lightning parameters, the main factors affecting the degree of timber damage are current peak, transferred charge, unit energy and waveform, of which peak current is the most important factor.

4. Conclusions

In this paper, a lightning simulation experiment was mainly used to study the lightning strike damage characteristics of ancient building timber, including the effect of the water content and thickness of the timber for the timber after lightning strike. The results show that the main ways of timber damage caused by lightning were lightning arc heat, gasification impact caused by lightning current on timber heating and lightning air shock wave effect, which played different roles under different conditions.

For the timber with low water content, the timber was easier to blacken or burnt after lightning strike, and the burning loss of lightning arc was the main reason. For the timber with high water content, the timber was more prone to pit crack damage, in which lightning mechanical energy was the main factor causing timber splitting damage. In essence, lightning current entered the timber, resulting in the increased heating and the rapid expansion of gas produced by high temperature. This effect produced a mechanical force similar to an explosion. The higher the water content, the greater the damage depth and the larger the damage area. In cases of timber with different thickness, the thicker the timber, the more difficult it was to ignite after lightning strike. When the timber was too thin, lightning shock waves would also cause damage. With the increase of the lightning current intensity, the timber damage will be more significant after lightning strike with the larger timber damage area and the greater damage depths. Therefore, it was concluded that the peak value of lightning current and the water content of timber were the main external and internal causes which influenced the damage degree of timber.

Generally speaking, these is always a series of complicated process within the timber after lightning strike. It is hoped that this study can provide some guidance and reference for the protection of ancient timber buildings from the possible damage caused by lightning.

Acknowledgements

This study is financially supported by the Beijing Natural Science Foundation (Grant Nos. 8192052, 8164072) and NSFC (Grants No. 61307065).

Author details

Jingxiao Li[1] and Jing Li[2*]

1 Disaster Prevention Center, Beijing Meteorological Service, Beijing, China

2 Technical Institute of Physics and Chemistry, Chinese Academy of Sciences, Beijing, China

*Address all correspondence to: lijingsp@mail.ipc.ac.cn

References

[1] S. Zhang, R. Lu. ICA3D – Intelligent computer-aided ancient Chinese architecture design. Advanced Engineering Informatics, 2012, (26): 705-715.

[2] X. Yang, C. Gao. Study on Lightning-Proof Mechanism in Buildings via Naturism Concept in Ancient China. Proceedings of Eco-Sophia 2011 Parallel Session-8th International Whitehead Conference-Creativity and Harmony. Japan, Tokyo, 2011:325-332.

[3] J. Li, P. Song, R. Li, M. Qian, F. Qi. Analysis and preventive measures on fire disaster from direct lightning stroke at the ancient timber buildings. Journal of Southwestern Normal University (Natural Science Edition), 2016, 41(10): 88-95.

[4] H. Zhang, Y. Liu, Y. Zhang. AHP-Based classification of lightning protection for heritage buildings [J]. Meteorological Science and Technology. 2015, 43(2):326-330.

[5] S. Wang. Lightning and fire protection of ancient buildings in gardens. Journal of Chinese garden, 1994,10(3): 40-45.

[6] Y. Hirano, S. Katsumata, Y. Iwahori, A. Todoroki, Artificial lightning testing on graphite/epoxy composite laminate [J].Composites Part A: Applied Science and Manufacturing, 2010, 41(10):1461-1470.

[7] F. Wang, N. Ding, Z. Liu, et al. Ablation damage characteristic and residual strength prediction of carbon fiber/epoxy composite suffered from lightning strike[J].Composite Structures, 2014,117:222-233.

[8] F. Wang, Y. Ji, Z. Liu. 3D numerical simulation of composite discharge effect under lightning strike. Journal of Vibration and Shock, 2014,33(22):13-20.

[9] D. Latham, J. Schlieter. Ignition probabilities of wildland fuels based on simulated lightning discharges. USDA, Forest Service Research Paper, INT-411. Washington DC: USDA, Intermountain Research Station, 1989.

[10] M. Darveniza, Y. Zhou. Lightning-initiated fires: Energy absorbed by fibrous materials from impulse current arcs. Journal of Geophysical Research, 1994,99:10663-10670.

[11] Y. Zhu, N. Liu, Z. Deng. Experimental study on the probability of lightning induced ignition of forest fuels. Fire Safety Science. 2012, 21(2):71-77.

[12] Q. Lin, J. Zhu, Q.Wang. Investigation on forest fire caused by lighting.Fire Safety Science. 1999, 8(1):15-21.

[13] I. Metwally, F. Heidler, W. Zischank. Measurement of the rear-face temperature of metals struck by lightning long duration currents [J]. European Transactions on Electrical Power, 2004, 14(4): 201-222.

[14] M. Paisios, C. Karagiannopoulos, P. Bourkas.Estimation of the temperature rise in cylindrical conductors subjected to heavy 10/350 μs lightning current impulses [J]. Electric Power Systems Research, 2007, 78(1): 80-87.

[15] Y. Liu, H. Xia, Y. He. Ablation Damage and Temperature Rise of Metal Oil Tanks Struck by Direct Lightning. High Voltage Engineering, 2016, 42(5): 1578-1585.

[16] X. Mi, Z. Xiao. Experimental study on round steel damaged caused by lightning stroke. Building Electricity, 2015, 34(4):45-50.

[17] N. Ding, B. Zhao. Analysis of damage influence factors on ablation damage of composite laminate subjected lightning strike. Transactions of

Materials and Heat Treatment, 2014, 35(2):186-192.

[18] X. Diao, P. Sun. The electrical properties of wood and the use of electric moisture content meter.Wood Processing Machinery, 1993, (3):18-20.

[19] W. Chen. Principles of Lightning Science, China Meteorological Press, Beijing, 2006.

[20] X. Zhao, J. Wu, N. Ma, R. Huang. A case study of a disastrous lightning stroke event caused by air shock wave. Meteorological Science and Technology. 2012, 40(6):1025-1029.

Chapter 20

Trends and Opportunities of Industry 4.0 in Wood Manufacturing Processes

Mario Ramos-Maldonado and Cristhian Aguilera-Carrasco

Abstract

Wood industry is key for sustainability and an important economic activity in many countries. In manufacturing plants, wood variability turns operation management more complex. In a competitive scenario, assets availability is critical to achieve higher productivity. In a new fourth industrial revolution, Industry 4.0, data engineering permits efficient decisions making. Phenomena difficult to model with conventional techniques are turned possible with algorithms based on artificial intelligence. Sensors and machine learning techniques allow intelligent analysis of data. However, algorithms are highly sensitive of the problem and his study to decide on which work is critical. For the manufacturing wood processes, Industry 4.0 is a great opportunity. Wood is a material of biological origin and generates variabilities over the manufacturing processes. For example, in the veneer drying, density and anatomical structure impact the product quality. Scanners have been developed to measure variables and outcomes, but decisions are made yet by humans. Today, robust sensors, computing capacity, communications and intelligent algorithms permit to manage wood variability. Real-time actions can be achieved by learning from data. This paper presents trends and opportunities provided by Industry 4.0 components. Sensors, decision support systems and intelligent algorithms use are reviewed. Some applications are presented.

Keywords: wood manufacturing, industry 4.0, multiple sensors, bigdata, machine learning

1. Introduction

In industrialized nations, manufacturing has become a key growth factor. "Great Britain was the first industrializer and became the technological leader of the world economy. Manufacturing became the main engine of the economic growth in the 19th century, spreading manufacturing production technologies to other countries" [1]. Digital manufacturing (Artificial Intelligence, bigdata analytics, cloud computing, among others) is changing the nature of manufacturing production. The adoption of these technologies (by developing countries) can foster inclusive and sustainable industrial development and the achievement of the Sustainable Development Goals [2].

Wood industry is key for sustainability and an important economic activity in many countries. In recent years, manufactured wood products for construction

 IntechOpen

have had special relevance. These represent 38.1% of wood-based products worldwide [3].

As a biological material, wood is variable in its physical and anatomical properties. Wood mechanical or chemical transformation processes are affected. In manufacturing processes, process variables must be managed to achieve quality and productivity standards. In industrial systems, productivity and product quality are affected by multiple variables. The "wood material" adds an additional complexity degree. Traditionally, human experience has been able to control many variables and maintain the operating system. However, the human capacity has limits and to use sensors and computers is needed to help online decision-making.

For decades, multiple sensors for generic physical variables such as pressure, speed, or temperature have been developed. Specific sensors capable of measuring and characterizing wood, including destructive and non-destructive testing, have been developed. For industrial use in manufacturing process control non-destructive characterization is the most important. This allows monitoring and control (usually by humans) the process in real time. In a competitive scenario, to achieve higher operating factors, lack assets availability is critical. To have in real time information on the operational behavior allows online decision-making. Lack of data and its analysis does not allow to forecast and prescribe operation behaviors and improve performance. Decision-making has become more complex, uncertain, and rapidly changing external conditions. However, the use of online data in industrial settings is still incipient. Computing power and robust algorithms capable of predicting behavior in complex environments using data are recent.

The fourth industrial revolution presents great opportunities for wood manufacturing processing. Sensor's development, high computing capacity, industrial internet (Internet of Things, IoT) and learning algorithms can allow a much better handling of uncertainty and material variability. Algorithms based on artificial intelligence make possible online decisions and prediction of phenomena that are difficult to model with conventional techniques. Sensor's availability and Machine Learning techniques allow the intelligent capture, display, and data analysis [4, 5]. Machine Learning (and Deep Learning) uses the history of data, of positive or negative experiences to model real processes and automatically conclude for other situations. The choice of the algorithms depends to the problem. Therefore, new opportunities are open for academia and industry to improve the industrial wood processing. Studying appropriate performant models and determining which variables and which sensors to use, among other considerations, are part of the data engineering and futures research work.

This work presents the industry 4.0 scopes, components, and opportunities for the wood industrial manufacturing. Data collection and engineering are the focus. Some author's examples are shown.

2. Industry 4.0 and components

In recent decades, the industry has evolved toward more intensive use of digital technologies. And this has passed from traditional automation to a new industrial revolution, the fourth industrial revolution or Industry 4.0. The first revolution was the introduction of mechanization and steam power. The second was the incorporation of mass production and division of labor. And the third revolution incorporated electronics, automation and CAX (Computer Aided X) technologies.

The 1980s changed the direction of industrial production policies. Taylorian mass production evolved toward a production of a variety of products, with costs like those of the production of large quantities. In 1990s, the focus was on integrating manufacturing automation technologies mainly robotics, CNC (Computer Numerical Control), CAD / CAM (Computer Aided Design/ Computer Aided Engineering), and automatic control (Advanced Manufacturing Technologies). Computer Integrated Manufacturing (CIM) emphasized data integration into the production cycle of the firm, it was the third industrial revolution [6]. Manufacturing evolved from the intensive labor use in traditional manufacturing to a sophisticated set of processes based on information technology [4].

In 2011, Industry 4.0 emerges as a focus of the competitiveness of Germany's industry [7]. It is defined as networks that incorporate Cyber-Physical Systems that handle bigdata and use Artificial Intelligence. Industry 4.0 is based on developments of the last 20 years, at least four: research in artificial intelligence, better computing capacity and speed, internet development and wireless communication. For example, the combination of distributed systems, self-organizing systems and artificial intelligence was the prelude to what today in industry 4.0 is known as Cyber-Physical Systems [8–11].

Data science and artificial intelligence would be the "core" of Industry 4.0 [2]. The consequences are the virtual factory (or digital twin) and autonomous machines (Cyber-physical Systems) capable of interacting "intelligently" with other machines and humans [12]. Data analysis (data science) makes it possible to make decisions and predict dynamic phenomena that are difficult to model with conventional techniques. Industry 4.0 is the data revolution, especially in the manufacturing industry [13]. Today object is about the massive use of data and analysis for the design and operation of industrial systems.

2.1 Components

Many authors have defined components part of this fourth revolution. Some place more emphasis on hardware devices and others on software elements. However, both data and automatic analysis are the base and more common denominator: the data science approach. Several methodologies exist to drive data projects, but in general that consists in fourth steps: to know the problem, to understand the data, to extract features and to model an analyze [14]. Data engineering is complementary and fundamental to achieve implementations: from data capture to the action over the physical system. First task in data engineering is to make available data: to select sensors, to process signals and to generate descriptors and data warehouse. Second, it is to know the physical processes, to understand and visualize data and to extract features. And finally, tasks are modeling and implementation for actions (**Figure 1**).

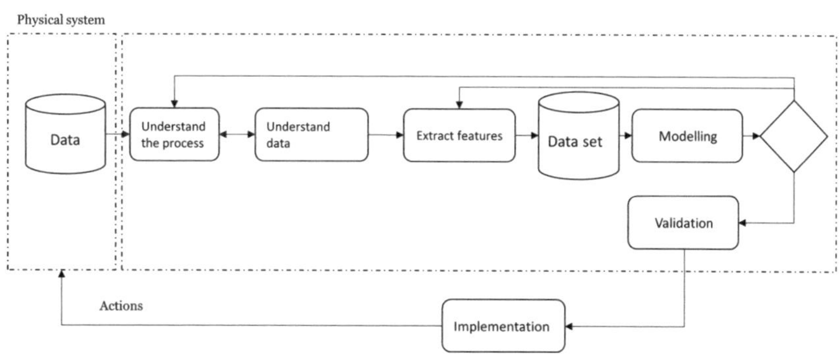

Figure 1.
Data engineering process.

Without data engineering 4.0 technologies would not be possible. In this work, four technological components are visited: Industrial Internet, Cloud Computing, Virtual Factory and Cyber-Physical Systems.

2.1.1 Industrial internet (II)

II Is to use the internet for industrial purposes. All is called Internet of Things (IoT). An IoT system consists of Industrial Wireless Networks (IWN) and Internet of Things (IoT) [15]. It includes machines and equipment, networks, the cloud, and terminals. "Things" and "objects" interact with each other and cooperate to achieve common goals. "IoT is capable of offering specific and personalized products. Users can customize products via web pages. Then, web servers transmit data to the industrial cloud and plants via wired or wireless networks" [5]. 5G technology will allow high speeds of communication and industrial internet feasible.

Also, it is possible to define Internet of Services (IoS). IoS allows providers to offer their services over the Internet. "IoS is emerging, based on the idea that services are made easily available through web technologies, allowing companies and private users to combine, create and offer new kind of value- added services" [16].

2.1.2 Cloud computing (CC)

CC is a set of resources, including physical servers, networks, storage, and user applications accessible from Internet [17]. CC is a new concept. it is a collection of configurable computing services to be made accessible and released as specified [18]. It also allows easy and on-demand network access. Different networks, servers, storage, applications, and services resources are disponible today. Services providers, e.g., Microsoft Azure, Alibaba Cloud, Amazon, and Google Cloud, provide access through the internet. Clients pay only for the resources they use. CC services are one crucial components of the Industry 4.0 including IoT and CPSs [19].

2.1.3 Cyber-physical system (CPS)

CPS comprise intelligent devices capable of exchanging information autonomously, causing actions and controlling each other independently. "CPS are systems of collaborating computational entities which are in intensive connection with the surrounding physical world and its on-going processes, providing and using, at the same time, data-accessing and data-processing services available on the Internet" [20]. CPS is an integration of computation with physical world. Computers monitor and control the physical processes. Feedback loops act where physical processes affect computations and vice versa [21, 22]. Software and hardware with sensor and action are integrated (**Figure 2**).

2.1.4 Virtual factory (VF)

VF is defined as a virtual model that assists people and machines in the execution of their tasks. They are systems that work in the background. In 1993, the VF concept was introduced by Onosato and Iwata [23]. VF It considers the actual context information such as the position and state of an object. In a virtual factory, the CPS perform tasks, communicate, and take those actions to the real world of the plant [9]. VF include virtual organization, emulation facility and integrated simulation. In [24]. VF is defined "as an integrated simulation model of major subsystems in a factory that considers the factory as a whole and provides an advanced decision

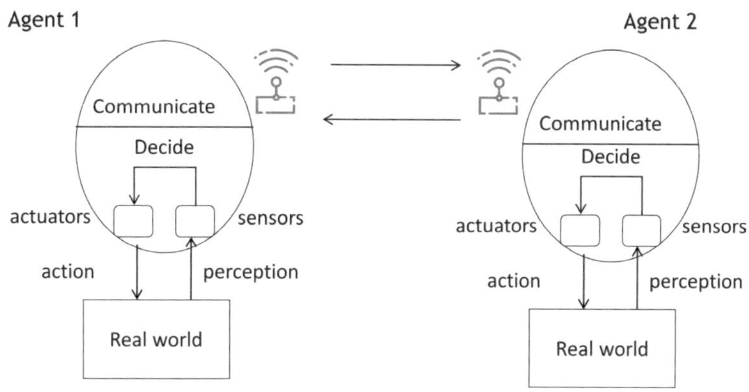

Figure 2.
CFS: Interaction of agents [10].

support capability." Virtual models can guide physical entities responding to the changes in their environment and to improve operations [25].

A similar concept, Digital Twin (DT) has been proposal. In [26] DT a production line is integrated with the real production processes using a simulation model. Real-time interaction between virtual and physical world allows DTs to respond to unexpected changes in manufacturing processes more rapidly [23].

3. Manufacturing processes in the wood industry

3.1 Manufacturing and process industry

It is "manufacturing, any industry that makes products from raw materials using manual labour or machinery and that is usually carried out systematically with a division of labour. In a more limited sense, manufacturing denotes the fabrication or assembly of components into finished products on a fairly large scale" [27]. Many authors difference manufacturing and process industry [28].

Manufacturing is a discrete system and uses machines or workstations to change forms, dimensions, or surfaces. Lines or cells assembly parts to obtain final products. Process industry is a continuous system and put emphasis over chemical processes, or batch like reaction, heat, cold, to generate final product liquid, gas or solid. In the forest industry, manufacturing is concerned to the "solid" wood transformation and process industry more with pulp and paper industry.

3.2 Wood transformation

The first wood transformation begins with the log after harvest. In the sawmill industry the main product is dry sawn wood. In the board industry products are veneers, flakes, particles, and fibers. Second transformation generates appearance and engineering products such as moldings, furniture parts, plywood, CLT (Cross Laminated Timber), OSB (Oriented Strands Board) and particle or fiberboard. Different operations can be considered: milling, molding, peeling, pressing, drying, gluing, painting, among others.

Main operations of the manufacturing industry are cutting operations that produce changes in shape, dimensions, and surfaces. Wood is an anisotropic material, but it is treated as an orthotropic material. Its mechanical properties change on the radial, tangential and longitudinal axes. This affects the "cutting" behavior according to the direction of the stress of the cutting tool [29]. Similarly, anatomy, density, singularities, and moisture content impact product quality and productivity. For example, well known is the effect of properties and species on the drying of lumber or veneer.

In cutting with or without chip, the tool interaction with the material produces cutting forces that release energy producing pressure waves and tool wear [29]. In the sawmill and remanufacturing industry, tool wear directly affects production costs due to its negative effects on dimensional and surface quality of the product. In sawmill, cutting forces wear the tooth on all faces, increasing friction. Friction changes heat and cutting angles producing inefficient cut over time. The surface quality increases its roughness.

Wear and heat of tool lead to loss of rigidity increasing kerf and dimensional inaccuracy.

In longitudinal sawing, working angles α, β and γ of the cutting tool, geometry, feed per tooth, the feed and cutting speed movement must be optimized (**Figure 3**). These and other variables depend on the properties of the wood and the cutting height. In high productivity sawmilling feeding speeds of over 120 m / min are driven, correct monitoring and control in real time is key. In [30], factors involved in the sawing process are classified into three categories: (1) workpiece, (2) feed, and (3) tool. Combined effects are analyzed showing a complete review of studies. Here, emphasis is to put the sensor to allow online prediction and intelligent monitoring systems and increase the productivity.

3.3 Main factors

For wood manufacturing processes, influencing factors are the material, the operation, and the transformation technology. In cutting processes, factors are combined, the cutting tool being important. These impact on assets, rotating mechanisms, landings, materials, motors, auxiliary systems, and other devices. Heat and mechanical power impact machine availability. In sawing, saws fatigue generates cracks and microcracks in the bottom of the blade throat. Vibrations impact on clamping and feeding systems acting on the products dimensional accuracy [31].

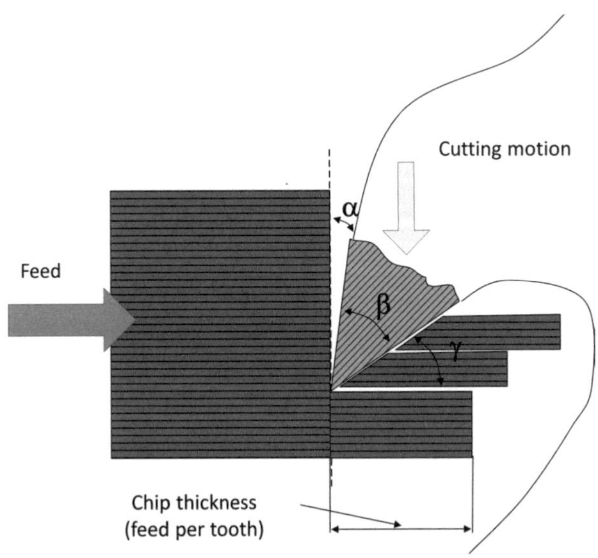

Figure 3.
A typical sawmill tool and material interaction.

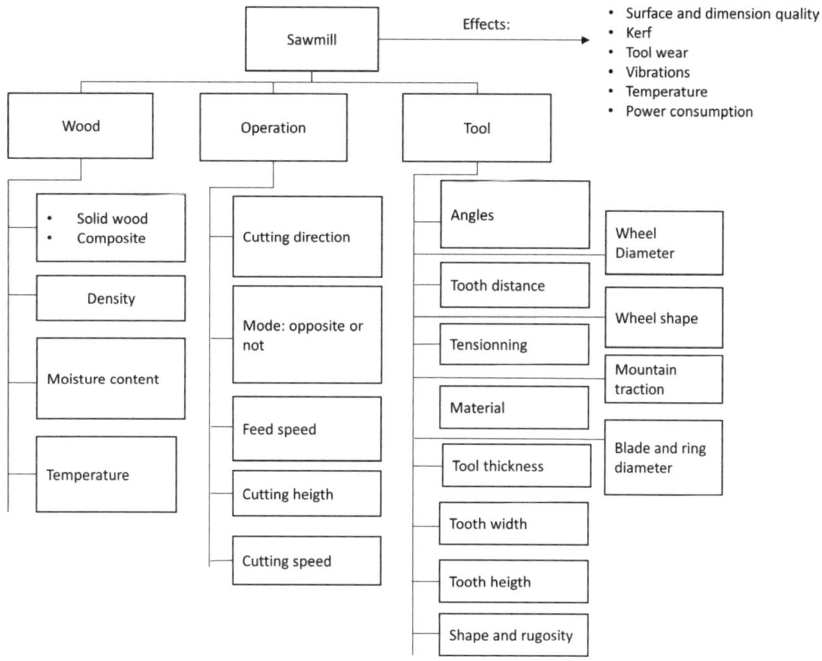

Figure 4.
Factors in a sawmill monitoring.

In other manufacturing operations such as drying, painting, gluing, or pressing, heat and mass transfer phenomena, adhesion and stress-deformation intervene. For example, in veneer continuous drying, air velocity, steam temperature and feed rate determine the cracks presence and the product moisture content. For years, for different species different transformation technologies have been studied. For sawmill, main factors are presented **Figure 4**.

4. Data engineering and wood manufacturing processes

4.1 Sensors and data in wood manufacturing

To make available data is a key task. Unfortunately, many industrial environmental have not yet all availability. Wood industries are not the exception. To measure power, temperatures, tool wear, pressions, velocities, vibrations and physical and wood anatomical characteristics requires robust sensors. Today, dimensions, moisture, density and many wood and panels defects can be tested online. Indirectly certain critical variables can be quantified. That is especially important when these variables depend on more conventional measurable physical phenomena (e.g., electrical variables, temperatures, vibrations, sound, etc.). In [32], cutting and feed per tooth are correlate with acoustic emission and saw temperature. In mechanical operations, cutting forces explain good machining behavior [33, 34]. In general, tool wear can be related indirectly with heat liberation, power consummation vibrations or acoustic emission.

In the plywood industry, peeling cutting forces with vibrations, acoustic emission and artificial vision can be correlated [33]. Cutting in particle panels can be explained by power consummation [34]. For milling, [35] show that sound and vibrations can be used to predict the online surface quality. For *P. radiata*, both acoustic emission and electric power to predict surface quality are showed in [36].

In [37], microwaves are used to detect knots and log defect using laser are shown in [38]. Others researches on non-destructive test exist.

In the same order, many superficial wood characteristics can be measured by artificial vision techniques [39, 40]. Today, industrial scanners can extract different knout types and singularities, colors, timber edger and in certain applications X-ray determines internal defects [41–43]. In the panel industry, dimensions, density, and panel moisture are captured online. To classify veneers, scanner test splits, discoloration, and holes. However mechanical properties are yet tested outline.

Several industrial applications are today available. For example, in sawmills, vision and laser are used to capture the logs true shape and the dimensions of boards (by companies like USNR, MPM and Microtec). Many modern sawmills around the word are users of these technologies. X-rays was yet developed for logs, probably Microtec is a company leader over this segment. To detects wood defects, both internal and external, to board in second transformation applications exist. Main suppliers are Weining, GreCon and Microtec with WoodEye©.

In the wood manufacturing industry, normally scanners aide to control specifical and local operation like parts classification, first cutting in sawmill or thickness mat on particle or fiber panels. Yet, data is non stocked for analysis or to create prediction models. World class wood producers are beginning to use and collect real-time data to extract information and add value (interviews and experiences of the authors).

4.2 Artificial intelligence

Artificial Intelligence (AI) or more specific Machine Learning (ML) is the core of industry 4.0 [44]. Artificial Intelligent has been defined by E. Rich like "the study of how to make computers do things at which, at the moment, people are better" [45]. If it is believed that intelligence is only a human property. Another Langton's definition of intelligence involves all living system [46]. A prominent AI area is Machine Learning (ML) consisting in the capacity to learn to solve problems. ML is the study of computer algorithms that improve automatically through experience [47]. And experience are historical data. Last years, Deep Learning (DL) is a new approach and area of ML. In DL, news algorithms using multi-layer artificial neural network work [48]. ML and DL permit today successful applications and an increase considerable research in many fields.

Complex structure of bigdata can be discovery using DL technics like Convolutional Neural Network (CNN) [49, 50]. Support Vector Machine, Random Forest or Bayes technics work on an important set of problems. However, CNN are advantageous to extract features of industrial bigdata [48]. Computing capacity and bigdata turn possible DL technics to industrial systems applications [51–53].

To wood industry several authors have showed advances using ML. In [54], plywood defects are classified by Support Vector Machine (SVM). In [55], wood quality is automatically classified. In [35], Neural Networks like cutting prediction is used. Recurrent Neural Networks (RNN) is special type of Neural Network. Pass knowledge can be used to learning and predict. In [56], by RNN productivity prediction of a high production sawmill is modeled.

In ML, practice and testing are keys. Data engineering methodologies are important to validate complex problems having many variables and non-lineal relations [52]. Choice of model's hyperparameters, learning and evaluation data size can become decisive to achieve good performances. Learning data sizes can be different according to the problem. Always, more data is better. Fortunately, in industrial environments data can be "bigdata". To validate models, data size can go from some miles to millions.

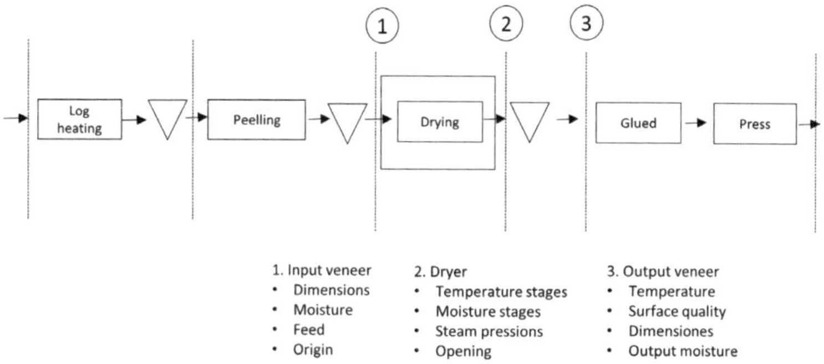

Figure 5.
Data collection in a drying veneer process.

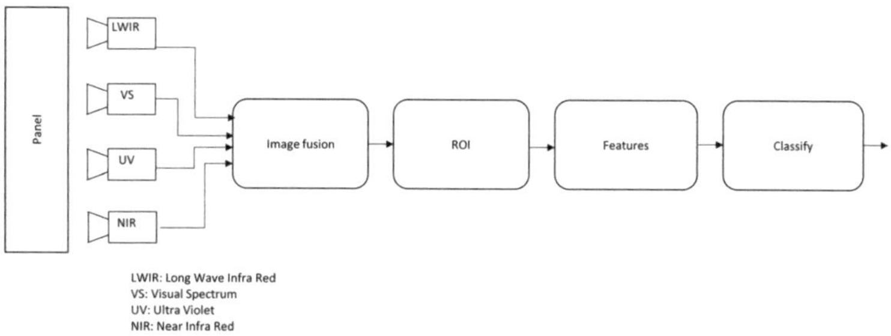

Figure 6.
Multispectral data process to classify industrial melamine panels [58, 59].

4.3 Cases: ML and data collection

Studies in sawmill industrial process show that RNN to predict productivity with 30,000 records 0.8 of coefficient of determination can be obtained [56]. To classify veneer quality in a plywood continuous industrial drying using Neural Network and Random Forest 6,000 records allowed accuracies over 0.8 [57]. In this case, online data collection was implemented to stock veneers, operation, and technology variables. Raw data was pre-process and filtered and a data set warehouse was generated. Material factors considered dimensions, moisture, and forest origin. Operations taken account feed and batch sizes. Drying technology variables were different important factors like steam temperature, pressions and opening. Response was veneer quality (**Figure 5**).

In a melamine particle panels industry too much money can be lost if the final product classification is not good. Using computer vision, multispectral sensors, cloud computing and ML algorithms it is possible to classify panels with 0.95 of accuracy (**Figure 6**). Multi sensor and data integration permit better performances. More of 14,000 records were used to learning and testing.

5. Conclusions

Benefices and components of Industry 4.0 was presented. Focus is on data engineering. Data analysis, Machine Learning and Deep Learning are in the core of Industry 4.0. Availability of sensors, better computing processor and wireless communication turn possible this new revolution and great opportunities for manufacturing industries. IoT is beginning. 5G technology will allow high speeds of communication and industrial internet feasible. Computing cloud represents opportunities for Small and Medium Enterprises too. Lower cost can be obtained when data processing and stockage is done in the cloud. CFS are still in growth and ML models to autonomous computing are showing auspicious and robust. Virtual factory (digital twin) is subject of a series of investigations. Prototypes of virtual reality and simulation models using real-time data is a reality.

In the wood manufacturing industry, last year, research contributions toward 4.0 techniques have been focuses in developing no-destructive sensors and models. Most of the investigations have been driven into the labs. But industry 4.0 woks with data, bigdata. Learning and testing ML models requires a lot of experiences. Industry to increase productivity and product quality need robust algorithms working within hazard environment to carry out intelligent actions. Either actions to aide decision making or automatic control. ML or DP models must be performants. The authors argue that it is necessary to approach academy and producers. Experience is fundamental to understanding data. The fourth revolution is the data revolution. In this context, researcher and practitioners should be overcome three factors: know the wood, understand the process, and use data engineering methodologies.

Acknowledgements

Authors thank to: Research Office of University of Bio-Bio, Chile: Research Project 2060360 IF/R, Machine Learning and Vision Computer Group and Intelligent Industry and Complex System Research Group; CORFO-Chile Agency; FONDEF-IDEA of R&D National Agency of Chile (ANID), Project ID14I20364; Project CYTED TICs4CI: Applications TICS for Intelligent Cities; Arauco Company; MASISA Company; Biobio Manufacturers Association (AGMET); Wood Engineering Department at the University of Bio-Bio.

Author details

Mario Ramos-Maldonado* and Cristhian Aguilera-Carrasco
University of Bio-Bio, Concepcion, Chile

*Address all correspondence to: mramos@ubiobio.cl

References

[1] Szirmai A. Industrialization as an engine of growth in developing countries. Structural Change and Economic Dynamics, 1950-2005. 2012; (23): 406-420

[2] United Nations Industrial Development Organization, 2019. Industrial Development Report 2020. Industrializing in the digital age. Vienna: 2019. 206 p.

[3] Ramage M, Burridge H, Wicher M, Fereday G, Thomas R, Shah D, Guanglu W, Li Y, Fleming P, & Densley D, Allwood J, Dupree P, Linden P.F, Scherman O. The wood from the trees: The use of timber in construction. Renewable and Sustainable Energy Reviews. 2017; 68. 333-359. DOI: 10.1016/j.rser.2016.09.107.

[4] Shipp S, Gupta N, Lal B, Scott J.A, Weber C. L, Finnin M. S, Blake M, Newsome S, Thomas S. Emerging Global Trends in Advanced Manufacturing. Virginia, USA: Institute for Defense Analyses; March 2012

[5] Lu Y. Industry 4.0: A survey on technologies, applications, and open research issues. Journal of Industrial Information Integration. 2017; 6: 1-10

[6] Rembold U, Nnaji B.O, Storr A. Computer integrated manufacturing and engineering. Addison-Wesley; 1996. 640 P.

[7] Hermann M, Pentek T, Boris O. Design Principles for Industrie 4.0 Scenarios: A Literature Review. Technische Universität Dortmund: Working Paper No. 01; 2015

[8] Van Brussel H, Bongaerts L, Wyns J. Valckenaers P, Van Ginderatcher T. A conceptual framework for holonic manufacturing: identification of manufacturing holons, Journal of Manufacturing Systems. 1999; Vol. 18 Issue 1: 35-52. DOI:10.1016/S0278-6125(99)80011-9

[9] Patriti V, Schäfer K, Ramos M, Charpentier P, Martin P, Veron M. Multi-agent and manufacturing: A multilevel point of view. In Plonka F, Olling G, editors. Computer Applications in Production and Engineering. CAPE 1997. IFIP — The International Federation for Information Processing, Boston, MA: Springer 1998. p. 47-57 DOI/10.1007/978-0-387-35291-6_50

[10] Ramos, M. Vers un pilotage auto-organisant de systèmes flexibles de fabrication [Thèse de Docteur]. Nancy, France: Université Henri Poincaré Nancy I; 1998.

[11] Ramos M, Maness T, Salinas D. Modelo de un sistema multi-agente para la optimización de la cadena de suministros en la industria de la madera de coníferas. Maderas: Ciencia y tecnología. 2015; 17(3): 613 – 624, DOI:10.4067/S0718-221X2015005000054

[12] Pranab K. Muhuri, Amit K. Shukla, Ajith Abraham. Industry 4.0: A bibliometric analysis and detailed overview. Engineering Applications of Artificial Intelligence. 2019; 78: 218-235

[13] McKinsey Global Institute, Manufacturing the future: The next era of global growth and innovation. Report, November 2012

[14] Martinez I, Viles G. Data Science Methodologies: Current Challenges and Future Approaches. Big Data Research. 2021; Volume 24, 100183. DOI: 10.1016/j.bdr.2020.100183.

[15] Lin F, Chen C, Zhang N, Guan X, Shen X. Autonomous channel switching: towards efficient spectrum sharing for industrial wireless sensor networks.

IEEE Internet Things J. 2016; 3 (2): 231-243

[16] . Wahlster W, Grallert H.J, Wess S, Friedrich H, Widenka T. (Eds.). Towards the Internet of Services. Switzerland: The THESEUS Research Program, Springer;2014

[17] Mohamed A, Hamdan M, Khan S, Abdelaziz A, Babiker S, Imran M, Marsono M.N. Software-Defined Networks for Resource Allocation. In: Cloud Computing: A Survey, Computer Networks; 2021. 108151, DOI: 10.1016/j.comnet.2021.108151

[18] Mell P, Grance T. The NIST definition of cloud computing. National Institute of Standards and Technology. Inf. Technol. Lab. Version; 2009, 915 (10.07)

[19] Arwa M, Mosab M, Khan S, Abdelaziz A, Babiker S, Imran M, Marsono M.N. Software-Defined Networks for Resource Allocation in Cloud Computing: A Survey, Computer Networks (2021), DOI: 10.1016/j.comnet.2021.108151

[20] Monostori L, Kadar B, Bauernhansl T, Kondoh S, Kumara S, Reinhart G, Ueda K. Cyber-physical systems in manufacturing, CIRP Ann.-Manuf. Technol; 2016; 65 (2): 621-641

[21] Lee E. A. Cyber physical systems: design challenges. In: Proceedings of 11th IEEE Symposium on Object Oriented Real-Time Distributed Computing (ISORC), May 5-7; 2008 Orlando, FL: IEEE; 2008: 363-369

[22] Ramos M, Duran-Faundez C, Monsalve D, Aguilera C. Towards an Agent-Based Control of a Cyber-Physical Production System using JADE: The CIMUBB Case. In: International Conference of Production Research, ICPR- Americas 2016 January 2016 [Internet]. 2020 Available from: https://www.researchgate.net/publication/314283737_Towards_an_Agent-Based_Control_of_a_Cyber-Physical_Production_System_using_JADE_The_CIMUBB_Case

[23] Yildiz E, Møller C, Bilberg A. Virtual Factory: Digital Twin Based Integrated Factory Simulations, Procedia CIRP; 2020 Volume 93: 216-221. DOI: 10.1016/j.procir.2020.04.043

[24] Jain S, Choong N. F, Aye K. M, Luo M. Virtual factory: an integrated approach to manufacturing systems modeling. Int. J. Oper. Prod. Manag. 2001; vol. 21, N° 5/6: 594-608

[25] Schluse M, Rossmann J. From simulation to experimentable digital twins: Simulation-based development and operation of complex technical systems. ISSE 2016 - 2016 Int. Symp. Syst. Eng. - Proc. Pap.; 2016

[26] Vachálek J, Bartalský L, Rovný O, Šišmišová D, Morháč M, Lokšík M. The Digital Twin of an Industrial Production Line Within the Industry 4.0 Concept. In: 21st International Conference on Process Control (PC). 2017: 258-262

[27] Britannica, The Editors of Encyclopedia. "Manufacturing". Encyclopedia Britannica, [Internet]. 2021. Available from: https://www.britannica.com/technology/manufacturing.

[28] Russell R, Taylor B. Operations Management. USA: Prentice Hall; 2011, 810 p

[29] Koch, P. Wood Machining Processes. USA: The Ronald Press Company; 1964. 530p

[30] Aguilera C, Ramos M, Roa G. An Automatic Grading System for Panels Surfaces Using Artificial Vision. International Journal of Computers,

Communications & Control. 2006; Vol. I, N° 2: 15-22

[31] Nasir V, Cool J. A review on wood machining: Characterization, optimization, and monitoring of the sawing process, Wood Material Science & Engineering. 2018; 15(3):1-16. DOI: 10.1080/17480272.2018.1465465

[32] Hutton S. G, Taylor J. Operating stresses in band saws and their effects on fatigue life. Journal of Forestry Production. 1999; 141, 7: 12-20.

[33] Zhao C, Tanaka C, Nakao T, Takahashi A. Adaptive-Control Optimization. In: Band Sawing. 1. Monitoring Methods of Band-Saw Deviation in Sawing. Mokuzai Gakkaishi. 1991; 37(10): 897-903

[34] Marchal R, Mothe F, Denaud L. Cutting forces in wood machining - Basics and applications in industrial processes. A review. Wood machining - micromechanics and fracture. 2009; 63 (2):157-167

[35] Mandic M, Svrzic S, Danon G. The Comparative Analysis of two Methods for the Power Consumption Measurement in Circular Saw Cutting of Laminated Particle Board. Wood Research. 2015; 60 (1): 125-136

[36] Iskra P, Hernández R. Towards a process monitoring of CNC wood router. Sensor selection and surface roughness prediction. Wood Sciences and Technology 2012; 46:115-112

[37] Aguilera A, Méausoone P.J, Rolleri A, Barros J.L, Burgos F, Aguilar C. Advances on indirect methods to evaluate tool wear for Radiata pine solid wood molding. Wear. 2016; 350-351: 27-34

[38] Baradit E, Aedo R, Correa J. Knot detection in Wood using microwaves, Wood Science and Technology. 2006; 40: 118-123

[39] Thomas L, Mili L. Defect Detection on Hardwood Logs Using Laser Scanning. Wood and Fiber Science. 2006; 38: 243-246

[40] Aguilera C, Ramos M, Sappa A. Simulated Annealing: A Novel Application of Image Processing in the Wood Area. In: Guerra M. Editor. Simulated Annealing - Advances, Applications and Hybridizations. London, UK: Intechopen; 2012. DOI: 10.5772/50635

[41] Oja J, Wallbäcks L, Grundberg S, Hägerdal E, Grönlund A. Automatic grading of Scots pine (*Pinus sylvestris* L.) sawlogs using an industrial X-ray log scanner. Computers and Electronics in Agriculture. 2003; Volume 41, Issues 1-3: 63-75

[42] Salinas D. Búsqueda del cilindro nudoso en trozas de pinus Radiata d. Don utilizando imágenes de CT y patrones de nudos [tesis]. Concepción, Chile: Universidad del Bío-Bío; 2004

[43] Rojas G, Condal A, Beauregard R, Verret D, Hernández R. Identification of internal defect of sugar maple logs from CT images using supervised classification methods. Holz als Roh- und Werkstoff. 2006; 64: 295-303. DOI:10.1007/s00107-006-0105-0.

[44] Schwab K. La cuarta revolución industrial. Foro Económico Mundial. Barcelona, España: Editorial Debate; 2016. 216 p.

[45] Rich, E. A. Artificial Intelligence. New York: McGraw-Hill; 1983. 806 p.

[46] Langton C. Artificial Life p. 37. In: Wilson R, Keil F, editors. The MIT Encyclopedia of the Cognitive Sciences. MA, USA: The MIT Press; 1999

[47] Mitchell T. Machine Learning. USA: McGraw Hill; 1997. 440 p.

[48] Sharp M, Ak R, Hedberg T. A survey of the advancing use and development

of machine learning in smart manufacturing. Journal of Manufacturing Systems. 2018; 48: 170-179

[49] Qingchen Z, Yang L, Chen Z, Li P. A survey on deep learning for big data. Information Fusion. 2018; 42: 146-157

[50] G.E. Hinton, R.R. Salakhutdinov. Reducing the dimensionality of data with neural networks, Science. 2006; 313 (5786): 504-507

[51] LeCun Y, Bengio Y, Hinton G. Deep Learning. Nature. 2015; 521(7553): 436-444

[52] Berzal, Fernando. Redes Neuronales & Deep Learning. Publicado independiente (Edición en español). 2018. 753 p.

[53] Diez-Olivan A, Del Ser J, Galar D, Sierra B. Data fusion and machine learning for industrial prognosis: Trends and perspectives towards Industry 4.0. Information Fusion. 2019; 50: 92-111

[54] Pham D. T, Muhamad Z, Mahmuddin M, Ghanbarzadeh A, Koc E, Otri S. Using the bees algorithm to optimize a support vector machine for wood defect classification. In: Innovative Production Machines and Systems Virtual Conference [Internet]. 2007. Available from: http://repo.uum.edu.my/154/

[55] Koskinen J, Vaarala T, Alatalo J, Heikkilä T. Automated Quality Classification of Wooden Parts for Flexible Manufacturing. Journal of Engineering Technology (JET). 2013; Vol. 2, N° 1: 239-243

[56] Reyes D. Análisis de datos a través de Machine Learning para predecir el comportamiento de máquinas de aserrío de alta productividad [tesis]. Concepción, Chile: Universidad del Bío-Bío. 2020

[57] Navarrete C. Evaluación de método predictivo para variables de secado de chapas en planta de paneles Arauco Nueva Aldea [tesis]. Concepción, Chile: Universidad del Bío-Bío. 2020

[58] Aguilera C, Ramos M. Project: multispectral sensors to detect defects and to control quality in the panels industry. FONDEF-IDEA Stage 2, Internal Report. ANID Chile, University of Bio-Bio: 2020

[59] Aguilera C. A, Aguilera C, Sappa, A. D. Melamine Faced Panels Defect Classification beyond the Visible Spectrum. Sensors. 2018; 18, 3644. DOI:10.3390/s18113644